QK
627
A1
C87
Cummins, G.B.
Rust fungi on legumes and composites in North America.
NOV 2 1 1988
MAY 1 0 1996
D0772373

RUST FUNGI

ON LEGUMES AND COMPOSITES IN NORTH AMERICA

RUST FUNGI

ON LEGUMES AND COMPOSITES IN NORTH AMERICA

GEORGE B. CUMMINS

author of
Illustrated Genera of Rust Fungi
and
The Rust Fungi of Cereals, Grasses, and Bamboos

University of Arizona Press
Tucson, Arizona

About the Author...

George B. Cummins, visiting research professor of plant pathology at the University of Arizona since 1970, has been engaged in teaching and research about rust fungi since 1930 when he joined the faculty of Purdue University. He retired from Purdue as professor emeritus of botany in 1970. While there he was curator of the Arthur Herbarium. He was for several years a collaborator with the National Fungus Collections, United States Department of Agriculture. He is author of two other books on rust fungi, *Illustrated Genera of Rust Fungi* and *The Rust Fungi of Cereals, Grasses, and Bamboos*.

THE UNIVERSITY OF ARIZONA PRESS

Library of Congress Cataloging in Publication Data

Cummins, George Baker, 1904–
Rust fungi on legumes and composites in North America.

Includes indexes.
1. Rusts (Fungi)—North America—Identification. 2. Legumes—Diseases and pests—North America—Identification. 3. Compositae—Diseases and pests—North America—Identification. 4. Fungi, Phytopathogenic—North America—Host plants. I. Title.
QK627.A1C87 589'.225'097 78-60541
ISBN 0-8165-0653-1

To distinguished predecessors, especially

Joseph Charles Arthur, 1850–1942

Paul Dietel, 1860–1947

Edward Willet Dorland Holway, 1853–1923

Paul Sydow, 1851–1925

Hans Sydow, 1879–1946

Contents

Preface

The decision was largely arbitrary to limit this study to the Uredinales (plant rust fungi) that parasitize the plant families Compositae and Leguminosae of the continental land mass of North America. Both families are large in numbers of genera and species. The Compositae are the largest family of flowering plants, with some 950 genera and 20,000 species (19) worldwide. Some of the tribes, e. g., Heliantheae, Eupatorieae and Vernonieae, are strikingly abundant, especially in the southern United States and in Mexico and Central America. Second in size are the Leguminosae with some 550 genera and 13,000 species (19). Three subfamilies: Mimosoideae, Caesalpinioideae and Papillionoideae, usually are recognized or sometimes are treated as families. Members of the Mimosoideae and Caesalpinioideae largely are inhabitants of the southern United States and southward, but the Papillionoideae have representatives more or less throughout the continent.

The Compositae include as crop plants globe artichoke, lettuce, endive, salsify, sunflower, safflower, and a wide variety of garden flowers. Members of the Leguminosae that commonly are grown as crop plants are alfalfa, the clovers, bird'sfoot trefoil, soybeans, common beans, garden peas, cowpeas, lentils, chickpeas, broadbeans, and peanuts. Ornamental legumes such as sweetpeas, lupines, wisteria, and redbud are common in temperate climates. Southward, the royal poinciana, orchid trees (*Bauhinia*), sennas, brooms, paloverdes, and some acacias are attractive trees and shrubs.

Both plant families are parasitized worldwide by the rust fungi. For the Compositae, species of *Puccinia* and *Uromyces* are the predominant parasites, totaling 166 for the continent and as delimited. The next most represented genus is *Coleosporium*, with 21 species. The Leguminosae support a quite different set of rust fungi. With only two species of *Puccinia*, the genus is conspicuous by its absence.

One looks askance at such species and wonders if they may not belong elsewhere. But *Uromyces*, despite a strong and justified tendency to consider that it is only a one-celled variant of *Puccinia*, has 39 species on the legumes, including many of the important pathogens. Slightly larger than *Uromyces* is *Ravenelia*, with 61 species. Its species are common in the south and especially southwestern area of the United States and increasingly common in Mexico and Central America. The species of *Ravenelia* are commonest on the Mimosoideae and Caesalpinioideae, whereas species of *Uromyces* predominantly are on the Papillionoideae.

The *North American Flora* (1) provides the only previous treatment of the Uredinales of North America but it included the offshore islands of the Caribbean and the West Indies. But recent monographic studies have made the task much easier. With respect to the species of *Puccinia* on the composites, Parmelee (25, 26, 27) studied those on the Heliantheae; Urban (30) those on the Vernonieae; Cummins, Britton, and Baxter (9) those on the Eupatorieae; Savile (29) those on the Cardueae; Hennen and Cummins (14) the species of *Puccinia* and *Uromyces* on the Senecioneae; and Lindquist (22) those on the genus *Baccharis*. Less work has been done with the rust fungi of legumes, but there is Baxter's (4) monograph of the genus *Uropyxis*.

Cummins and Stevenson (10) brought the records of the North American rust flora up to 1956 and for the most part corrected the nomenclature. Unfortunately, in a few cases the Code was not followed or was not correctly interpreted. It is my intention that the names used here be in accord with the Code of Nomenclature.

Since 1956 there has been extensive field work in the southwestern United States and the northern two-thirds of Mexico by Cummins, and by Hennen in Mexico. This has significantly increased the numbers of species of the Uredinales, as well as known distributions. The majority of the records have been of rust fungi on the composites and legumes. Obviously, the record presented here will be stable only until more collecting is done, especially in Mexico and Central America. *Uromyces ciceris-arietinus* Jacz. was only recenty introduced into North America (20) and it can be very destructive. *Phakopsora pachyrhizi* Syd. now occurs on soybean and some other legumes in Puerto Rico, but it may not reach or be destructive in the principal soybean acreages.

The delimitation of species is relatively conservative by intention, in the hope that plant pathologists and mycologists generally may be able to identify collections with some certainty. The keys use characters that are fairly obvious, and the sequence of species places together those that have common characteristics. Presumably, such an arrangement is somewhat phylogenetic, but an attempt to build a phylogenetic system was not the goal. Neither have families been used, although there probably are four represented, as follows: (1) Coleosporiaceae, if properly restricted, (2) Melampsoraceae, (3) Pucciniaceae, and (4) Raveneliaceae. Leppik (21) recognized these four and two others that have no parasites of the composites or legumes. The family Coleosporiaceae, is not a new concept but it cannot include all genera that have "internal basidia." The Raveneliaceae of Leppik is comprised of *Ravenelia* and a considerable number of small genera, e. g. *Uropyxis, Phragmopyxis, Diabole, Dicheirinia, Spumula,* etc., that parasitize legumes. It doubtless is a valid concept but in need of refinement. At present its limits are a bit diffuse, but spermogonia are type 7 (16) in all genera. The only melamsoraceous genera are *Phakopsora* and *Baeodromus*. *Puccinia* and *Uromyces* are the principal representatives of the Pucciniaceae.

Generally, not all species of host plants are listed for each species of fungus. Thus *Baccharis* spp., for example, indicates that more than one species of *Baccharis* is parasitized by that fungus. The host of the type collection always is given. Neither states nor countries are listed for the species of rust fungi but the overall distribution is stated. If a species is known to occur in southern Mexico and in Costa Rica, it almost certainly also occurs in Guatemala even if there is no record. There has been relatively little collecting in vast area.

The *illustrations* for the volume were drawn on graph paper having 100 squares per square inch and with one square representing one space (2.2 µm) of the ocular scale. All drawings were made at this magnification and reduced to 570 diameters. In most genera, teliospores, urediniospores and paraphyses are illustrated but in *Coleosporium* urediniospores and basidiospores are figured. For *Ravenelia*, because of the size of the teliospore heads, only the central cells of the head were drawn together with a silhoutte-like sketch of the marginal cells. *Spore measurements* are given as (40-)43-55 (-67) x (18-)20-25(-30) µm. The figures 43-55 x 20-25 describe the majority of the population. Figures in parentheses are outside of the typical range yet not uncommon. Clearly unusual or freak spores were disregarded.

ACKNOWLEDGEMENTS

Many institutions and individuals have loaned specimens, and this is greatly appreciated. Because of the large numbers of type specimens held in the Stockholm Museum (S) and the

Arthur Herbarium (PUR) of Purdue University, the cooperation of the curators has been indispensable. I would be remiss if I did not acknowledge the backlog of experience gained during 40 years of work in the Arthur Herbarium, but the book in its entirety was written at the University of Arizona, where the Department of Plant Pathology provided space, optical equipment, some financial support, and much encouragement, all of which are appreciated.

George B. Cummins

Key to Genera of Rust Fungi

1. Teliospores united in radially arranged discoid heads 11
1. Teliospores not in discoid heads 2

   2. Teliospores sessile (may be in chains) 14
   2. Teliospores pedicellate 3

3. Teliospores 1 celled 4
3. Teliospores more than 1 celled 6

   4. Teliospores producing an internal basidium .... *Chrysella* (12)
   4. Teliospores producing an external basidium 5

5. Spermogonia subcuticular; teliospores sculpture coarse and block like .... *Pileolaria* (5)
5. Spermogonia subepidermal; teliospores smooth or sculpture not block like .... *Uromyces* (2)

   6. Teliospores 2 celled 7
   6. Teliospores more than 2 celled 9

7. Teliospores horizontally septate 8
7. Teliospores vertically septate 10

   8. Teliospores with 1 germ pore per cell *Puccinia* (1)
   8. Teliospores with 2 germ pores per cell .... *Uropyxis* (3)

9. Teliospores horizontally septate only *Phragmopyxis* (4)
9. Teliospores horizontally and vertically septate .... *Sphaerophragmium* (11)

10. Teliospores single on each pedicel *Dicheirinia* (9)
10. Teliospores one or more pairs on each pedicel .... *Diabole* (10)

11. Teliospore pedicels fascicled, of more than one hypha .... *Ravenelia* (6)
11. Teliospore pedicels simple, one hypha 12

12. Teliospore pedicel with apical but not hygroscopic cells .... *Dicheirinia* (9)
12. Teliospores with subtending hygroscopic cells 13

13. Teliospore heads with pedicel attached to the spore; cysts separate .... *Spumula* (7)
13. Teliospore heads with pedicel attached to fused cysts .... *Cystomyces* (8)

14. Teliospores producing an internal basidium .... *Coleosporium* (15)
14. Teliospores producing an external basidium 15

15. Teliospores in a single layer 16
15. Teliospores in more than one layer 17

16. Teliospores in groups on basal cells; spermogonia subcuticular; aecia uredinoid .... *Chaconia* (13)
16. Teliospores borne singly; spermogonia subepidermal; aecia caeomoid ...... *Chrysocelis* (14)

17. Telia 2, 3, or more spores deep; not aecidium-like .. 18
17. Telia several spores deep; aecidium-like 21

18. Teliospores in short chains 19
18. Teliospores irregularly arranged ... *Phakopsora* (16)

19. Microcyclic species; spermogonia subepidermal 20
19. Macrocyclic species; spermogonia subcuticular .... *Cerotelium* (18)

20. Teliospores not extruded in columns *Baeodromus* (17)
20. Teliospores extruded in hair like columns .... *Cionothrix* (19)

21. Spores only loosely or not adherent; distinguisable from *Aecidium* only by germination ..... *Endophyllum* (20)
21. Spores firmly adherent, the spore mass compact 22

22. Spores 1 celled .... *Endophylloides* (21)
22. Spores 2 celled .... *Pucciniosira* (22)

(The form genera *Aecidium* and *Uredo* not in this key; see pages 398 and 403)

1. PUCCINIA Persoon

Syn. Method. Fung. p. 225. 1801.

Spermogonia subepidermal, globoid, type 4 (16). Aecia supepidermal in origin, erumpent; aecidioid with catenulate spores, or uredinoid with spores borne singly on pedicels. Uredinia subepidermal in origin, erumpent, with or without paraphyses; spores borne singly on pedicels. Telia subepidermal in origin, erumpent in most species; spores borne singly on pedicels, typically 2 celled by horizontal septum, germ pores 1 in each cell, wall mostly pigmented; basidium external.

Type species: *Puccinia graminis* Pers.

KEY TO SPECIES OF *PUCCINIA* ON COMPOSITAE
(See p. 181 for the species on Leguminosae)

Teliospore wall smooth — Section I
Teliospore wall sculptured — Section II (p. 12)

SECTION I

Uredinia produced; macrocyclic or potentially so — Section IA
Uredinia lacking; demicyclic or microcyclic — Section IB (p.10)

Section IA

1. Teliospores germinating without dormancy; wall pale ... 2
1. Teliospores requiring dormancy; wall brown 20

    2. Urediniospores with scattered pores 3
    2. Urediniospores with equatorial or basal pores ... 8

3. Urediniospore wall cinnamon brown or darker 4
3. Urediniospore wall golden or paler 6

4. Uredinial paraphyses lacking ....*inaudita* (1)
4. Uredinial paraphyses present 5

5. Teliospores mostly 60-90 µm long, apical wall 8-15 µm .... *holwayula* (2)
5. Teliospores mostly 70-115 µm long, apical wall 2-4 µm .... *seorsa* (3)

6. Teliospores mostly 15-20 µm wide; aecia uredinoid .... *arthuriana* (4)
6. Teliospores mostly 24-28 µm wide, aecia aecidioid 7

7. Teliospores mostly 50-68 µm long, pedicel mostly broad .... *baccharidis* (5)
7. Teliospores mostly 54-74 µm long, pedicel narrow .... *evadens* (6)

8. Urediniospores with 3 pores typical 9
8. Urediniospores with 2 pores typical 13

9. Urediniospore pores with large lens like caps .... *baccharidis-multiflorae* (7)
9. Urediniospore pores with inconspicuous caps 10

10. Teliospore wall 8-12 µm thick apically ... *enixa* (8)
10. Teliospore wall 5 µm or less apically 11

11. Urediniospores brown; teliospores mostly 44-66 µm long .... *inaudita* (1)
11. Urediniospores yellowish 12

12. Teliospores mostly 44-55 µm long ...... *exornata* (9)
12. Teliospores mostly 55-85 µm long ..... *erratica* (10)

13. Urediniospores colorless or yellowish, pores obscure 14
13. Urediniospores brown; pores obvious 15

14. Urediniospores 26-34 µm long; teliospores 44-60 µm long .... *alia* (11)
14. Urediniospores 22-28 µm long; teliospores 35-48 µm long .... *oaxacana* (12)

15. Urediniospore pores with large lens like caps 16
15. Urediniospore pores with small or no caps 17

16. Teliospores 28-37 µm long .... *trixitis* (13)
16. Teliospores 43-55 µm long .... *ocellifera* (14)

17. Teliospores mostly 40-60 µm long .... *guardiolae* (15)
17. Teliospores mostly less than 40 µm long 18

18. Pore in lower teliospore cell at septum
.... *potosina* (16)
18. Pore in lower teliospore cell midway to hilum ... 19

19. Urediniospores 23-26 x 19-20 µm, pores in equator
.... *brachytela* (17)
19. Urediniospores 22-25 x 30-34 µm, pores toward
base .... *inermis* (18)

20. Urediniospore pores scattered, 5-9 21
20. Urediniospore pores equatorial or below 22

21. Telia covered, loculate; teliospores 13-17 µm wide
.... *pistorica* (19)
21. Telia exposed; teliospores mostly 24-27 µm wide
.... *porophylli* (20)

22. Urediniospore pores 3 or 4 23
22. Urediniospore pores typically 2 27

23. Urediniospore wall essentially colorless *sphenica* (21)
23. Urediniospore wall brown 24

24. Urediniospores 19-22 x 23-26 µm *spegazziniana* (22)
24. Urediniospores 28-35+ x 22-28 µm 25

25. Urediniospore pores all equatorial, caps
conspicuous .... *similis* (23)
25. Urediniospore pores usually 2 equatorial,
1 apical, caps slight or none 26

26. Urediniospores mostly 28-40 µm long, wall
2-2.5 µm .... *parthenii* (24)
26. Urediniospores mostly 24-33 µm long, wall
1.5 µm .... *affinis* var. *triporosa* (29b)

27. Telia tardily exposed; teliospores mostly clavate ... 28
27. Telia early exposed; teliospores ellipsoid or
broader 30

28. Urediniospore wall pale yellow .... *axiniphylli* (25)
28. Urediniospore wall brown 29

29. Urediniospores mostly 24-29 µm long; teliospores mostly 40-53 µm long .... *irregularis* (26)
29. Urediniospores mostly 25-37 µm long; teliospores mostly 43-66 µm long .... *senecionicola* (27)

30. Pore of lower teliospore cell depressed 1/3 or more 31
30. Pore of lower teliospore cell next to septum 38

31. Urediniospores yellowish; teliospore side wall 2-2.5 µm .... *punctoidea* (28)
31. Urediniospores brown; teliospore side wall 3 µm or more 32

32. Teliospore wall bilaminate nearly throughout .... *affinis* (29)
32. Teliospores with differentiated umbos only 33

33. Teliospore pedicel thick walled, not collapsing 34
33. Teliospore pedicel thin walled, mostly collapsing ... 36

34. Urediniospore wall 1-2 µm thick; aecia aecidioid 35
34. Urediniospore wall 2-3 µm thick; aecia uredinoid .... *kuhniae* var. *robusta* (32d)

35. Urediniospore wall 1-1.5 µm thick .... *massalis* (30)
35. Urediniospore wall 1.5-2 µm thick .... *chloracae* (31)

36. Urediniospore wall 2-3 µm thick; teliospores 30-37 µm wide .... *kuhniae* var. *decora* (32c)
36. Urediniospore wall 1-1.5 µm thick; teliospores narrower 37

37. Teliospores mostly 42-53 x 26-33 µm .... *kuhniae* var. *kuhniae* (32a)
37. Teliospores mostly 34-46 x 22-28 µm *calanticariae* (33)

38. Urediniospores typically wider than high 39
38. Urediniospores typically higher than wide 40

39. Urediniospores 32-38 µm between the pores; teliospores 42-50 x 33-38 µm .... *espinosarum* (34)
39. Urediniospores 24-28 µm between the pores; teliospores 36-42 x 33-37 µm .... *inanipes* (35)

40. Urediniospores strongly triangular with pores lateral 41
40. Urediniospores ellipsoid or slightly obovoid with pores lateral 46

41. Teliospore side wall 1-2 µm thick .... *enceliae* (36)
41. Teliospore side wall more than 2 µm thick 42

42. Teliospores mostly 38 µm or less long, pedicels mostly exceeding 100 µm .... *abrupta* var. *partheniicola* (38b)
42. Teliospores commonly more than 40 µm long, pedicel mostly less than 100 µm 43

43. Urediniospores mostly 22-28 µm long .... *noccae* (37)
43. Urediniospores mostly 20-24 µm long 44

44. Teliospore side wall 3-4 µm; telia commonly on stems .... *abrupta* var. *abrupta* (38a)
44. Teliospore side wall 2-3.5 µm; telia on leaves 45

45. Teliospore pedicel rugose basally ..... *gymnolomiae* (39)
45. Teliospore pedicel not rugose .... *verbesinae* (40)

46. Urediniospores mostly less than 24 µm long...... 47
46. Urediniospores mostly more than 24 µm long 49

47. Teliospores mostly 35-50 µm long .... *gnaphaliicola* (41)
47. Teliospores mostly 38 µm or less long 48

48. Teliospores 30-36 x 24-28 µm, side wall 2-2.5 µm .... *subglobosa* (42)
48. Teliospores 30-38 x 20-24 µm, side wall 1-2 µm .... *sonorae* (43)

49. Teliospore pedicel typically broken near hilum 50
49. Teliospore pedicel long, persistent 51

50. Teliospores 33-40 µm long; aecia uredinoid, systemic .... *helianthellae* (44)
50. Teliospores mostly 47-55 µm long; aecia unknown .... *redempta* (45)

51. Urediniospore pores with conspicuous caps 52
51. Urediniospore pores with thin or no caps 54

52. Teliospore pedicel thin walled, collapsing; aecia aecidioid .... *sinaloana* (46)
52. Teliospore pedicel thick walled, mostly not collapsing 53

53. Teliospore apical wall broadly thickened; aecia uredinoid .... *franseriae* (47)
53. Teliospore apical wall thickened by a low umbo; aecia aecidioid .... *caleae* (48)

54. Teliospore side wall less than 2.5 µm thick 55
54. Teliospore side wall more than 2.5 µm thick 61

55. Teliospores obovate clavate or oblong ellipsoid; aecia aecidioid or unknown 56
55. Teliospores ellipsoid or broadly so; aecia uredinoid 60

56. Urediniospore wall 2-3 µm, echinulate over pores .... *nuda* (49)
56. Urediniospore wall 2 µm or less thick 57

57. Urediniospores mostly 26-33 µm long; teliospores mostly 38-60 µm long .... *helianthi* (50)
57. Urediniospores mostly 22-28 µm long; teliospores shorter 58

58. Teliospores mostly 38-55 µm long, apical wall mostly 8-11 µm thick .... *cognata* (51)
58. Teliospores mostly 35-50 µm long, apical wall mostly less than 8 µm thick 59

59. Urediniospore wall smooth over pores .... *invelata* var. *invelata* (52a)
59. Urediniospore wall echinulate over pores .... *invelata* var. *echinulata* (52b)

60. Teliospore apex with a small pale umbo .... *eupatorii* (53)
60. Teliospore apex broadly thickened ... *abramsii* (54)

61. Teliospore apical wall essentially concolorous 62
61. Teliospore apical wall a differentiated pale umbo .. 63

62. Teliospores mostly 38-46 x 24-28 µm, not forming galls .... *vaga* (55)
62. Teliospores mostly 38-65 x 29-35 µm, forming stem galls .... *splendens* (56)

63. Urediniospore wall echinulate over pores; teliospores mostly 36-39 µm wide .... *solidipes* (57)
63. Urediniospore wall smooth over pores 64

64. Teliospore pedicels enlarged basally *turgidipes* (28)
64. Teliospore pedicels terete 65

65. Teliospores dimorphic, dark chestnut and golden brown .... *ximenesiae* (59)
65. Teliospores all dark chestnut brown 66

66. Teliospores mostly 42-50 µm long; aecia unknown .... *tageticola* (60)
66. Teliospores mostly 45-55 µm long; aecia uredinoid .... *kuhniae* var. *brickelliae* (32b)

Section IB

1. Demicyclic species, spermogonia, aecia and telia formed (Also see No. 77, *P. grindeliae*) 2
1. Microcyclic species, telia and sometimes spermogonia 10

2. Telia covered, loculate with brownish paraphyses 3
2. Telia exposed, without paraphyses 5

3. Teliospores mostly 32-43 µm long .... *tenuis* (61)
3. Teliospores mostly 40-55 µm long 4

4. Aeciospore wall 1 µm thick, minutely verrucose .... *batesiana* (62)
4. Aeciospore wall finely verrucosely rugose .... *desmanthodii* (63)

5. Teliospores germinate without dormancy, wall yellowish 6
5. Teliospores germinate after dormancy, wall brown ... 8

6. Teliospore wall uniformly thin, spores mostly 15-19 µm wide .... *inaudita* (1)
6. Teliospore wall thicker at apex; spores more than 19 µm wide 7

7. Teliospores mostly 62-78 x 23-27 µm .... *interjecta* (64)
7. Teliospores mostly 50-66 x 19-24 µm .. *vallartensis* (65)

8. Teliospore apical wall broadly thickened to 8-10 µm .... *investita* (66)
8. Teliospore apical wall thickened by a small papilla 9

9. Pore of lower teliospore cell in lower half of cell; aecia systemic .... *intermixta* (67)
9. Pore of lower teliospore cell at septum; aecia localized .... *senecionis* (68)

10. Telia covered by epidermis 11
10. Telia early exposed 13

11. Sori discrete, bounded by compacted hyphae .... *virgae-aureae* (69)
11. Sori compound, loculate with brown paraphyses 12

12. Sori encircling stems .... *gallula* (70)
12. Sori on leaves .... *stromatifera* (71)

13. Telia systemic .... *fraseri* (72)
13. Telia localized 14

14. Telia pulverulent; spore pedicels fragile 15
14. Telia compact; spore pedicels persistent 17

15. Apex of spore with a broad pale umbo .. *excursionis* (73)
15. Apex of spore with only a small umbo or papilla 16

16. Teliospores of relatively uniform size .... *conglomerata* (74)
16. Teliospores dimorphic, small and large .... *glomerata* (75)

17. Spore pedicels usually 100 µm or longer 18
17. Spore pedicels usually 60 µm or shorter 19

18. Spores mostly 22-30 µm wide, side wall 3-4 µm thick .... *marianae* (76)
18. Spores mostly 20-26 µm wide, side wall 1.5-2.5 µm .... *grindeliae* (77)

19. Spore wall less than 3 µm thick at apex *spegazzinii* (78)
19. Spore wall more than 3 µm thick at apex 20

20. Spore wall essentially colorless, apex very abruptly thickened .... *schistocarphae* (79)
20. Spores wall golden or darker, apex usually gradually thickened 21

21. Sporogenous layer chestnut brown, conspicuous 22
21. Sporogenous layer pale, inconspicuous 23

22. Teliospores mostly 13-19 µm wide ...... *xanthii* (80)
22. Teliospores mostly 18-23 µm wide .... *dyssodiae* (81)

23. Spore wall chestnut brown apically, only slightly paler basally 24
23. Spore wall golden or yellowish, usually paler basally 26

24. Spore pedicels mostly 20 µm or shorter .... *praemorsa* (82)
24. Spore pedicels mostly 30 µm or longer 25

25. Spores mostly elongately ellipsoid .... *silphii* (83)
25. Spores mostly elongately obovoid or clavate .... *cnici-oleracei* (84)

26. Spore pedicels mostly 25 µm or longer 27
26. Spore pedicels 20 µm or shorter 28

27. Spore side wall mostly 1 µm thick; spores mostly 35-58 µm long .... *melampodii** (85)
27. Spore side wall mostly 1.5-2 µm thick; spores mostly 36-48 µm long .... *recedens* (86)

28. Spores mostly 19-22 µm wide, apical wall 3-7 µm .... *tolimensis* (87)
28. Spores mostly 13-17 µm wide, apical wall 6-10 µm .... *semota* (88)

*Note: Numerous binomials have been consigned to the synonymy of *P. melampodii* and others keyed here are scarcely separable on morphological bases.

SECTION II

Uredinia produced; macrocyclic or potentially so Section IIA
Uredinia lacking; demicyclic or microcyclic Section IIB (p. 18)

Section IIA

1. Teliospore side wall smooth, apex striated .... *archibaccharidis* (89)
1. Teliospore side wall sculptured 2

   2. Urediniospore pores scattered, 4 or more 3
   2. Urediniospore pores zonate, 2 or 3 7

3. Paraphyses abundant around uredinia, to 25 µm wide .... *rata* (90)
3. Paraphyses lacking 4

   4. Lower teliospore pore at septum; urediniospore wall pale yellow 5
   4. Lower teliospore pore in lower half of cell; urediniospore wall brown 6

5. Teliospore wall uniformly 2.5-3 µm thick, finely rugosely reticulate .... *viatica* (91)
5. Teliospore wall 4-6 µm over pores, verrucose with plate like warts .... *jaliscana* (92)

   6. Urediniospore pores with conspicuous caps; aecia systemic .... *minussensis* (93)
   6. Urediniospore pores with inconspicuous caps; aecia unknown .... *arthurella* (94)

7. Urediniospore pores 2, below the equator 8
7. Urediniospore pores 2 or 3, in or above the equator 11

   8. Urediniospores longer than wide 9
   8. Urediniospores globoid or shorter than wide 10

9. Urediniospores obovoid with pores lateral, pores slightly subequatorial .... *praetermissa* (95)
9. Urediniospores oblong ellipsoid with pores lateral, pores near hilum .... *hogsoniana* (96)

   10. Teliospores mostly 34-39 x 25-28 µm *basiporula* (97)
   10. Teliospores mostly 43-48 x 33-40 µm .... *obesiseptata* (98)

11. Urediniospore pores 2 (rarely 3) above the equator; lower teliospore pore depressed 12
11. Urediniospore pores 2 or 3 at the equator; lower teliospore pore various 16

12. Aecia systemic, uredinoid .... *mirifica* (99)
12. Aecia localized, uredinoid where known 13

13. Teliospores mostly 35-46 μm long; urediniospores mostly 27-36 μm long 14
13. Teliospores mostly 30-40 μm long; urediniospores mostly 24-30 μm long 15

14. Teliospore verrucae spaced 0.5-1.5 μm, pedicel fragile .... *balsamorhizae* (100)
14. Teliospore verrucae spaced 2-2.5 μm, pedicel semipersistent *hieracii* var. *stephanomeriae* (101c)

15. Teliospore pedicel more or less persistent, teliospore wall 2-2.5 μm thick .... *hieracii* var. *harknessii* (101b)
15. Teliospore pedicel always broken short, teliospore wall 1.5-2 μm or less .... *hieracii* var. *hieracii* (101a)

16. Urediniospore pores typically 3 17
16. Urediniospore pores typically 2 38

17. Uredinia with large thin walled paraphyses 18
17. Uredinia without paraphyses 19

18. Lower teliospore pore at septum *ludovicianae* (102)
18. Lower teliospore pore much depressed *egregia* (103a)

19. Teliospore side wall 5-7 μm thick, spores mostly 30-40 μm wide .... *semiinsculpta* (104)
19. Teliospore side wall less than 5 μm thick, spores mostly less than 30 μm wide 20

20. Lower teliospore pore at septum or nearly so ... 21
20. Lower teliospore pore mostly midway or below ... 28

21. Urediniospores uncertain; aecia systemic *egressa* (105)
21. Urediniospores common; aecia localized 22

22. Urediniospore pores with conspicuous smooth caps 23
22. Urediniospore pores with echinulate caps, conspicuous or not 24

23. Teliospore wall rugose or pseudoreticulate .... *artemisiae-norvegicae* (106)
23. Teliospore wall finely punctately verrucose .... *tanaceti* (107)

24. Teliospore apical wall broadly thickened, mostly 4-7 µm thick 25
24. Teliospore apical wall only slightly thickened . 26

25. Teliospore wall closely punctately verrucose, spores mostly 36-44 µm long .... *longipes* (108)
25. Teliospore wall with discrete wartlets, spores mostly 32-37 µm long .... *oblata* (109)

26. Urediniospore pores with conspicuous caps .... *cnici* (110)
26. Urediniospore pores with only slight caps 27

27. Teliospores mostly 35-46 µm long .... *inclusa* (111)
27. Teliospores mostly 37-53 µm long ..... *californica* (112)

28. Infections systemic; aecia uredinoid .... *punctiformis* (113)
28. Infections localized 29

29. Urediniospore wall brown or brownish, pores obvious 30
29. Urediniospore wall colorless, pores obscure 35

30. Teliospores rarely more than 32 µm long; aecia aecidioid .. *variabilis* var. *variabilis* (137a)
30. Teliospores mostly more than 32 µm long 31

31. Urediniospores echinulate to the hilum 32
31. Urediniospores smooth on lower 1/4-1/3 .... *calcitrapae* var. *centaureae* (114b)

32. Teliospore apical wall 1/2 to 2 times thicker than the side wall .... *acroptili* (115)
32. Teliospore apical wall essentially as side wall 33

33. Aecia uredinoid; teliospore wall 2-2.5 µm thick; urediniospore wall 1.5-2 µm thick .... *calcitrapae* var. *bardanae* (114a)
33. Aecia aecidioid; teliospore wall 1.5-2 µm thick; urediniospore wall 2-2.5 µm thick 34

34. Aecia with peridium developed, obvious .... *crepidis-montanae* (116)
34. Aecia with no or rudimentary peridium .... *orbicula* (117)

35. Teliospore wall rugose or pseudoreticulate .... *inaequata* (118)
35. Teliospore wall with discrete verrucae 36

36. Sori deep seated, in crowded groups *praealta* (119)
36. Sori merely subepidermal, scattered 37

37. Teliospores echinulate verrucose with conical verrucae spaced 4-6 µm .... *notha* (120)
37. Teliospores verrucose with low rounded verrucae spaced 1.5-2 µm .... *idoneae* (121)

38. Teliospore side wall smooth, apex striated .... *archibaccharidis* (89)
38. Teliospore side wall sculptured 39

39. Pore of lower teliospore cell at septum 40
39. Pore of lower teliospore cell depressed 1/3 or more 49

40. Teliospores prominently verrucose with conical verrucae 41
40. Teliospore sculpture not prominent conical verrucae 42

41. Teliospores with an umbo 6-9 µm thick .... *conoclinii* var. *conoclinii* (122a)
41. Teliospores with no distinct umbo ..... *echinulata* (123)

42. Teliospore wall obviously thickened at apex 43
42. Teliospore wall uniformly thick or essentially so, punctate verrucose .... *altissimorum* (124)

43. Teliospores with discrete punctae spaced 1 µm or more 44
43. Teliospores rugose or reticulate 45

44. Teliospores 32-37 x 22-25 µm .... *oblata* (109)
44. Teliospores 38-46 x 25-30 µm .... *baccharidis-hirtellae* (125)

45. Teliospore prominently reticulate apically, decreasing basally 46
45. Teliospore finely rugose or rugosely reticulate throughout 47

46. Teliospores all alike, all germinating; aecia uredinoid .... *guatemalensis* (126)
46. Teliospores dimorphic, pale ones germinating; aecia aecidioid .... *proba* (127)

47. Teliospores dimorphic, mostly 40-46 µm or 29-35 µm long .... *concinna* var. *duranii* (128b)
47. Teliospores all more or less alike 48

48. Teliospores mostly 42-56 x 25-32 µm, appearing smooth in silhouette .... *concinna* (128a)
48. Teliospores mostly 34-48 x 21-28 µm, appearing slightly undulate in silhouette ... *zaluzaniae* (129)

49. Teliospores reticulate or rugose 50
49. Teliospores verrucose with discrete verrucae 51

50. Teliospores reticulate; aecia aecidioid .... *zexmeniae* (130)
50. Teliospores minutely rugose reticulate; aecia uredinoid .... *subdecora* (131)

51. Aecia systemic, uredinoid .... *cyani* (132)
51. Aecia localized where known 52

52. Teliospore wall bilaminate nearly throughout; apical umbo to 9 µm thick .... *iostephanes* (133)
52. Teliospore wall if differentiated at all, then only over pores 53

53. Teliospore wall with prominent cones spaced 2 µm or more 54
53. Teliospore wall punctate with small verrucae spaced 2µ m or less 57

54. Urediniospore wall smooth around the pores .... *electrae* (134)
54. Urediniospore wall uniformly echinulate 55

55. Teliospores with prominent umbos over each pore .... *conoclinii* var. *depressipora* (122b)
55. Teliospores with small or no umbos 56

56. Teliospore wall 2-2.5 µm thick; uredinia aparaphysate .... *globulifera* (135)
56. Teliospore wall 4-5 µm thick; uredinia paraphysate .... *egregia* var. *cumminsiana* (103b)

57. Urediniospore wall uniformly echinulate 58
57. Urediniospore wall smooth around pores; aecia uredinoid .... *balsamorhizae* (100)

58. Urediniospore wall 2-2.5 µm thick; teliospore wall 2-3 µm thick; aecia unknown .... *pinaropappi* (136)
58. Urediniospore wall 1.5 µm thick; teliospore wall 1.5-2 µm thick; aecia aecidioid 59

59. Urediniospore pores 3, rarely 2, spores mostly 22-26 x 19-23 µm ..... *variabilis* var. *variabilis* (137a)
59. Urediniospore pore 2, rarely 3 60

60. Urediniospores mostly 22-26 x 19-23 µm .... *variabilis* var. *insperata* (137b)
60. Urediniospores mostly 19-22 x 17-20 µm .... *variabilis* var. *lapsanae* (137c)

Section IIB

1. Spermogonia, aecia and telia produced, demicyclic .. 2
1. Spermogonia or usually only telia produced, microcyclic 5

2. Aecia systemic; teliospores punctately verrucose or with flat, spaced warts 3
2. Aecia localized; teliospores with conical verrucae 4

3. Lower teliospore pore at septum, wall punctately verrucose .... *egressa* (105)
3. Lower teliospore depressed, wall verrucose with flat warts .... *hysterium* (138)

4. Lower teliospore pore at septum .... *mcvaughii* (139)
4. Lower teliospore pore depressed *otopappicola* (140)

5. Lower teliospore pore at septum 6
5. Lower teliospore pore depressed 7

6. Umbos over pores pale and differentiated; spores dimorphic .... *ghiesbrechtii* (141)
6. Umbos over pores not differentiated; spores all alike .... *dovrensis* (142)

7. Spores echinulate verrucose with prominent cones spaced 2-3 µm 8
7. Spores minutely verrucose or rugose 9

8. Infections systemic; pedicels fragile .... *suksdorfii* (143)
8. Infections localized; pedicels thick walled .... *annulatipes* (144)

9. Spore wall minutely punctately verrucose, 1-1.5 µm thick .... *ferox* (145)
9. Spore wall rugose or rugosely reticulate 10

10. Apex of spore with a 4-6 µm thick umbo *absicca* (146)
10. Apex of spore without a differentiated umbo 11

11. Spores minutely rugose, sharply constricted at septum .... *discreta* (147)
11. Spores rugose reticulate, slightly or not constricted .... *neorotundata* (148)

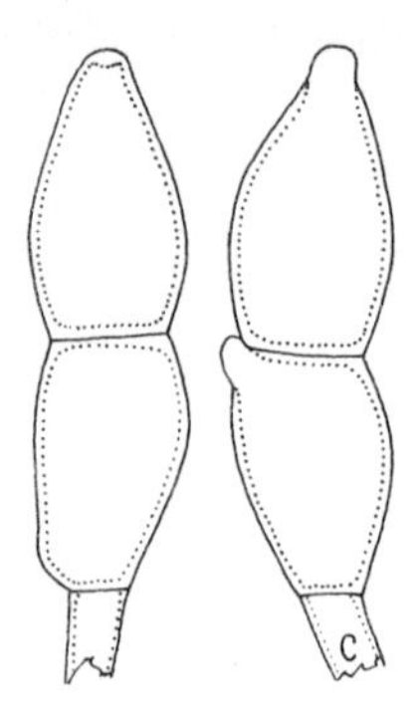

1. *PUCCINIA INAUDITA* H. S. Jack. & Holw. in Arthur, Amer. J. Bot. 5:535. 1918.

   Aecidium collapsum Mains, Contrib. Univ. Michigan Herb. 1:15. 1939.

Spermogonia mostly on adaxial leaf surface. Aecia in small groups and usually on slightly swollen areas, amphigenous, peridium whitish, cylindrical but fragmenting variously; spores (21-)25-34(-40) x (16-)19-27(-32) µm, mostly globoid or ellipsoid, wall including verrucae (1.5-)2-3(-4) µm thick, sometimes thicker apically, brownish, coarsely verrucose with irregularly shaped, straight sided, flat verrucae. Uredinia not or only rarely formed; spores occasionally in the telia (after Parmelee) 24-32 x 19-24(-27) µm, ellipsoid, globoid or obovoid, wall 1.4 µm thick, yellow brown, evenly echinulate, spines 0.7-1.4 µm high, 1.7-3.5 µm apart, pores 3-4, scattered. Telia on abaxial surface, exposed, whitish when old and dry, probably bright yellow when fresh, compact; spores (35-)44-66(-80) x (13-)15-19(-22) µm, narrowly ellipsoid or more or less cylindrical, wall 0.5-1.5 µm, essentially colorless, smooth, pore of upper cell apical, of lower cell at septum; pedicels colorless, to 50 µm long but usually shorter; spores germinate without dormancy.

Hosts and distribution: species of *Wedelia* and *Zexmenia*: northeastern Mexico to Guatemala and Honduras.

Type: on *Z. leucactis* Blake, San Felipe, Guatemala, Holway No. 693 (PUR 34492).

Parmelee (25) suggests that the species "... is in the process of losing its uredinia". Only by careful inspection of slides is one apt to see the spores.

2. *PUCCINIA HOLWAYULA* H. S. Jack. Mycologia 24:163. 1932.

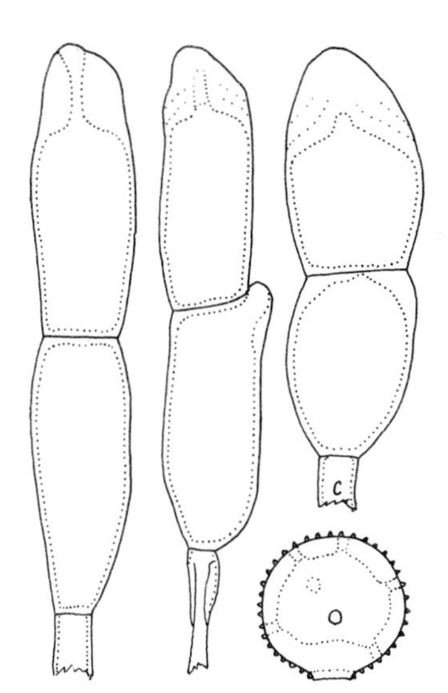

Spermogonia on adaxial leaf surface. Aecia mostly on adaxial surface on small galls, peridium short cylindric, lacerate, pale brownish or whitish; spores 25-32 x 21-27 µm, globoid or broadly ellipsoid, wall 2-3 µm thick, sometimes thicker above, verrucose with columnar warts, yellowish. Uredinia on abaxial surface, cinnamon brown, with peripheral, colorless, thin walled, mostly 1 septate, cylindrical to capitate paraphyses to about 50 µm long; spores (21-)23-26(-28) x (24-)26-29 µm, mostly slightly depressed globoid, wall (3-)3.5-4(-5) µm thick including verrucae, sordid brown, verrucose echinulate with cones spaced 1.5-2(-3) µm, connected basally by fine lines (ridges?) in a netlike pattern, pores 6-10, scattered, obscure, without caps. Telia on abaxial surface, exposed, about chestnut brown becoming gray from germination, compact; spores (48-)60-95(-120) x (15-)19-26(-29) µm, cylindrical or cylindrical clavate, wall (0.5-)1(-1.5) µm thick at sides, 8-15 µm at apex, clear golden brown, the apical thickening progressively paler, smooth; pedicels colorless, to 60 µm long, often shorter.

Type: on *Oyedaea verbesinoides* DC. (as *O. acuminata* (Benth.) Benth. & Hook.), San José, Costa Rica, Holway No. 356 (PUR 34498). Two other collections, both from Costa Rica, are known.

This fungus is listed in N. Amer. Flora (1) as *P. oyedaeae* Mayor, a species not known to occur in North America.

3. *PUCCINIA SEORSA* H. S. Jack. & Holw. in Jackson, Mycologia 24:103-104. 1932.

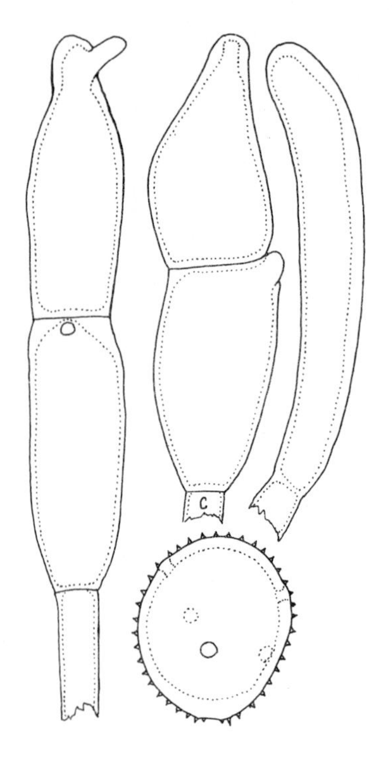

Spermogonia on adaxial leaf surface in small groups. Aecia on abaxial surface, few in a group, peridium short cylindrical, yellowish; spores (28-)30-36(-41) x (24-)27-30(-33) µm broadly ellipsoid, globoid or sometimes lemon shape, wall 2-2.5(-3) µm thick, colorless or pale yellowish, finely verrucose with rod like verrucae, discrete or fusing in striae. Uredinia on abaxial surface, chestnut brown, with long, peripheral, incurved, dorsally thick walled, pale golden paraphyses, 12-18 µm wide; spores (28-)30-35(-38) x (26-)29-33(-34) µm, mostly globoid, wall 2-2.5(-3) µm thick, nearly chestnut brown, echinulate, pores (3)4 or 5(6), scattered or sometimes nearly equatorial, with slight or no caps. Telia on abaxial surface, exposed, chestnut brown becoming grayish from germination, relatively compact, with paraphyses as in uredinia; spores (50-)70-115(-122) x (17-)20-25(-31) µm, mostly fusiform cylindrical, wall 1 µm thick at sides, 2-4 µm thick apically, pale chestnut brown or deep golden brown, smooth, pore of upper cell apical, of lower cell at the septum; pedicels colorless, thin walled and collapsing, 115 µm long.

Hosts and distribution: *Piptocarpha chontalensis* Baker: Guatemala (but needs confirmation because only aecia are recorded). Also in Brazil.

Type: on *Piptocarpha axillaris* (Less.) Baker, Taipas, São Paulo, Brazil, Holway No. 1540 (PUR F7980).

4a. *PUCCINIA ARTHURIANA* H. S. Jack. Bot. Gaz. 65:295. 1918 var. *ARTHURIANA*.
Argomyces vernoniae Arth. N. Amer. Flora 8(3):218. 1912, not *Puccinia vernoniae* Schw. 1832.

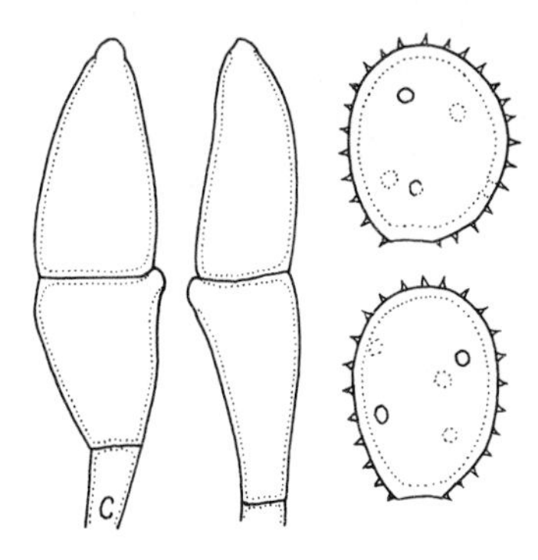

Spermogonia on adaxial leaf surface, few. Aecia on abaxial surface, uredinioid, in small groups; spores (20-)22-26(-29) x (18-)20-23(-25) µm, mostly globoid or broadly ellipsoid, wall 1.5-2(-2.5) µm thick, pale yellowish, uniformly echinulate, pores obscure but apparently (4) 5 or 6, scattered or bizonate, rarely equatorial. Uredinia not seen, perhaps lacking. Telia on abaxial surface, exposed, pulvinate, usually with the aecia, cinnamon brown becoming gray from germination; spores (45-)55-70(-78) x (13-)15-20(-22) µm, mostly fusiform or narrowly ellipsoid, wall 0.5-1 µm thick, pale cinnamon brown or golden, smooth, pore apical in upper cell, at the septum in lower cell, not clearly differentiated; pedicels colorless, to 75 µm long but usually broken shorter.

Hosts and distribution: *Vernonia canescens* H.B.K.: southern Mexico to Costa Rica; also in the Caribbean Islands and South America.

Type: on *Vernonia borinquensis* Urban, Cayey, Puerto Rico, Holway No. 3 (PUR 37200).

4b. *PUCCINIA ARTHURIANA* var. *TABASCANA* Urban , Acta Univ. Carolinae Biol. 1971:16. 1973.

Urediniospores 29-36 x 27.5-33.5 µm, wall 3(-3.5) µm thick, teliospores 50.5-66 x 21.5-26.5(-29) µm; occasional colorless, clavate paraphyses occur.

Type: on *Vernonia ctenophora* Gleason, Balancán, Tab., Mexico, Matuda No. 3112 (PUR 62821).

5. *PUCCINIA BACCHARIDIS* Diet. & Holw. in Dietel, Erythea 1:250-251. 1893.

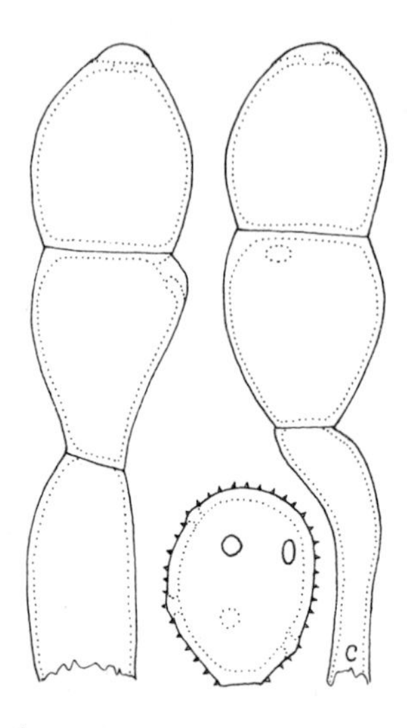

Spermogonia amphigenous. Aecia on abaxial leaf surface and on stems, in groups, without peridium, rupturing the epidermis, bright yellow but fading when dry; spores (27-)30-48 (-60) x (20-)22-26(-30) µm, varying from globoid to fusiform, mostly oblong ellipsoid, the apex often acute, wall (1.5-)2-2.5 µm thick or the apex slightly thicker especially when acute, verrucose with bead like warts usually in more or less lineal arrangement, colorless or pale yellowish. Uredinia amphigenous, yellowish to cinnamon brown; spores (28-)32-43 (-46) x (17-)20-24(-27) µm, oblong ellipsoid, ellipsoid or obovoid, wall 1.5-2(-2.5) µm thick, echinulate with fine spines spaced (1-)2(-3) µm, yellowish or golden brown, pores 5-8, scattered or tending to be bizonate. Telia mostly on abaxial leaf surface, exposed, in groups, pale cinnamon brown becoming gray from germination, compact; spores (45-)50-68(-72) x (22-)24-28(-31) µm, elongately ellipsoid or oblong ellipsoid, wall 1-1.5 µm thick at sides, 2.5-5(-6) µm over pores, about golden brown except paler over pores, smooth, pore apical in each cell; pedicel colorless, usually or often to 25 µm wide, to 140 µm long but usually 100 µm or less; germinating without dormancy.

Hosts and distribution: *Baccharis* spp., especially *B. glutinosa* Pers.: the southwestern United States to Mexico and Guatemala; also in South America.

Type: on *Baccharis viminea* DC., Pasadena, California, 1893, McClatchie No. 359 (S).

6. *PUCCINIA EVADENS* Hark. Bull. Calif. Acad. 1:34-35. 1884.

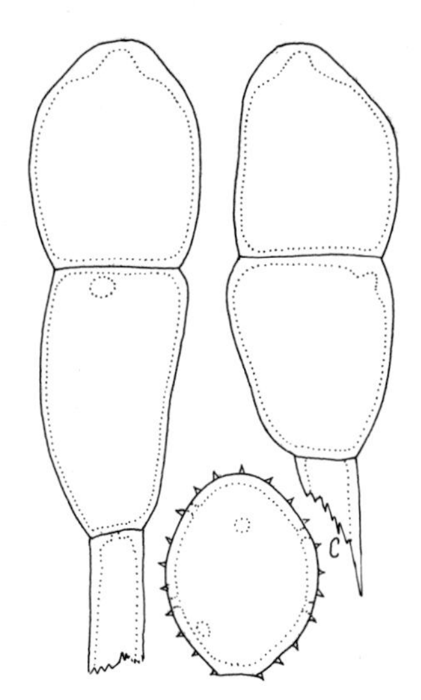

Spermogonia and aecia mostly on stems causing fusiform swellings and often witches' brooms; aecia pustulate, caeomoid but often columnar, bright yellow when fresh, tan colored when dry; spores (32-)36-55(-60) x (17-)23-25(-28) μm, variable but mostly ellipsoid or fusiformly ellipsoid, wall (1.5-) 2-3 μm thick at sides, (4-)6-12(-15) μm at apex, sometimes thickened basally, rugosely verrucose with irregular verrucae that often unite in labyrinthiform patterns, often striate apically, colorless or pale yellowish. Uredinia on abaxial leaf surface and on stems and branchlets, colored like the aecia; spores (25-)30-38(-42) x (19-)22-27 (-30) μm, ellipsoid, broadly ellipsoid or obovoid, wall 1.5-2 μm thick, colorless or pale yellowish, echinulate, pores scattered or tending to be bizonate, obscure, about 6. Telia on abaxial surface and on branchlets, exposed, pulvinate, dark cinnamon brown becoming gray from germination; spores (48-)54-74(-80) x (22-)26-30(-33) μm, mostly oblong ellipsoid, wall (1-)1.5(-2) μm thick at sides, 3-5(-6) μm at apex, about golden brown, smooth, pore of upper cell apical, of lower cell at septum; pedicels colorless, to 160 μm long.

Hosts and distribution: *Baccharis* spp.: southern United States south to southern Mexico; also in South America;

Type: on *Baccharis pilularis* DC., San Francisco, California, Nov. 1883, Harkness No. 3384 (BPI; probable isotypes Ell. & Ever. N. Amer. F. 1843).

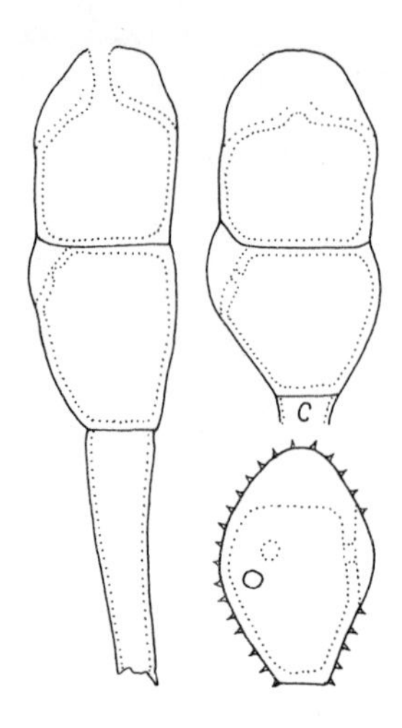

7. *PUCCINIA BACCHARIDIS-MULTIFLORAE* Diet. & Holw. in Holway, Bot. Gaz. 31:331. 1901.

Spermogonia on the adaxial leaf surface, few in a group. Aecia on adaxial leaf surface in small groups with the spermogonia, uredinoid, pale yellowish brown; spores similar to urediniospores. Uredinia on abaxial surface, cinnamon brown; spores (27-)32-38(-42) x (21-)23-27(-29) μm, mostly obovoid, wall 1.5-2 μm thick at sides, abruptly thickened apically (5-)6-8(-9) μm, echinulate, golden to cinnamon brown, pores 3, equatorial, with caps. Telia on abaxial surface, exposed, about cinnamon brown becoming gray from germination, relatively compact; spores (40-)44-52(-58) x (20-)22-27(-29) μm, mostly more or less oblong ellipsoid, wall 1(-1.5) μm thick at sides, (6-)7-9(-13) μm over pores, golden brown but the thickened areas paler, smooth, pore apical in each cell; pedicel colorless, to 75 μm long but usually shorter.

Hosts and distribution: *Baccharis* spp.: southern Mexico to Guatemala; also reported in Brazil.

Type: on *Baccharis multiflora* H.B.K., Amecameca, Mex., Mexico, Holway No. 3757 (S; isotype PUR 37519).

Teliospores are known only in Mexican specimens.

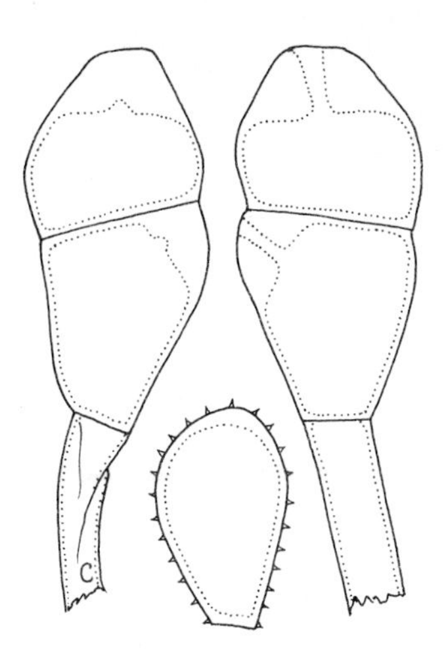

8. *PUCCINIA ENIXA* Cumm. Bull. Torrey Bot. Club 79:220. 1952.

Spermogonia on adaxial leaf surface, few in groups. Aecia amphigenous or mostly on abaxial surface, without peridium, opening by irregular rupture of the epidermis, yellowish (dry); spores 35-48 x 22-32 µm, oblong ellipsoid, ellipsoid or broadly ellipsoid, wall 3-4 µm thick except apex which is 8-16 µm, tuberculate with irregularly shaped, flattopped warts or short ridges tending to be striate, colorless. Uredinia on abaxial surface, yellowish; spores (26-)33-42(-46) x (16-)18-24 µm, mostly ellipsoid or obovoid, wall 1.5(-2) µm thick except apex usually is about 3 µm thick, echinulate, colorless, pores obscure. Telia on abaxial surface, exposed, yellowish becoming gray from germination, compact; spores (43-)50-68 x 27-33(-38) µm, mostly more or less oblong or obovoid, wall 1 µm thick at sides, 8-12 µm at the pores which are apical in each cell, colorless, smooth; pedicel colorless, wide, to 82 µm long.

Type: on *Baccharis braunii* (Polak.) Standl. (as *Baccharis* sp.), Turrialba, Costa Rica, Müller No. 1974 (PUR 51956). Two other Costa Rican collections are known.

9. *PUCCINIA EXORNATA* Arth. Bull. Torrey Bot. Club 38:370. 1911.

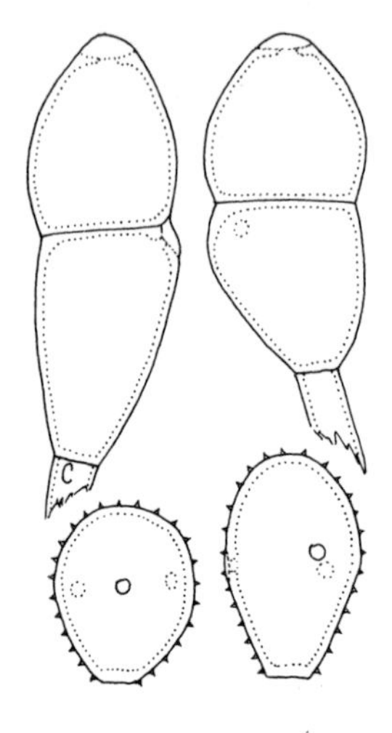

Spermogonia amphigenous in small groups. Aecia yellowish when dry, bright orange when fresh, without a peridium, opening by irregular rupture of the host, on abaxial leaf surface; spores (25-)27-32(-35) x (22-)24-27 μm, broadly ellipsoid or obovoid, wall 2-3(-5) μm thick or thicker apically, colorless, ornamented with warts or more typically with ridges of varying lengths arranged longitudinally or spirally. Uredinia on abaxial surface, yellowish when dry, bright yellow when fresh; spores (24-)26-34(-40) x (19-) 21-24 μm, mostly broadly ellipsoid or obovoid, wall 1-1.5 μm thick, pale yellowish, echinulate, pores 3, equatorial, obscure. Telia on abaxial surface, exposed, cinnamon brown becoming gray from germination, compact; spores variable in size in different collections, (40-)44-55(-66;-74) x (17-)20-26(-29) μm, mostly elongately ellipsoid or more or less oblong ellipsoid, wall 1(-1.5) μm thick at sides, pale golden brown, (2-)2.5-4(-4.5) μm over pores by a nearly colorless, low umbo, smooth, pore apical in each cell; pedicel colorless, to 90 μm long but usually shorter.

Hosts and distribution: *B. trinervis* (Lamb.) Pers. and var. *rhexioides* (H.B.K.) Baker: Veracruz, Mexico to Costa Rica; also in Brazil.

Type: on *B. thesioides* H.B.K. (= error; the host is *B. trinervis*), Guatemala, Dept. Guatemala, Kellerman No. 5368 (PUR 33882).

10. *PUCCINIA ERRATICA* H. S. Jack. & Holw. in Jackson, Bot. Gaz. 65:294. 1918.
Dietelia vernoniae Arth. Bot. Gaz. 40:198. 1905, not *Puccinia vernoniae* Schweinitz 1832.

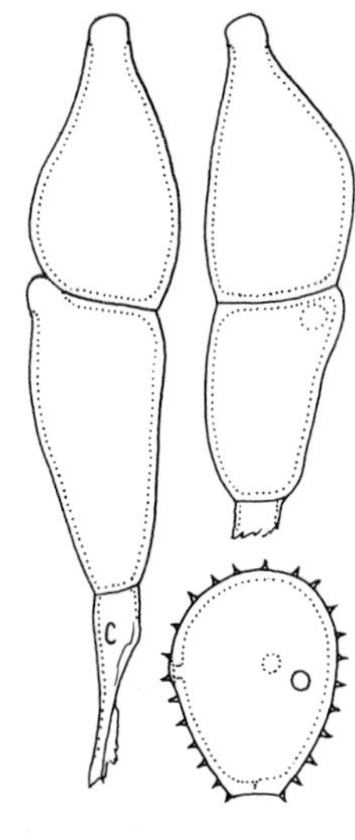

Spermogonia on adaxial leaf surface in close groups. Aecia on abaxial surface opposite the spermogonia, blister like, pale yellowish when dry, peridium doubtful, if present the cells readily separable and similar to aeciospores; spores (29-)32-38(-40) x (23-)24-28(-30) μm, variable but mostly broadly ellipsoid or ovoid, wall 2-2.5 μm thick, often slightly thicker at ends, essentially colorless, verrucose with rod like verrucae of various shapes, these discrete or united, or forming ridges at ends of spore. Uredinia on abaxial surface, pale cinnamon brown; spores (27-)29-33(-35) x (20-)23-27(-28) μm, mostly broadly ellipsoid or obovoid, wall 1(-1.5) μm thick, yellowish or pale golden, uniformly echinulate, pores obscure, 3 or 4, equatorial, with small caps. Telia on abaxial surface, exposed, dark cinnamon or chestnut brown becoming gray from germination, compact, spores (46-)55-85(-100) x (11-)16-22 (-24) μm, cylindrically fusiform or narrowly ellipsoid, wall 1 μm thick, about cinnamon brown or golden, colorless at the pores, which are apical, smooth; pedicels colorless, to 110 μm long but commonly broken short.

Hosts and distribution: *Vernonia tortuosa* (L.) Blake, *V.* sp.: southern Mexico to El Salvador and Guatemala.

Type: on *Vernonia schiediana* Less. (=*V. tortuosa*), Jalapa, Veracruz, Mexico, Holway No. 3111 (PUR 33779).

Urban (30) consigned Guatemalan records of *P. insulana* (Arth.) H. S. Jack. to this species.

11. *PUCCINIA ALIA* H. S. Jack. & Holw. in Jackson, Mycologia 24:137. 1932.

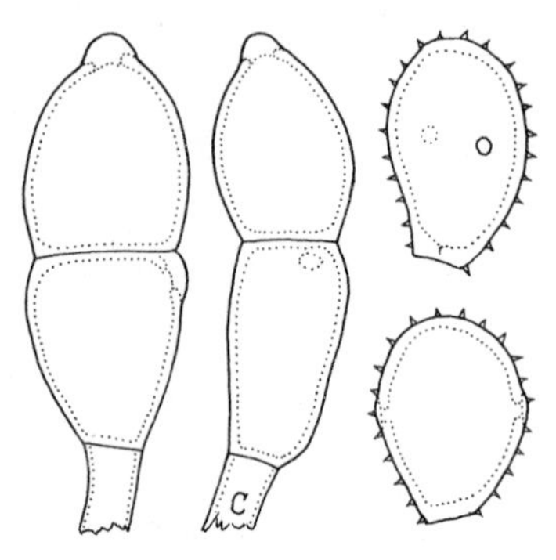

Spermogonia amphigenous. Aecia amphigenous, deep-seated, without peridium, opening by a pore in the epidermis; spores (23-)26-35(-40) x (16-)18-23(-25) µm, but mostly ellipsoid or obovoid, wall 2-2.5(-3) µm thick at sides, often slightly thicker at one or both ends, echinulate with spines spaced (2-)3-5 µm. Uredinia on abaxial leaf surface, rather slowly exposed, pale yellowish (dry), probably bright yellow when fresh; spores (22-)26-35(-40) x (17-)18-22(-24) µm, mostly obovoid or ellipsoid, wall 1.5 (-2) µm thick, sometimes slightly thicker apically, yellowish or nearly colorless, echinulate, pores obscure, equatorial, probably 2, perhaps sometimes 3. Telia on abaxial surface, exposed, about cinnamon brown, becoming gray from germination, compact; spores 44-60(-68) x (15-)17-22(-23) µm, mostly elongately ellipsoid, wall 1 µm thick at sides, golden brown, 2-4 µm thick over pores by a small, nearly colorless umbo, smooth, pore of upper cell apical, of lower cell at septum; pedicels colorless, to 60 µm long but often shorter.

Hosts and distribution: *Baccharis trinervis* (Lam.) Pers.: Guatemala; also in Brazil.

Type: on *B. trinervis*, Rio de Janeiro, Brazil, Holway No. 1007 (PUR F8156; isotypes Reliq. Holw. 599).

One specimen is known in North America (7) and, because it bears only aecia and uredinia, its identity is open to question.

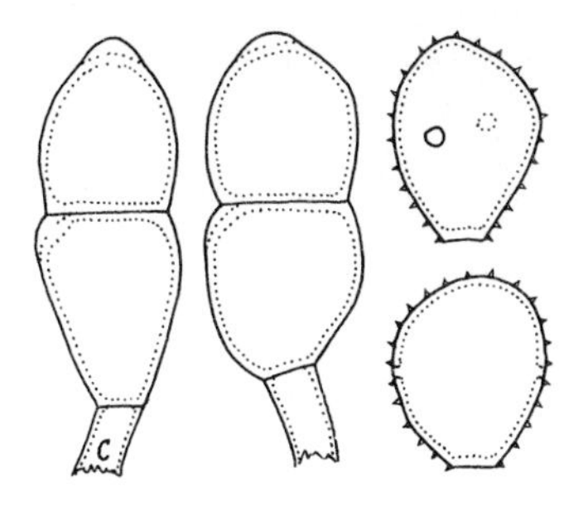

12. *PUCCINIA OAXACANA* Diet. & Holw. in Holway, Bot. Gaz. 31: 331. 1901.

Spermogonia few in a group, amphigenous. Aecia amphigenous in small groups or singly, or on hypertrophied twigs, peridium whitish, short cylindrical; spores (20-)25-35(-40; -44) x (14-)17-23(-25) µm, variable but mostly more or less ellipsoid, wall including warts 2-3 µm thick or sometimes thicker apically, verrucose with warts about 1-2 µm high and 1-3 µm wide, the outline irregular, the sides parallel, the warts discrete or sometimes pseudoreticulately joined. Uredinia on abaxial leaf surface, not abundant, pale brownish; spores (20-)23-28(-30) x (17-)19-21(-23) µm, mostly obovoid, wall 0.5-1 µm thick, pale brownish or golden, echinulate, pores 2, equatorial, obscure with slight or no caps. Telia on abaxial surface, exposed, compact, about cinnamon brown becoming gray from germination; spores (32-)35-48(-53) x (18-)19-24(-26) µm, mostly ellipsoid or elongately obovoid, wall (0.5)1(-1.5) µm thick at sides, about golden brown, 2.5-3.5(-4) µm thick and paler over pores, pore apical in each cell, smooth; pedicels colorless, to 90 µm long.

Hosts and distribution: *Archibaccharis torquis* Blake: southern Mexico to Costa Rica.

Type: on *Baccharis hirtella* (now considered to be *A. torquis*), Oaxaca, Oax., Mexico, Holway No. 3673 (S: isotype PUR 33888).

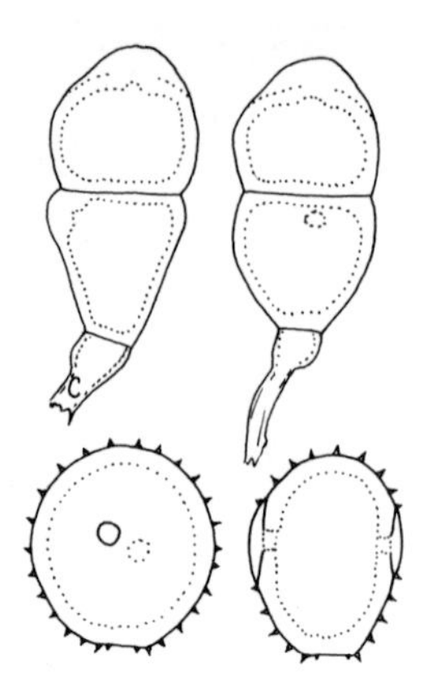

13. *PUCCINIA TRIXITIS* Arth. N. Amer. Flora 7:604. 1922.

Spermogonia and aecia unknown. Uredinia on abaxial leaf surface, dark chestnut brown; spores (24-)26-32(-35) x (20-)21-26(-28) µm, mostly broadly ellipsoid, broadly obovoid or globoid, wall (1.5-)2-3 µm thick, near chestnut brown, echinulate except over the pores, pores 2, equatorial in the somewhat flattened sides, with conspicuous caps. Telia on abaxial surface, exposed, about chestnut brown, compact; spores (24-)28-37(-42) x (17-)18-20(-22) µm, mostly ellipsoid or obovoid, wall (1-)1.5-2 µm thick at sides, golden brown, (3.5-)4.5-6 µm over the pores as a pale, differentiated umbo, smooth, pore apical in each cell; pedicel colorless, to 50 µm long but usually broken shorter.

Hosts and distribution: *Trixis radialis* (L.) Kuntze: Guatemala.

Lectotype: on *Trixis frutescens* (=*T. radialis*), Antigua, Guatemala, Holway No. 71 (PUR 42673).

14. *PUCCINIA OCELLIFERA* Cumm. Mycotaxon 5:405. 1977.
Puccinia biocellata Vest. Microm. rar. sel. Nos. 1267, 1368. 1908, illegit.
Puccinia plucheae Arth. Bull. Torrey Bot. Club 49:194. 1922, illegit.
Puccinia biocellata Cumm. Mycologia 48:606. 1956, illegit.

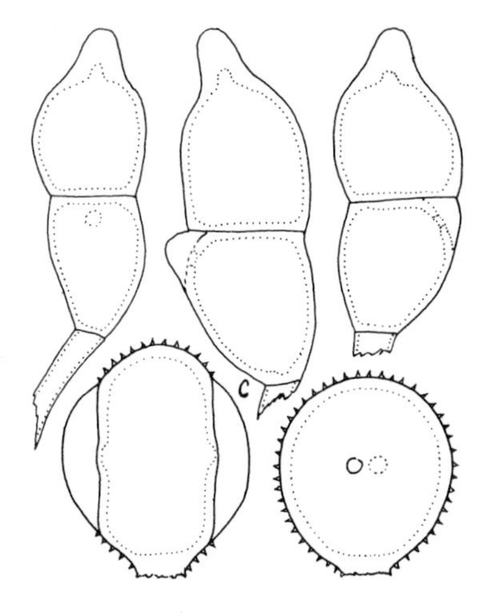

Spermogonia not seen. Aecia on abaxial surface in small groups, peridium fragile; spores 13-17 µm diam, angularly globoid, wall 1 µm thick, colorless, minutely verrucose. Uredinia amphigenous, dark brown; spores (24-)28-33(-35) x (22-)24-27 µm, obovoid or broadly ellipsoid, wall 1.5-2 µm thick, echinulate except over pores, cinnamon brown, pores 2, equatorial in the strongly flattened sides of spore, overlaid by large caps. Telia on abaxial leaf surface, exposed, compact, about cinnamon brown becoming gray from germination; spores (40-)43-55(-67) x (16-)18-24(-26) µm, ellipsoid or oblong ellipsoid, wall 1(-1.5) µm thick at sides, golden brown, 5-8 µm thick apically by an abrupt, pale, umbo, smooth, pore of upper cell apical, of lower cell at the septum; pedicels colorless, to 65 µm long but often broken short.

Hosts and distribution: *Pluchea* spp.: Florida, U.S.A., Veracruz and Baja California, Mexico and in Guatemala; also in South America and Puerto Rico.

Type: on *Pluchea fastigiata* Griseb., Prov. Jujuy, Argentina, Vestergren, (holotype in BPI bound set; isotypes Vest. Microm. Rar. Sel. 1368).

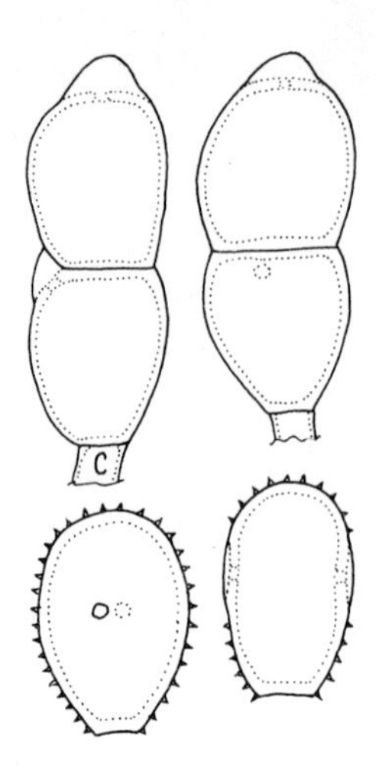

15. *PUCCINIA GUARDIOLAE* Diet. & Holw. in Holway, Bot. Gaz. 31:334. 1901.

Spermogonia and aecia unknown. Uredinia amphigenous, about cinnamon brown; spores (24-)26-30(-33) x (17-)19-26 (-27) µm, broadly ellipsoid or obovoid with pores face view, wall (1.5-)2-2.5 µm thick, dark cinnamon brown, echinulate except around the pores, pores 2, equatorial in flattened sides. Telia on abaxial leaf surface, exposed, compact, pale cinnamon brown becoming gray from germination; spores (37-)42-60(-64) x (16-)19-23(-26) µm, mostly elongately ellipsoid, wall (1-)1.5(-2) µm thick at sides, pale golden brown, (3.5-)4-6(-7) µm over pores as pale umbos which disappear with germination, smooth, pore of upper cell apical, of lower cell at septum; pedicels colorless to 80 µm long but usually broken shorter; the spores germinate without dormancy.

Hosts and distribution: *Guardiola* spp.: Nayarit and Durango to Morelos and Guerrero, Mexico.

Type: on *Guardiola mexicana* Humb. & Bonpl., Cuernavaca, Mor., Mexico, Holway No. 3513 (S; isotype PUR 42592). Not otherwise known.

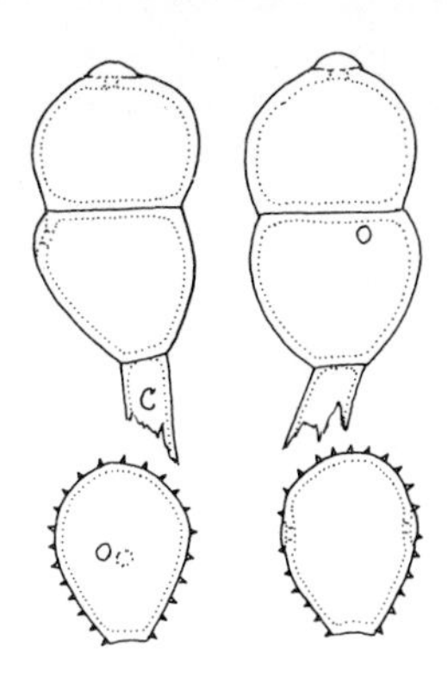

16. *PUCCINIA POTOSINA* Cumm., Brit. & Baxt. Mycologia 61: 926. 1969.

Spermogonia on adaxial leaf surface. Aecia circinately grouped around the spermogonia, uredinoid, cinnamon brown; spores (21-)24-27(-30) x (17-)19-22(-24) µm, broadly ellipsoid or obovoid, wall 1-1.5 µm thick, pale cinnamon brown or golden, echinulate, pores 2, equatorial. Uredinia similar to aecia except scattered and few; spores like the aeciospores. Telia on abaxial leaf surface, often with the aecia, exposed, dark brown; spores (30-)33-40(-45) x (18-)21-25(-28) µm, mostly ellipsoid or slightly obovoid, wall uniformly 1.5 µm thick and pale chestnut brown except for a small, colorless, papilla over each pore which disappears during germination, smooth, pore apical in upper cell, at septum in lower cell; pedicels colorless, to 60 µm long but usually broken short; the spores germinate without dormancy.

Hosts and distribution: *Eupatorium longifolium* B. L. Rob.: mountains of northeastern Mexico.

Type: west of Cd. Mante in San Luis Potosí state, Cummins No. 63-115 (PUR 61875).

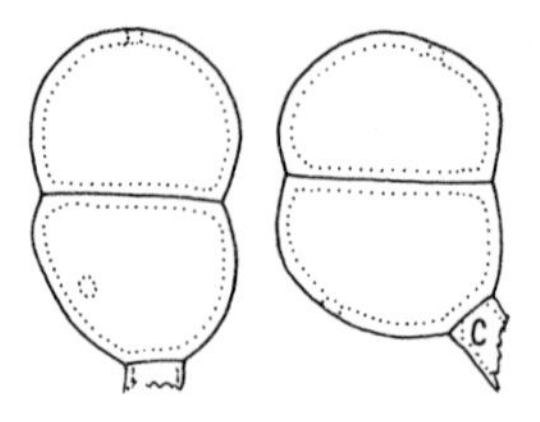

17. *PUCCINIA BRACHYTELA* H. Syd. Ann. Mycol. 23:315. 1925.

Spermogonia and aecia unknown. Uredinia not seen; urediniospores in the telia 23-26 x 19-20 µm, ellipsoid or ovate ellipsoid, wall 1-1.5 µm thick, echinulate, pale yellowish brown, pores 2, equatorial. Telia on abaxial leaf surface, about dark cinnamon brown; spores (25-)28-35(-37) x (19-)22-26(-29) µm, ellipsoid, broadly ellipsoid or oblong ellipsoid, variable, wall uniformly 1-1.5 µm thick, about golden brown, smooth, germinating without dormancy, pore at or near apex of upper cell, about midway in lower cell; pedicels fragile, broken near the hilum.

Type: on *Otopappus verbesinoides* Benth., near Grecia, Costa Rica, Sydow (holotype destroyed; isotype PUR 48933 = Sydow F. exot. exsic. No. 565).

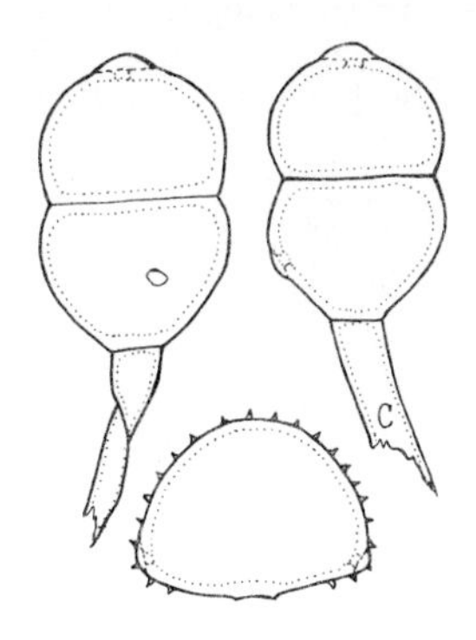

18. *PUCCINIA INERMIS* H. S. Jack. & Holw. in Arthur, Mycologia 10:142. 1918.

Spermogonia and aecia unknown. Uredinia mostly on abaxial leaf surface, cinnamon brown; spores 30-34(-36) µm between the pores, (21-)22-25(-27) µm hilum to apex, (25-)26-29(-32) µm wide with hilum in optical axis, strongly asymmetrical, depressed ovoid with hilum basal, transversely ellipsoid with hilum in optical axis, wall 1-1.5(-2) µm thick, cinnamon brown, echinulate except around hilum, pores 2, subequatorial in ends of spore. Telia not seen; teliospores in the uredinia, (29-)31-37(-40) x (22-)24-26(-29) µm, oblong ellipsoid, wall uniformly 1.5-2 µm thick except for a low umbo over the pore, chestnut brown, smooth, pore apical in upper cell, about midway to pedicel in lower cell; pedicels colorless, broken near the spore.

Type: on *Eupatorium* sp., near Cartago, Costa Rica, Holway No. 434A (PUR 37464). Not otherwise known.

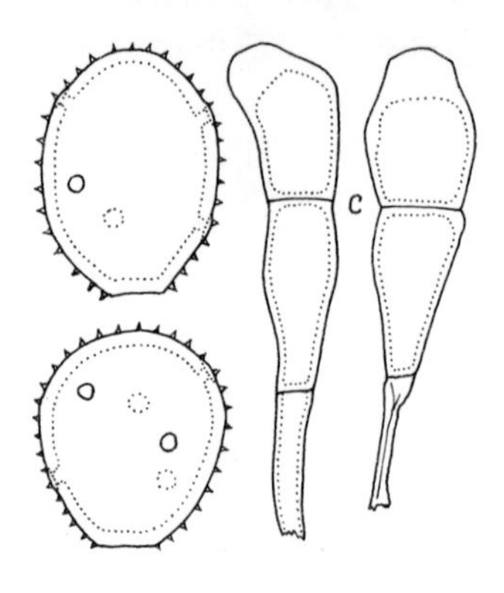

19. *PUCCINIA PISTORICA* Arth. Bull. Torrey Bot. Club 38:372. 1911.

Spermogonia and aecia unknown. Uredinia on abaxial leaf surface, large, pale yellow when old, probably bright yellow when fresh, long covered by epidermis; spores (28-) 30-38(-42) x (24-)27-30(-32) µm, mostly broadly ellipsoid, wall (1.5-)2(-2.5) µm thick, pale yellowish, echinulate, pores scattered, 5 or 6, obscure. Telia amphigenous, blackish brown, covered by the epidermis, divided into locules by golden to chestnut brown, stromatic paraphyses; spores (37-)40-53(-56) x (11-)13-17(-19) µm, variable but mostly narrowly oblong ellipsoid or more or less cylindrical, wall 1-1.5 µm thick at sides, (5-)7-10(-12) µm apically, clear chestnut brown, smooth; pore apical in each cell but obscure; pedicels yellowish, to about 25 µm long.

Type: on *Baccharis glomeruliflora* Pers., Mt. Dora, Florida (PUR) 33873). Not otherwise known.

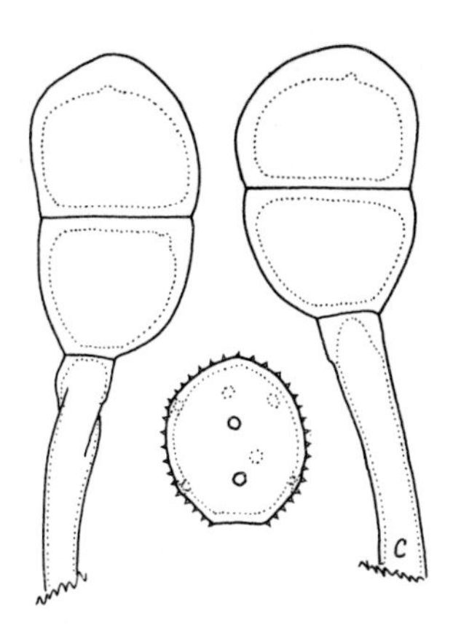

20. *PUCCINIA POROPHYLLI* P. Henn. Hedwigia 39(Beibl.):153. 1900.

Spermogonia and delicate peridiate aecia have been reported (Davidson, Mycologia 24:227. 1932) but without a description. Uredinia mostly on the abaxial leaf surface, pale cinnamon brown; spores (20-)22-25(-27) x (17-)19-22 µm, mostly broadly ellipsoid or nearly globoid, wall 1.5 µm thick, pale yellowish, echinulate, pores about 8, scattered, with slight caps, obscure. Telia mostly abaxial and on stems, exposed, blackish brown, compact; spores (32-)35-44 (-48) x (22-)24-27(-30) µm, mostly ellipsoid or obovoid, wall 2-3 µm thick at sides, (3.5-)4-7 µm at apex, uniformly deep chestnut brown, smooth, pore in each cell apical; pedicels brownish near hilum, colorless below, to 130 µm long but usually broken shorter.

Hosts and distribution: *Porophyllum* spp.: Baja California Sur to Jalisco and Chiapis, Mexico and in Guatemala; also in South America.

Type: on *Porophyllum ellipticum* Cass., Caracas, Venezuela, Urban No. 255 (B).

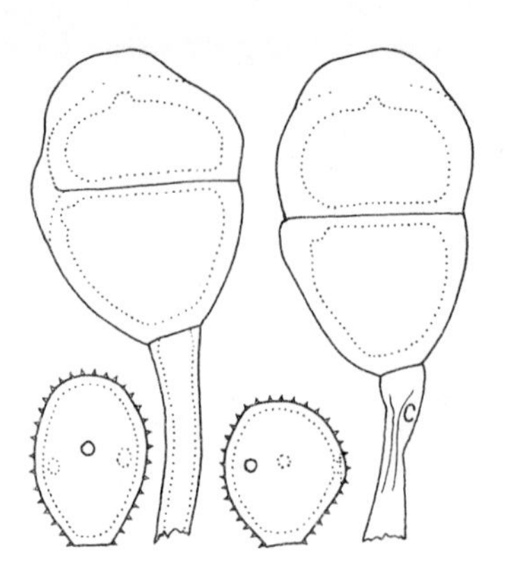

21. *PUCCINIA SPHENICA* Arth. Bull. Torrey Bot. Club 38:371. 1911.

Spermogonia and aecia unknown. Uredinia mostly on abaxial leaf surface, pale yellowish when old, probably bright yellow when fresh; spores (21-)23-26(-28) x (17-)18-22 µm, mostly obovoid, wall 1.5(-2) µm thick, yellowish or essentially colorless, echinulate, pores 3(4?), equatorial, without caps. Telia on abaxial surface, exposed, dark cinnamon brown, compact; spores (36-)40-46(-48) x (26-)28-34 (-36) µm, mostly broadly ellipsoid, wall (2-)2.5-4 µm thick at sides, clear chestnut brown or deep golden brown, (6-)7-9 (-11) µm thick apically and progressively paler but not as a clearly differentiated umbo, smooth, pore apical in each cell; pedicels colorless, mostly thin walled, to 90 µm long but usually shorter.

Type: *Baccharis sordescens* DC., Cuernavaca, Mor., Mexico, Holway No. 5266 (PUR 33874). Not otherwise known.

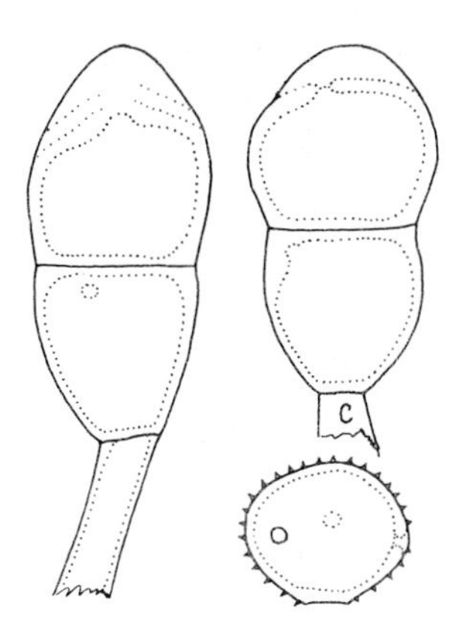

22. *PUCCINIA SPEGAZZINIANA* De-Toni in Sacc. Syll. Fung. 7: 644. 1888.

Spermogonia and aecia unknown. Uredinia on abaxial leaf surface, cinnamon brown; spores 19-22 µm high, 23-26 µm wide, slightly transversely broadly ellipsoid, wall 1-1.5 µm thick, cinnamon brown, uniformly echinulate but more sparsely so basally, pores 3 or 4, equatorial or slightly subequatorial. Telia on abaxial surface, exposed, chocolate brown, pulvinate; spores (41-)45-64(-72) x (23-)24-28(-30) µm, mostly obovoid, sometimes nearly ellipsoid, wall (1-)1.5-2 (-3) µm thick at sides, (6-)7-10(-11) µm at apex, clear chestnut or deep golden brown, smooth, the apical thickening pale and differentiated, pore of upper cell apical, of lower cell at septum; pedicel colorless, to 70 µm long, often broken short.

Hosts and distribution: *Eleutheranthera ruderalis* (Sw.) Sch., *Wedelia acapulcensis* H.B.K., *Xexmenia hispida* (H.B.K.) A. Gray: Honduras, Guatemala, and El Salvador; also in South America.

Type: on *Aspilia montevidensis* (Spreng.) O. Kuntze (as *Verbesina m.*), near Boca del Riachuela, Argentina, Schnyder (LPS).

Central American hosts, previously assigned to *Puccinia subaquila* H. S. Jack., are included here in the absence of telia.

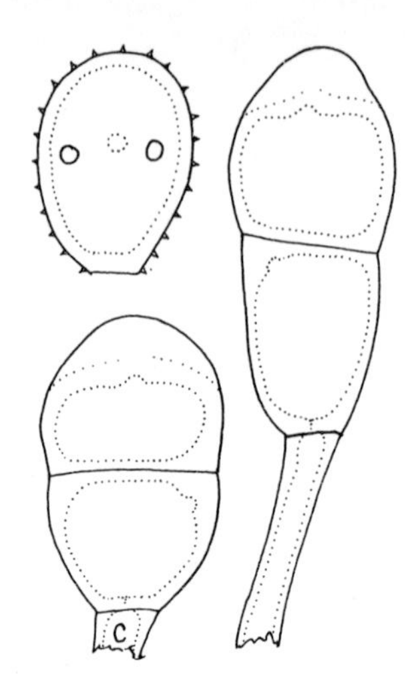

23. *PUCCINIA SIMILIS* Ellis & Ever. Bull. Torr. Bot. Club 25: 508. 1898.
Puccinia seriphidii Fahr. Ann. Mycol. 39:182. 1941.
Puccinia sphaeromeriae Fahr. Ann. Mycol. 39:182. 1941.

Spermogonia and aecia unknown. Uredinia amphigenous, cinnamon brown; spores (26-)28-35(-38) x (20-)23-26(-29) µm, broadly ellipsoid or obovoid, wall 1.5-2(-2.5) µm thick, pale cinnamon or golden brown, echinulate except around pores, pores 3, equatorial, with conspicuous caps. Telia amphigenous and on stems, exposed, blackish brown, compact; spores (40-)44-56(-60) x (20-)23-28(-31) µm, mostly oblong ellipsoid, wall (1-)1.5-2(-2.5) µm thick at sides, deep golden or chestnut brown, (5-)6-9(-12) µm at apex and progressively paler but not as a defined umbo, smooth, pore apical in upper cell, at septum in lower cell; pedicels nearly colorless, to 175 µm long.

Hosts and distribution: *Artemisia* spp., especially *A. tridentata* Nutt., *Tanacetum* spp.: North Dakota to New Mexico and the Pacific Coast.

Type: on *Artemisia tridentata*, Albany County, Wyoming, Elias Nelson No. 3309 (NY).

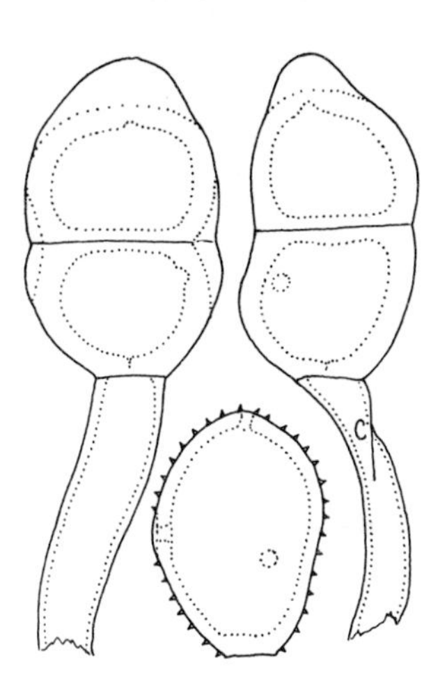

24. *PUCCINIA PARTHENII* Arth. Bull. Torrey Bot. Club 37:570. 1910.

Spermogonia and aecia unknown. Uredinia amphigenous, dark cinnamon brown; spores (24-)28-40(-50) x (18-)22-28 (-30) µm, typically broadly ellipsoid or obovoid but variable, often misshapen, wall (1.5-)2-2.5 µm thick, about cinnamon brown, echinulate except for a small area over pores, pores 1-4 variously distributed but commonly 2 in the equator or 2 equatorial and one apical, with indistinct caps. Telia amphigenous, blackish brown, exposed, compact or rather pulverulent; spores (40-)44-56(-60) x (24-)27-35(-40) µm, variable but mostly broadly ellipsoid, wall (2.5-)3.5-4(-5) µm thick at sides, chestnut brown, 8-11(-13) µm over pores as defined umbos, smooth, pore apical in upper cell, at or near septum in lower cell; pedicels colorless, to 26 µm wide, 85 µm long; 1 celled spores frequent.

Hosts and distribution: *Parthenium argentatum* Gray, *P. incanum* H.B.K.: southern Texas to central Mexico; also in South America.

Type: on *P. argentatum*, Mazapil, Zac., Mexico, Lloyd (PUR 42602).

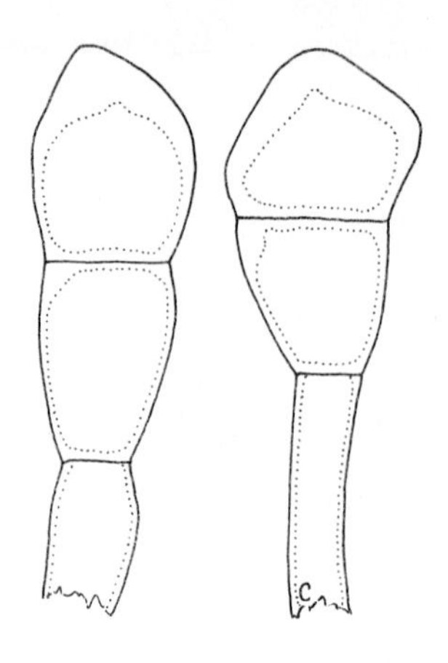

25. *PUCCINIA AXINIPHYLLI* Arth. Bot. Gaz. 40:201. 1905.

Spermogonia and aecia unknown. Uredinia on abaxial leaf surface, yellowish, tardily rupturing the epidermis; spores 24-31 x 22-24 μm, obovoid, ellipsoid or globoid, wall 1-1.5 μm thick, pale yellowish, echinulate with spines spaced 2-3.5 μm, pores obscure, perhaps equatorial. Telia on abaxial surface, tardily exposed, blackish, compact; spores (40-)45-62(-65) x (19-)23-30(-33) μm, variable but mostly oblong ellipsoid or obovoid, wall 1.5-2(-2.5) μm thick at sides, progressively thicker apically to (4-)6-10 (-14) μm at apex of upper cell, clear chestnut or golden brown, the apex progressively paler externally, smooth, pore apical in each cell; pedicel nearly colorless, often broad, to 70 μm long but usually broken shorter.

Hosts and distribution: *Axiniphyllum tomentosum* Benth.: southern Mexico.

Type: Oaxaca, Oax., Holway No. 3710 (PUR 42572); isotypes Barth. N. Amer. Ured. 1529). One other collection is known.

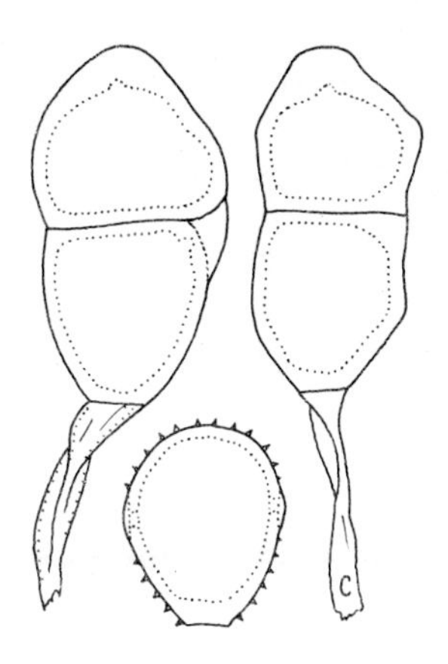

26. *PUCCINIA IRREGULARIS* Diet. Hedwigia 36:33. 1897 (Feb.), not Ellis & Tracy, June 1897.

Spermogonia and aecia unknown. Uredinia on abaxial leaf surface, pale cinnamon brown; spores (21-)24-29(-32) x (17-)20-24(-27) µm, mostly obovoid or ellipsoid, wall (1-) 1.5-2(-2.5) µm thick, about cinnamon brown, echinulate except a small area around the pores, pores 2, equatorial in scarcely or not flattened sides. Telia on abaxial leaf surface, covered by the epidermis but tardily exposed, blackish brown; spores (35-)40-53(-57) x (18-)22-29(-32) µm, irregular but mostly ellipsoid or elongately obovoid, wall 1.5-2.5 (-3) µm thick at sides, (3-)4-7(-8) µm at apex, clear chestnut brown or slightly paler apically, smooth or minutely punctate, pore apical in each cell; pedicels golden, to 70 µm long but usually less than 55 µm.

Hosts and distribution: *Verbesina* spp.: central Mexico to Nicaragua; also in South America.

Type: on *Verbesina subcordata* DC., Serra Geral, Brazil, Ule No. 1691 (S; isotype PUR F8315).

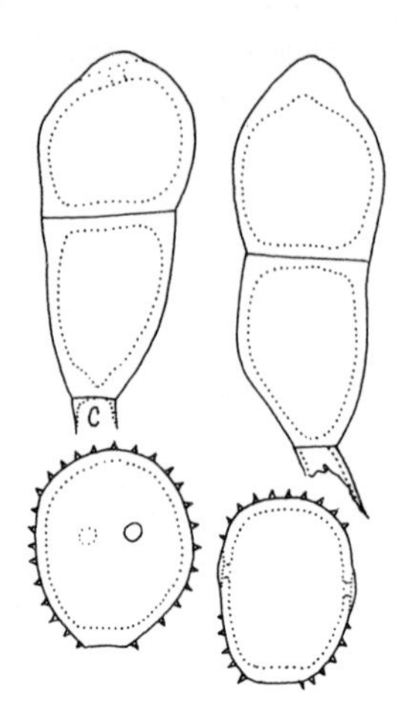

27. *PUCCINIA SENECIONICOLA* Arth. Bot. Gaz. 40:199. 1905.

Spermogonia on adaxial leaf surface. Aecia on abaxial surface, in groups, peridium white, fragile; spores 26-36 x 22-30 µm, irregularly globoid or broadly ellipsoid, wall 2-3.5 µm thick, rugose with irregular warts, these sometimes merging. Uredinia amphigenous or only on abaxial surface, cinnamon brown; spores 25-37(-40) x 21-28(-30) µm, broadly ellipsoid or obovoid, wall 1.5-2 µm thick at apex and base thinner around pores, chestnut brown, echinulate, pores 2 (rarely 3 or 4), in flattened, smooth sides. Telia mostly on abaxial surface, usually covered by the epidermis, sometimes with peripheral, stromatic paraphyses, blackish brown; spores (39-)43-66(-85) x (18-)22-28(-31) µm, mostly oblong ellipsoid or elongately obovoid, wall 1.5-2.5(-4) µm thick at sides, (4.5-)5.5-8(-10) µm at apex, from pale golden to chestnut brown, the apical thickening paler externally, smooth; pedicels colorless or yellowish, to 55 µm long but usually broken near the hilum; 1 celled spores sometimes common.

Hosts and distribution: species of *Cacalia* and *Senecio*: Durango, Mexico south to Honduras and El Salvador.

Type: on *Senecio angulifolius* DC., Amecameca, Mex., Mexico, Holway No. 5189 (PUR 34636).

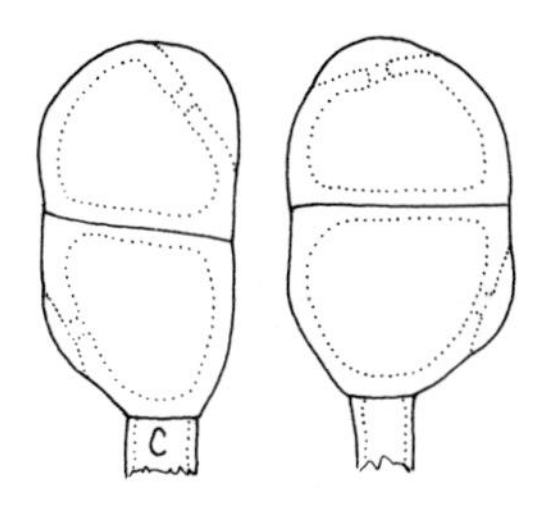

28. *PUCCINIA PUNCTOIDEA* P. Syd. & H. Syd. Monogr. Ured. 1: 182. 1902.

Spermogonia, aecia and uredinia unknown. Urediniospores in telia 24-29 µm diam, globoid, wall 1-1.5 µm thick, golden yellow, echinulate, pores 2(3?), equatorial. Telia amphigenous, exposed, dark chocolate brown, pulvinate; spores 33-43 x 21-27 µm, mostly ellipsoid or broadly ellipsoid, wall 2-2.5(-3) µm thick at sides and clear chestnut brown, 5-7 µm thick over each pore as a pale umbo, smooth, pore of upper cell apical or somewhat displaced laterally, pore of lower cell about midway to pedicel; pedicels colorless, to 75 µm long or often broken short.

Type: on *Viguiera pringlei* B. L. Rob. & Greenm., near Zapotlán, Jal., Mexico, Pringle (S). Not otherwise known.

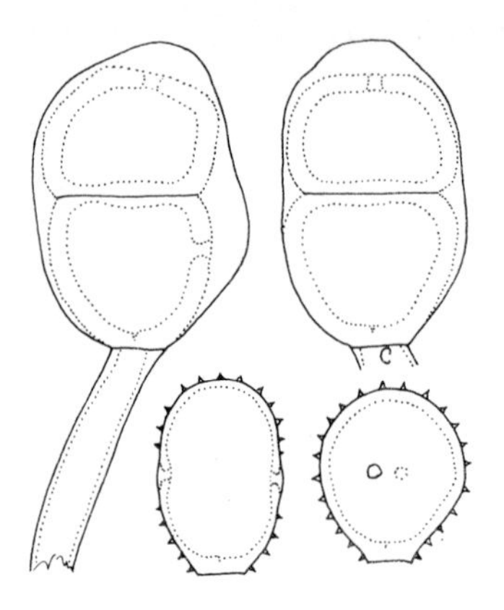

29. *PUCCINIA AFFINIS* P. Syd. & H. Syd. Monogr. Ured. 1:174. 1902 var. *AFFINIS*.
Puccinia otopappi P. Syd. & H. Syd. Monogr. Ured. 1:129. 1902.

Spermogonia and aecia unknown. Uredinia mostly on adaxial leaf surface, cinnamon brown; spores (22-)24-29(-32) x (17-)19-24(-26) µm, obovoid with pores face view, oblong or oblong obovoid with pores lateral, wall 1.5-2(-2.5) µm thick or 1-1.5 µm on pore bearing sides, cinnamon or golden brown, echinulate with spines spaced (2-)2.5-4 µm, pores 2, equatorial or slightly below, in smooth areas. Telia mostly on adaxial surface, blackish brown, pulverulent; spores (32-) 37-48(-54) x (25-)29-35(-38) µm, broadly ellipsoid, wall usually obviously bilaminate, (2-)3-4(-4.5) µm thick at sides, dark chestnut brown, (4-)7-9(-11) µm over the pores as a golden brown umbo, smooth, pore apical in upper cell, 1/4-1/2 toward hilum in lower cell; pedicel colorless, to 130 µm long but usually 75-100 µm.

Hosts and distribution: *Verbesina* spp.: central Mexico southward to Guatemala.

Type: on *Verbesina triloba* B. L. Rob. & Greenm., Oaxaca, Oax., Mexico, Holway (S; isotype PUR 34579; probable isotypes Sydow Ured. 1514; Barth. F. Columb. 3831).

The host of *P. otopappi* was given as *Otopappus alternifolius* B. L. Rob., collected by Pringle in San José Pass, San Luis Potosí. Pringle's collection is listed as the type of *O. alternifolius* which now is a synonym of *Verbesina robinsonii* (Klatt) Fern.

29b. *PUCCINIA AFFINIS* var. *TRIPOROSA* J. Parm. Can. J. Bot. 45:2283. 1967.

Urediniospores 24-33 x 21-29 µm, ellipsoid or obovoid with pores face view, mostly triangularly obovoid with pores lateral, wall 1.5 µm thick, pale cinnamon brown, uniformly echinulate, pores 2, equatorial or slightly below and 1 apical. Teliospores 37-48 x 29-37 µm, ellipsoid or broadly so, wall 4-5.5 µm thick at sides, dark chestnut brown, (7-)8-10(-11) µm over pores with pale umbos, smooth, pore of upper cell apical, of lower cell about midway to pedicel.

Type: on *Verbesina montanoifolia* B. L. Rob. & Greenm., Patzcuaro, Mich., Mexico, Holway (PUR 34573). Not otherwise known.

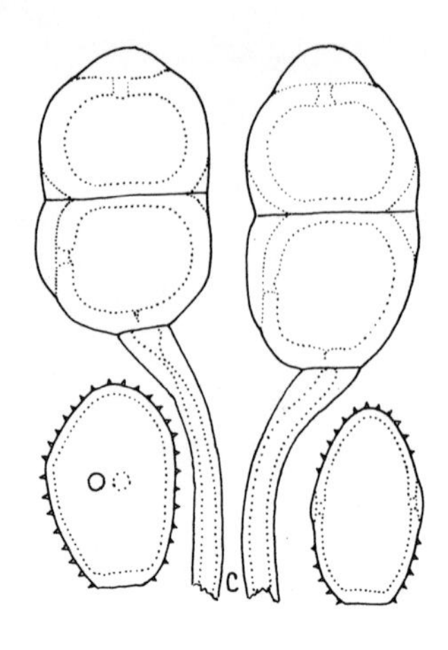

30. *PUCCINIA MASSALIS* Arth. Bull. Torrey Bot. Club 46:119. 1919.

Spermogonia amphigenous and on stems. Aecia amphigenous, mostly associated with veins, and on stems, peridium cylindrical, whitish or yellowish; spores (19-)22-30(-35) x (15-)18-22(-23) μm, from oblong to globoid, wall 1 μm thick, colorless, minutely verrucose. Uredinia amphigenous, cinnamon brown; spores (23-)26-35(-40) x (15-)19-23(-25) μm, mostly ellipsoid or obovoid with pores in face view, wall 1-1.5 μm thick, cinnamon brown, echinulate, pores 2, equatorial in smooth, flattened sides, with low caps. Telia amphigenous, exposed, blackish brown, pulvinate; spores (36-)42-48(-54) x (23-)25-30(-32) μm, mostly ellipsoid or broadly ellipsoid, wall (2.5-)3-4(-4.5) μm thick at sides, chestnut brown, 6-10(-11) μm over pores with pale umbos, smooth, pore apical in upper cell, midway to pedicel in lower cell; pedicels colorless, to 175 μm long but usually shorter.

Hosts and distribution: *Helianthus ciliaris* DC.: the Rio Grande Valley from Albuquerque, New Mexico to the Big Bend National Park, Texas.

Type: Brazito, New Mexico, W. A. Archer (PUR 34457).

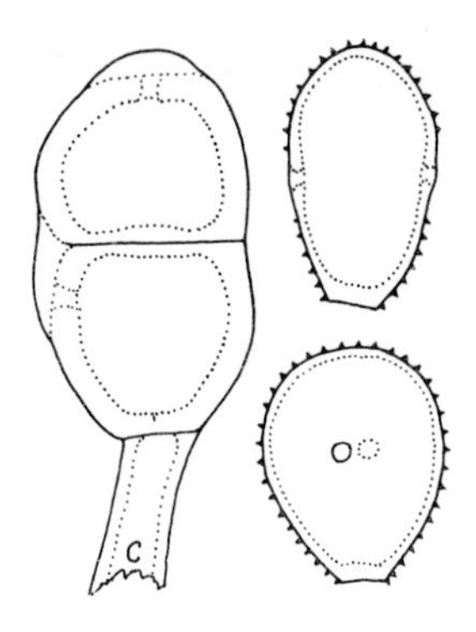

31. *PUCCINIA CHLORACAE* J. Parm. Can. J. Bot. 45:2300. 1967.

Spermogonia amphigenous. Aecia few, amphigenous, peridium cupulate, whitish; spores 22-37 x 16-22 µm, globoid or ellipsoid, wall colorless, 1 µm thick, densely verrucose. Uredinia amphigenous, cinnamon brown; spores (24-)26-32(-36) x (17-)19-24(-27) µm, obovoid with pores face view, oblong ellipsoid or elongately obovoid with pores lateral, wall 1.5-2 µm thick, cinnamon brown, echinulate with spines spaced (1-)2-3(-3.5) µm, pores 2, equatorial in smooth flattened sides. Telia amphigenous, becoming pulverulent, blackish brown; spores (32-)36-48(-53) x (24-)26-31(-33) µm, ellipsoid or broadly ellipsoid, wall (2.5-)3-4 µm thick at sides, dark chestnut brown, (6-)7-9(-10) µm over pores as a golden brown umbo, smooth, pore of upper cell apical, of lower 1/4 to 1/2 toward hilum; pedicel colorless or yellowish near hilum, to 250 µm long but often about 100 µm.

Hosts and distribution: *Viguiera deltoidea* Gray, *V. laciniata* Gray, *V. stenoloba* Blake: the southwestern United States and adjacent Mexico.

Type: on *Viguiera stenoloba*, Big Bend Natl. Park, Texas, Cummins No. 61-311 (PUR 58347).

Only uredinia are known on *V. deltoidea* and *V. laciniata*.

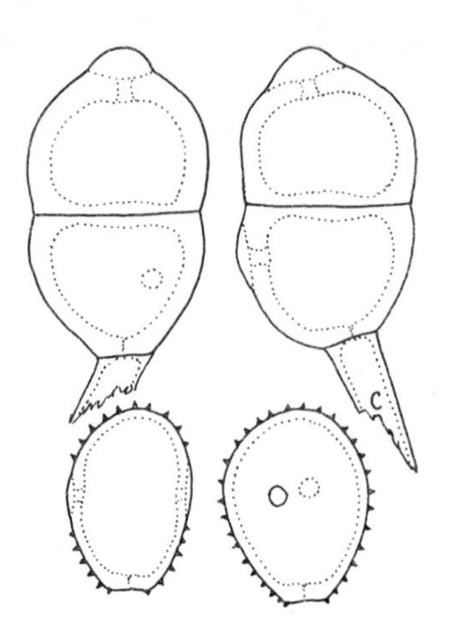

32a. *PUCCINIA KUHNIAE* Schw. Trans. Amer. Phil. Soc. 4:296. 1832 var. *KUHNIAE*.

Spermogonia amphigenous. Aecia amphigenous, uredinoid, in groups with the spermogonia, dark brown; spores (23-)25-33(-38) x 20-25(-27) μm pores face view, (15-)17-20(-22) μm wide with pores lateral, oblong ellipsoid, broadly ellipsoid or obovoid, wall (1-)1.5-2 μm thick, cinnamon brown or golden, echinulate with spines spaced 2-3(-3.5) μm, pores 2, equatorial in smooth areas of flattened sides, with slight or no caps. Uredinia mostly on abaxial surface; spores about as the aeciospores. Telia mostly on abaxial surface, occasionally on stems, exposed, blackish brown; spores (30-)42-53(-60) x (22-)26-33(-40) μm, ellipsoid or occasionally oblong, wall (2-)3-4(-6) μm thick at sides, chestnut brown, (5-)7-8(-10) μm over pores with a pale brownish to nearly colorless umbo, smooth, pore in upper cell apical, in lower cell midway to hilum or near septum; pedicels colorless, to 160(-200) μm long, sometimes rugose basally.

Hosts and distribution: *Barroetia subuliger* (Schauer) Gray and species of *Brickellia* and *Kuhnia*: the Great Lakes region of the United States to Baja California and Jalisco, Mexico.

Type: on *Kuhnia* sp., locality uncertain, perhaps Indiana, despite Bethlehem, Pa. on the label, Schweinitz (PH).

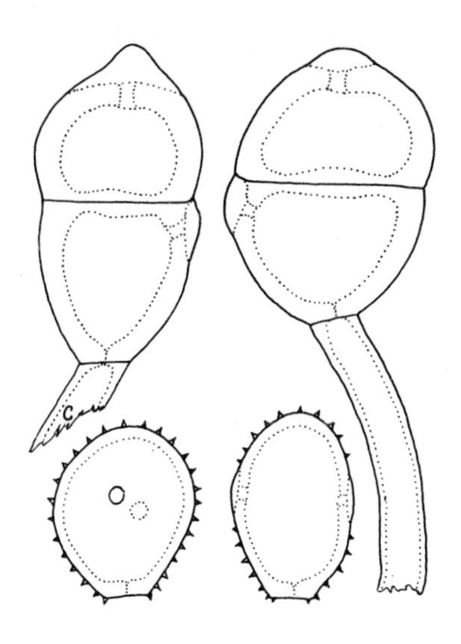

32b. *PUCCINIA KUHNIAE* var. *BRICKELLIAE* (Peck) Cumm. Mycotaxon 5:405. 1977.
Puccinia brickelliae Peck, Bull. Torrey Bot. Club 12: 34-35. 1885.

Aecia uredinoid; spores (24-)26-33(-35) x (20-)22-26 (-28) µm, wall 1.5-2 µm thick, cinnamon brown, echinulate, pores 2. Urediniospores similar to the aeciospores, dark cinnamon brown. Teliospores (42-)45-55(-66) x (24-)28-35 (-38) µm, broadly ellipsoid or obovoid, variable, wall 4-7 µm thick at sides, to 11 µm over pores, chestnut brown, paler over pores, pore apical in each cell; pedicel, sometimes with sterile branches, to at least 200 µm long.

Hosts and distribution: *Brickellia*, especially *coulteri* Gray, *Kuhnia chlorolepis* Woot. & Standl.: southern Arizona to Sinaloa and western Chihuahua, Mexico.

Type: on *Brickellia*, "Arizona, September", Jones. (The holotype in NYS bears data as follows: Bowie, Arizona, 9/16/84, M. E. Jones No. 541; isotype PUR 37467). The host plant unquestionably is *B. coulteri*. Robinson (28) cites a Jones specimen from Bowie, but without other data.

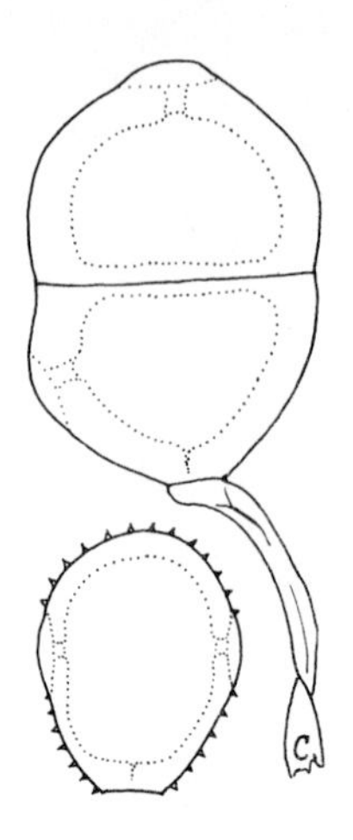

32c. *PUCCINIA KUHNIAE* var. *DECORA* (Diet. & Holw.) Cumm., Brit. & Baxt. Mycologia 61:936. 1969.
Puccinia pinguis Diet. & Holw. in Holway, Bot. Gaz. 24: 34. 1897. (July) not Dietel (Feb.).
Puccinia decora Diet. & Holw. in Dietel, Hedwigia 37: 202. 1898.

Spermogonia and aecia unknown. Uredinia cinnamon brown or darker; spores (23-)26-32(-35) x 23-29 µm, broadly ellipsoid or obovoid, 18-23(-25) µm wide, ellipsoid or oblong ellipsoid with pores lateral, wall 2-3(-4) µm thick, thinner on pore bearing sides, dark cinnamon brown, echinulate, pores 2. Teliospores (35-)40-52(-60) x 30-37 µm, broadly ellipsoid, wall 4-6(-7) µm thick at sides, 8-10 µm over pores by a paler umbo, chestnut brown, pore of upper cell apical or subapical, of lower cell midway to hilum, smooth; pedicels to 75 µm long but usually shorter.

Hosts and distribution: *Brickellia secundiflora* (Lag.) Kuntze, *B. tomentella* Gray, *B.* sp.: southern half of Mexico.

Type: on *Brickellia* sp., Rio Hondo near Mexico City, Holway (S; isotype PUR 37517).

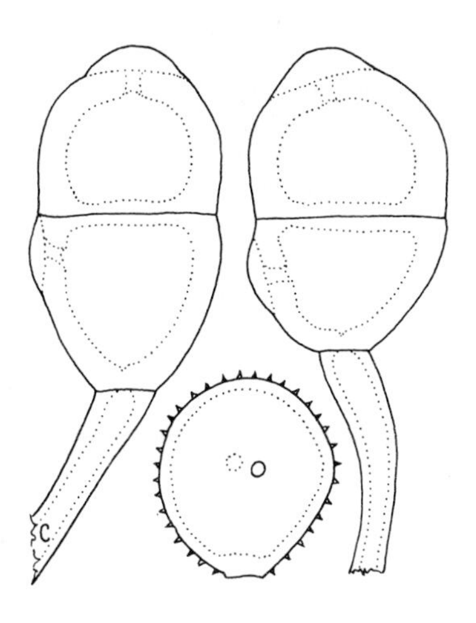

32d. *PUCCINIA KUHNIAE* var. *ROBUSTA* Cumm., Britt. & Baxt.
Mycologia 61:935. 1969.

Spermogonia, uredinoid aecia and uredinia about as in var. *kuhniae* except the spores mostly 30-36 μm long and the wall (1.5-)2-3(-4) μm thick, dark cinnamon brown (some amphisporic ?), sometimes thinner and paler. Teliospores (35-)40-50(-52) x (28-)30-36(-38) μm, broadly ellipsoid, wall (3-)4-5(-6) μm thick at sides, (5-)7-10(-12) μm apically, chestnut brown, pore of upper cell apical, of lower cell at septum or midway to hilum, each with a pale umbo; pedicel colorless, to 200 μm long.

Hosts and distribution: species of *Brickellia* and *Kuhnia*: Montana to Texas and Arizona, Durango, Mexico and Guatemala.

Type: on *Brickellia lemmonii* Gray, Chiricahua Mts., Arizona, Cummins No. 61-258 (PUR 59958).

Arizona specimens, especially, have large, chestnut brown urediniospores which perhaps are amphisporic.

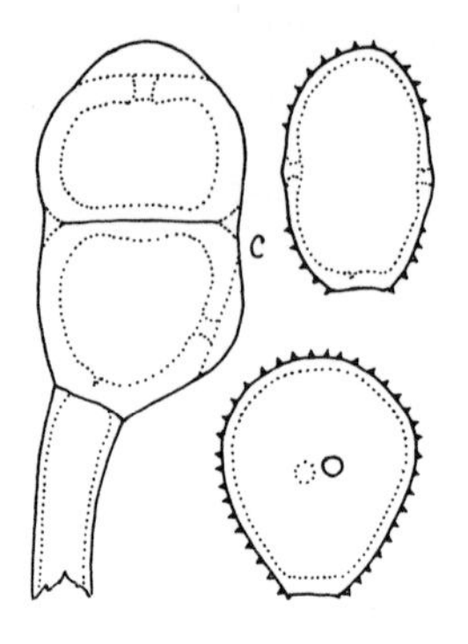

33. *PUCCINIA CALANTICARIAE* J. Parm. Can. J. Bot. 45:2298. 1967.

Spermogonia amphigenous. Aecia amphigenous, uredinoid, around the spermogonia, cinnamon brown; spores as in uredinia. Uredinia amphigenous, scattered, cinnamon brown; spores (24-)27-30(-34) x (17-)19-25(-28) µm, mostly obovoid with pores face view, ellipsoid with pores lateral, wall 1-1.5 (-2) µm thick, cinnamon brown, echinulate with fine spines spaced (1.5-)2(-2.5) µm, pores 2, equatorial, in flattened, smooth sides. Telia amphigenous and on swellings of stems, blackish brown, pulverulent; spores (30-)34-46(-50) x (19-) 22-28(-32) µm, mostly broadly ellipsoid, wall (2-)2.5-4 µm thick at sides, 6-8(-10) µm over pores, deep chestnut brown but umbo over pores pale brown, smooth, pore apical in upper cell, usually midway to hilum in lower cell; pedicels colorless or nearly so, to 185 µm long, usually about 100 µm.

Hosts and distribution: *Viguiera* spp., especially *V. cordifolia* Gray and *V. linearis* (Cav.) Sch. Bip.: southern Arizona south at least to Mexico City.

Type: on *Viguiera budleiaeformis* (DC.) Benth. & Hook., Rio Hondo, near Mexico City, Holway No. 3563 (PUR 34586; isotypes Barth. N. Amer. Ured. 1215; Barth. F. Columb. 3452, 4460).

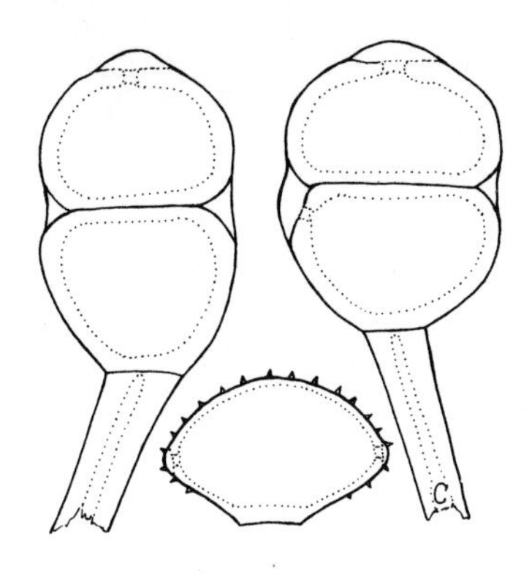

34. *PUCCINIA ESPINOSARUM* Diet. & Holw. in Holway, Bot. Gaz. 31:332. 1901.

Spermogonia and aecia unknown. Uredinia amphigenous, cinnamon brown; spores (30-)32-38(-40) µm between the pores, (18-)20-24(-27) µm hilum to apex, 24-29(-31) µm wide with hilum in optical axis, strongly asymmetrical, depressed ovoid with hilum basal, transversely ellipsoid with hilum in optical axis, wall 1-1.5(-2) µm thick, cinnamon brown, echinulate except around the hilum, pores 2, slightly below the equator in ends of spore. Telia amphigenous, exposed, blackish brown, pulverulent; spores (37-)42-50(-55) x (31-) 33-38(-40) µm, mostly broadly ellipsoid, wall 2-3(-4) µm thick at sides, 5-8 µm over each pore, dark chestnut brown except the yellowish umbos over the pores, pore of upper cell apical, of lower cell near septum, smooth; pedicels colorless, to 160 µm long but often 100 µm or less, lower portion rugose and swelling to 12-18(-30) µm.

Hosts and distribution: *Eupatorium* spp.: central Mexico to Guatemala.

Type: on *Eupatorium espinosarum* Gray, Oaxaca, Oax., Mexico, Holway No. 3651 (S; isotypes Barth. N. Amer. Ured. 1241).

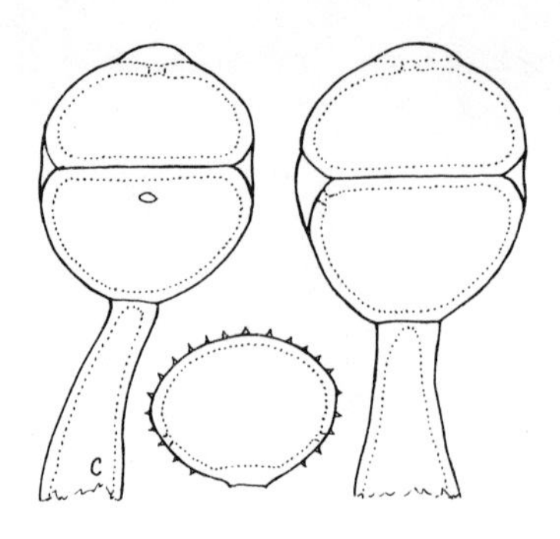

35. *PUCCINIA INANIPES* Diet. & Holw. in Holway, Bot. Gaz. 31: 332. 1901.

Spermogonia mostly on adaxial leaf surface. Aecia mostly abaxial in groups of 4 to 8, bullate, peridium scarcely exserted, fragmenting; spores (22-)24-30 x (17-)20-24(-26) µm, mostly ellipsoid or globoid, wall 1.5-2(-2.5) µm thick, yellowish, finely and often striolately verrucose. Uredinia amphigenous, cinnamon brown; spores (22-)24-28(-30) µm between pores, 18-24 µm hilum to apex, 21-25 µm wide with hilum in optical axis, depressed ovoid with hilum basal, transversely ellipsoid with hilum in optical axis, wall 1-1.5 µm thick, cinnamon brown, echinulate except around hilum, pores 2, subequatorial. Telia amphigenous, exposed, blackish brown, pulverulent; spores (34-)36-42(-45) x (30-)33-37(-39) µm, broadly ellipsoid, wall (1.5-)2-3(-4) µm thick at sides, 5-8 µm over pores, dark chestnut brown except the paler umbos over pores, smooth, pore apical in upper cell, near septum in lower cell; pedicels to 160 µm long but usually 30-100 µm, enlarged and rugose below.

Hosts and distribution: *Eupatorium* spp.: southern Arizona to southern Mexico.

Type: on *Eupatorium brevipes* DC., Oaxaca, Oax., Mexico, Holway No. 3677 (S; isotypes Barth. N. Amer. Ured. 150).

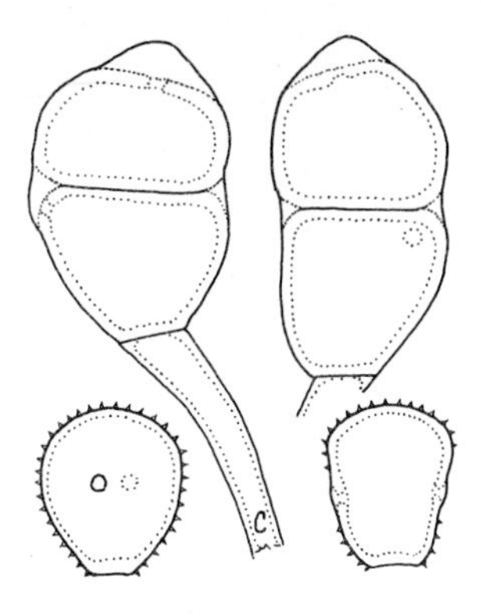

36. *PUCCINIA ENCELIAE* Diet. & Holw. in Holway, Bot. Gaz. 24:32. 1897.
Puccinia tithoniae Diet. & Holw. in Holway, Bot. Gaz. 24:31. 1897.
Puccinia aemulans P. Syd. & H. Syd. Ann. Mycol. 4:31. 1906.

Spermogonia on adaxial leaf surface. Aecia grouped on abaxial surface, peridium short; spores 18-24(-27) x (13-) 16-21(-23) µm, ellipsoid or globoid, wall 1-1.5 µm thick, colorless, verrucose. Uredinia mostly on abaxial surface, cinnamon brown; spores (18-)20-24(-28) x (17-)19-23(-25) µm, obovoid or globoid, triangularly obovoid with pores lateral, wall 1-1.5 µm thick, cinnamon brown, echinulate except around pores, pores 2, subequatorial in flattened sides. Telia mostly on abaxial surface, exposed, blackish brown, compact; spores (32-)36-46(-52) x (19-)22-26(-29) µm, ellipsoid or obovoid, wall 1-2(-2.5) µm thick at sides, (5-)6-9(-11) µm over pores, chestnut brown, the umbo over each pore pale, smooth, pore in each cell apical; pedicels colorless, to 125 µm long but commonly less than 100 µm.

Hosts and distribution: *Enceliopsis nudicaulis* (Gray) A. Nels., *Helianthus argophyllus* Torr. & Gray, species of *Simsia*, *Tithonia* and *Viguiera*: the Rocky Mt. region from Wyoming south to Panama; also in South America.

Type: on *Encelia mexicana* (=*Simsia foetida* (Cav.) Blake, Cuernavaca, Mor., Mexico, Holway (S; isotypes Barth. N. Amer. Ured. No. 1240).

Aecia have been found only on *Viguiera* and *Simsia*.

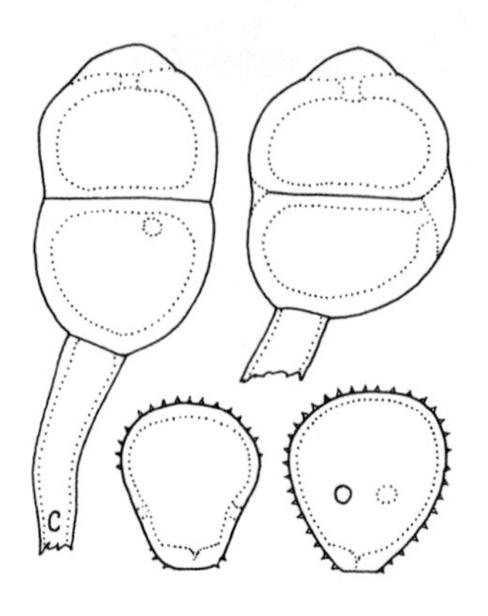

37. *PUCCINIA NOCCAE* Arth. Bot. Gaz. 40:202. 1905.

Spermogonia and aecia unknown. Uredinia mostly amphigenous, cinnamon brown; spores (19-)22-28(-30) x (18-)20-26 (-30) µm, obovoid with pores lateral, wall 1-1.5 µm thick at sides, usually slightly thicker at hilum and often at apex, echinulate, cinnamon brown or slightly darker at apex and base, pores 2, subequatorial in smooth flattened sides. Telia mostly amphigenous, exposed, blackish brown, pulverulent; spores (32-)35-44(-48) x (20-)24-29(-32) µm, ellipsoid or broadly so, wall (1.5-)2-3 µm thick at sides, (5-)6-8 (-10) µm at apex, deep chestnut brown except the yellowish umbos over pores, smooth, pore apical in each cell; pedicels colorless, to 85 µm long but usually broken shorter, sometimes rugose basally.

Hosts and distribution: *Lagascea* spp.: the southern half of Mexico south to Honduras and El Salvador.

Type: on *Lagascea decipiens* Hemsl. (as *Nocca decipiens*), Sayula, Jal., Mexico, Holway No. 5122 (PUR 42593).

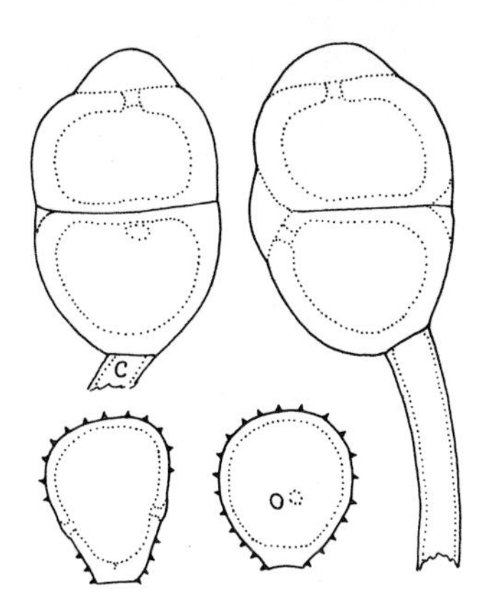

38. *PUCCINIA ABRUPTA* Diet. & Holw. in Dietel, Hedwigia 37: 208. 1898, var. *ABRUPTA*.

Spermogonia and aecia unknown. Uredinia amphigenous or mainly on adaxial leaf surface and stems, dark cinnamon brown; spores (18-)20-25(-28) x (16-)18-22(-24) µm, strongly obovoid or triangular, wall (1-)1.5(-2) µm thick, thicker at hilum, cinnamon brown, some spores (amphisporic?) chestnut brown with wall 2.5-3 µm thick, echinulate with spines spaced 2-2.5(-3) µm, pores 2, subequatorial and rarely with 1 apical. Telia amphigenous and on fusiform stem galls, blackish brown, pulverulent; spores (30-)35-44(-50) x (21-) 26-31(-35) µm, broadly ellipsoid, wall (2.5-)3-4(-5) µm thick at sides, chestnut brown or some spores deep golden brown, (6-)7-9(-11) µm over pores by yellowish umbos, smooth, pore apical in each cell; pedicel colorless, to 130 µm but usually less than 100 µm long.

Hosts and distribution: *Viguiera dentata* (Cav.) Spreng. and vars., *V. sylvatica* Klatt, *V. tenuis* Gray?: southern Arizona and southern Texas to Costa Rica; also in South America.

Type: on *Viguiera helianthoides* (now = *V. dentata* var. *helianthoides* (H.B.K.) Blake), Tula, Mexico, Holway (S; isotypes Barth. N. Amer. Ured. 118).

The teliospores are dimorphic in some collections.

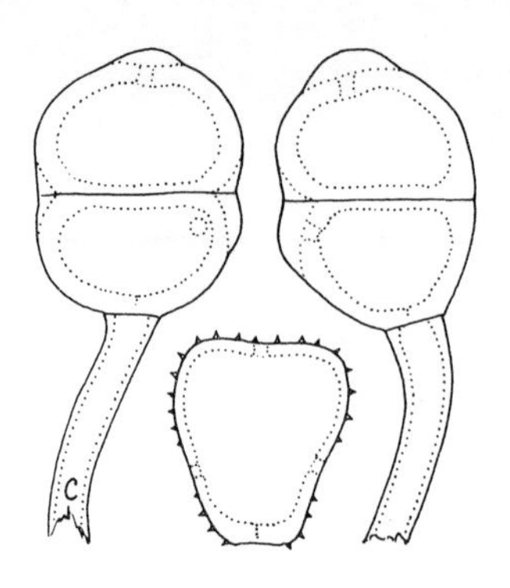

38b. *PUCCINIA ABRUPTA* var. *PARTHENIICOLA* (H. S. Jack) J. Parm. Can. J. Bot. 45:2293. 1967.
Puccinia partheniicola H. S. Jack. Mycologia 24:166. 1932.

Spermogonia and aecia unknown. Uredinia amphigenous and on stems, dark cinnamon brown; spores (21-)24-28(-31) x (18-)22-27(-28) µm, mostly obovoid or triangular, wall 1-1.5 µm thick, cinnamon brown, echinulate, pores mostly 2 subequatorial and 1 apical. Telia amphigenous and on stems, exposed, compact, blackish brown; spores (31-)33-38(-42) x (23-)26-30(-33) µm, broadly ellipsoid, wall 2.5-3.5 µm thick at sides, (4-)5-7(-8) µm over pores, chestnut brown except pale umbos over pores, smooth, pore apical in upper cell, at or near septum in lower cell; pedicels colorless, to 160 µm long.

Hosts and distribution: *Parthenium confertum* Gray, *P. hysterophorus* L.: the northern one-third of Mexico; also in South America.

Type: on *P. hysterophorus*, Cochabamba, Bolivia, Holway No. 349 (PUR F8298; isotypes Reliq. Holw. 639).

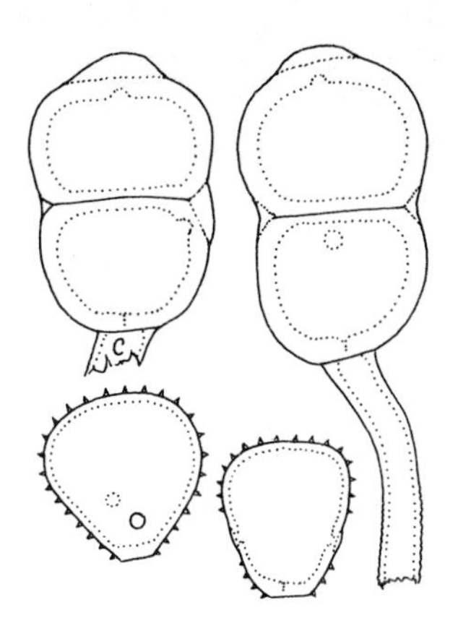

39. *PUCCINIA GYMNOLOMIAE* Arth. Bot. Gaz. 40:200. 1905.

Spermogonia and aecia unknown. Uredinia mostly on the abaxial surface, pale cinnamon brown; spores (19-)21-24(-26) x (18-)20-24(-25) µm, mostly obovoid with pore face view, triangularly obovoid with pores lateral, wall (0.5-)1(-1.5) µm thick, about cinnamon brown, echinulate except around pores, pores 2, below equator. Telia on abaxial surface, exposed, blackish brown, more or less pulverulent; spores (35-)38-44(-46) x (23-)26-29(-32) µm, mostly ellipsoid, wall (1.5-)2.5-3.5 µm thick at sides, chestnut brown, 4.5-7 µm over pores with nearly colorless umbos, smooth, pore of upper cell apical, of lower cell at septum; pedicels colorless, rugose basally, to 70 µm long.

Hosts and distribution: *Hymenostephium* spp.: southern Mexico to Costa Rica.

Type: on *Gymnolomia subflexuosa* (= *Hymenostephium cordatum* (Hook. & Arn.) Blake, Oaxaca, Oax., Mexico, Holway No. 3645 (PUR 33934; isotypes Barth. N. Amer. Ured. 1247, 1248).

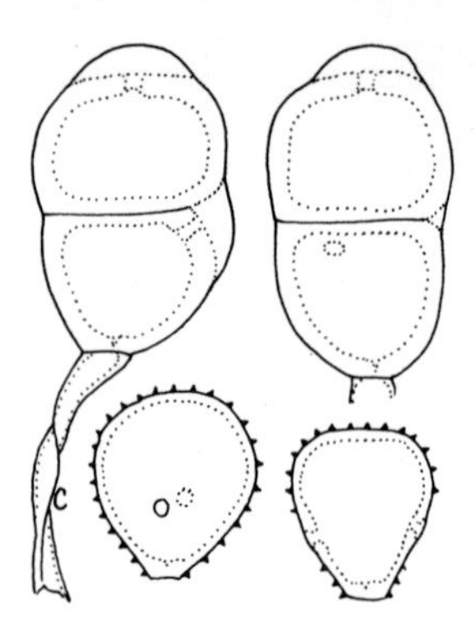

40. *PUCCINIA VERBESINAE* Schw. Schr. Nat. Ges. Leipzig 1:73. 1822.

Spermogonia on adaxial leaf surface. Aecia on abaxial surface in groups, cylindric, erose; spores (17-)20-25(-27) x (15-)17-20(-21) µm, globoid or broadly ellipsoid, wall 1-1.5 µm thick, colorless, finely and closely verrucose. Uredinia mostly on abaxial surface, cinnamon brown; spores (18-)20-24(-27) x (18-)19-23(-25) µm, mostly slightly higher than wide, mostly obovoid with pores face view, obovoid or triangularly obovoid with pores lateral, wall 1-1.5 µm thick, cinnamon brown, echinulate except around pores, pores 2, subequatorial. Telia mostly on abaxial surface, exposed, more or less pulverulent, blackish brown; spores (33-)36-42 (-45) x (22-)24-28(-30) µm, broadly ellipsoid, wall (1.5-) 2-3(-3.5) µm thick at sides, (5-)6-8(-10) µm at pores, chestnut brown except a pale umbo over each pore, smooth, pore apical in each cell; pedicel to 65 µm long but usually broken shorter.

Hosts and distribution: *Verbesina alternifolia* (L.) Britt., *V. occidentalis* (L.) Wats.: the southeastern United States.

Neotype: on *Verbesina occidentalis*, Advance, North Carolina, Higgins (PUR 34508 ; isotypes Barth. F. Columb. 2969).

According to Arthur and Bisby (3) there is only an empty packet of the original (PH).

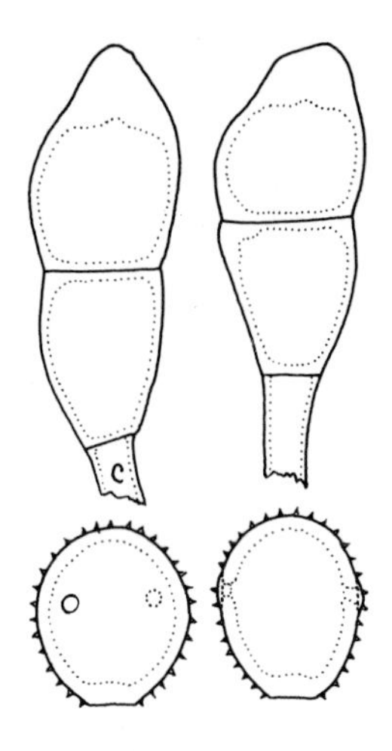

41. *PUCCINIA GNAPHALIICOLA* P. Henn. Hedwigia 38(Beibl.):68. 1899.
Puccinia gnaphalii P. Henn. Hedwigia 41(Beibl.):66. 1902.

Spermogonia and aecia unknown. Uredinia mostly on abaxial leaf surface and stems; cinnamon brown; spores (21-)22-25(-27) x (19-)21-24 µm, broadly obovoid or globoid, wall (1.5-)2(-2.5) µm thick, uniformly echinulate, yellowish brown, pores 2, rarely 3, equatorial or slightly above, in slightly or not flattened sides, with slight or no caps. Telia like the uredinia but blackish brown, exposed, compact; spores (32-)35-50(-55) x (17-)19-23(-25) µm, oblong ellipsoid or elongately obovoid, wall 1-1.5(-2) µm thick at sides, (4-)5-7(-10) µm at apex, uniformly golden or clear chestnut brown, smooth, pore apical in each cell; pedicels colorless, to about 65 µm long.

Hosts and distribution: *Facelis retusa* (Lam.) Sch.-Bip. and species of *Gnaphalium*: the southeastern United States to Durango, Mexico and in Guatemala; also in South America, Australia and New Zealand.

Type: on *Gnaphalium* sp., Rio de Janiero, Brazil, Ule (B).

The date (9 Jan. 1896) and collector's number (2162) as published differ from the date (9 Sept. 1896) and collector's number (2126) on the type sheet.

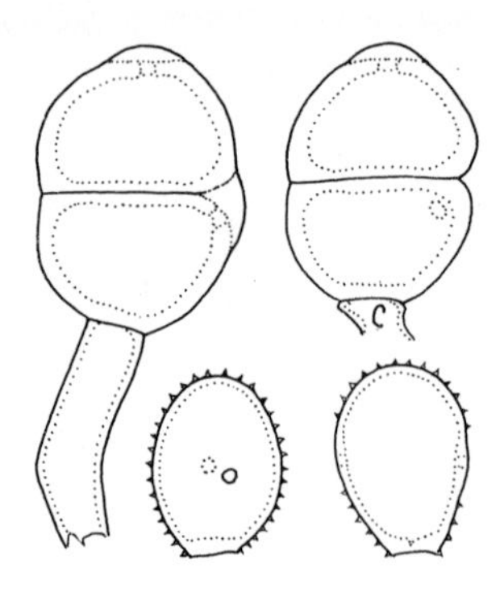

42. *PUCCINIA SUBGLOBOSA* Diet. & Holw. in Holway, Bot. Gaz. 31:332. 1901.

Spermogonia and aecia unknown. Uredinia amphigenous, dark cinnamon brown; spores (19-)21-24(-27) x 16-19 µm, globoid, broadly ellipsoid or obovoid, wall 1-1.5 µm thick or to 2 µm at apex and base, dark cinnamon brown or near chestnut brown, echinulate except around pores, pores 2, equatorial. Telia amphigenous, exposed, blackish brown, more or less pulverulent; spores (27-)30-36(-40) x (21-)24-28(-30) µm, mostly broadly ellipsoid, wall (1.5-)2-2.5 µm thick at sides, (4-)5-7 µm over the pores, deep chestnut brown but the umbo over each pore slightly paler, smooth, pore in upper cell apical, next to septum in lower cell; pedicels colorless, to 65 µm long.

Hosts and distribution: *Rhysolepis palmeri* (Gray) Blake: the area around Guadalajara, Jal., Mexico.

Type: Chapala, Holway No. 3488 (S; isotype PUR 62558).

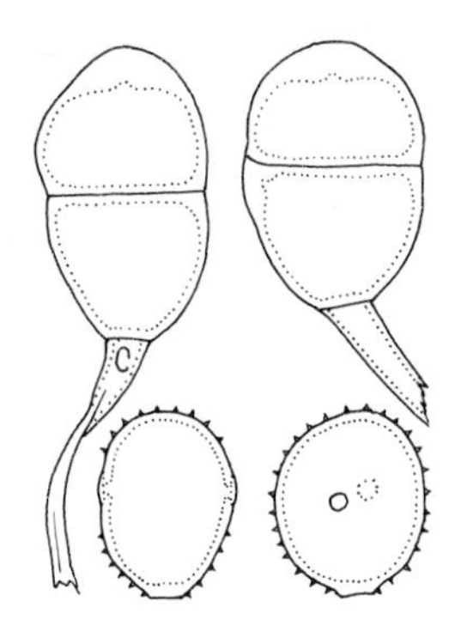

43. *PUCCINIA SONORAE* J. Parm. Can. J. Bot. 47:1395. 1969.

Spermogonia on adaxial leaf surface. Aecia on abaxial surface, cupulate to short cylindric, in groups on slightly hypertrophied areas; spores 24-29 x 17-22 µm, globoid or ellipsoid, wall 1-2.5 µm thick, verrucose. Uredinia on abaxial surface, cinnamon brown; spores (17-)18-22(-24) x (15-)17-20(-22) µm, globoid or broadly ellipsoid, wall 1-1.5 µm thick, cinnamon brown, echinulate with spines spaced about 2 µm, pores 2, equatorial, in smooth areas of the not or slightly flattened sides. Telia on abaxial surface, exposed, compact, blackish brown; spores (26-)30-38(-42) x (18-)20-24(-26) µm, broadly ellipsoid or sometimes broadly obovoid, wall 1-2 µm thick at sides, (2.5-)3.5-7(-8) µm thick at apex, uniformly clear chestnut brown, smooth, pore apical in each cell; pedicel hyaline, to 115 µm long but usually broken shorter.

Hosts and distribution: *Ambrosia carduacea* (Greene) Payne, *A. cordifolia* (Gray) Payne: Sonora, Sinaloa and Baja California, Mexico.

Type: on *Ambrosia cordifolia*, Alamos, Son., Cummins No. 63-762 (PUR 61984).

The single collection on *A. carduacea* has longer teliospores than the type.

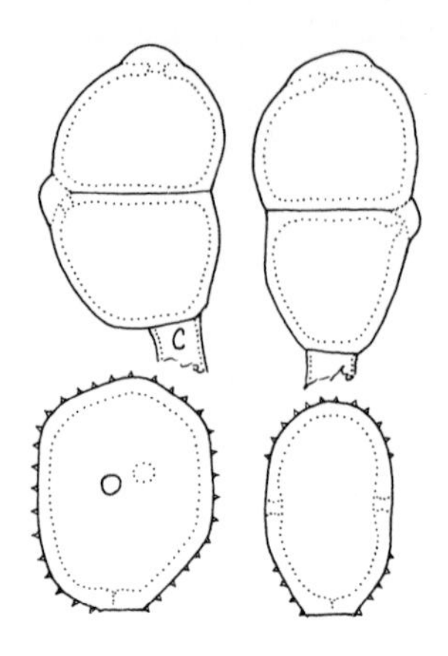

44. *PUCCINIA HELIANTHELLAE* Arth. Bull. Torrey Bot. Club 31: 4. 1904.
Uredo gaillardiae Diet. & Holw. in Dietel, Erythea 7:98. 1899.

Spermogonia amphigenous in elongate chlorotic areas (locally systemic) or absent. Aecia in chlorotic areas, mostly on abaxial leaf surface, locally systemic, uredinoid, chocolate brown; spores (25-)28-34(-37) x (16-)20-29(-32) µm, mostly broadly obovoid or broadly ellipsoid, wall (1.5-) 2-2.5(-3) µm thick, dark cinnamon brown, echinulate, pores 2, equatorial in flattened smooth sides. Uredinia scattered, chocolate brown; spores as the aeciospores. Telia associated with the spermogonia or usually scattered, amphigenous, exposed, chocolate brown, pulverulent; spores (28-)33-40 (-43) x (18-)21-26(-29) µm, mostly ellipsoid, wall (1-)1.5-2(-2.5) µm thick at sides, clear chestnut brown, smooth, 4-6(-7) µm over pores, pore apical in each cell, each covered by a pale umbo; pedicels colorless, to 50 µm long but usually broken at or near hilum.

Hosts and distribution: *Helianthella* spp.: Wyoming and Colorado to California.

Type: on *Helianthella californica* (as *H. nevadensis*), Nevada County, California, Heller No. 7072 (PUR 34465).

45. *PUCCINIA REDEMPTA* H. S. Jack. Mycologia 14:107. 1922.

Spermogonia and aecia unknown. Uredinia on abaxial leaf surface, cinnamon brown; spores (24-)25-32 x (23-)25-28(-30) µm, mostly broadly obovoid or nearly globoid, wall 1.5-2 µm thick or slightly thicker at pores, cinnamon brown, echinulate except around pores, pores 2, equatorial in slightly flattened sides. Telia on abaxial surface, exposed, blackish brown, pulverulent; spores (42-)47-55(-58) x (30-)32-39(-43) µm, variable but mostly oblong ellipsoid, often angularly so, wall 2-2.5(-3.5) µm thick at sides, 4-6 µm over pores, chestnut brown except paler umbos over pores, smooth, pore apical in upper cell, near septum in lower cell; pedicel hyaline, broken near spore.

Hosts and distribution: *Eupatorium hebebotryum* (DC.) Hemsl.: central Mexico and in the West Indies.

Type: on *Eupatorium atriplicifolium* Lam., Tortola, W. J. Britton and Shafer ex phanerogam No. 795 (= PUR 37465).

The Mexican fungus (uredinia only) may not be this species.

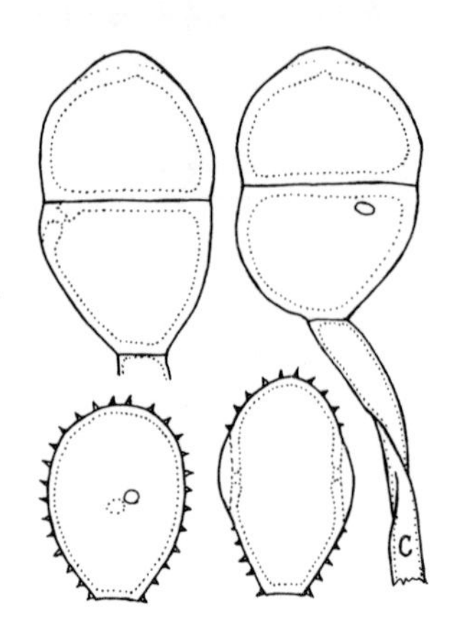

46. *PUCCINIA SINALOANA* Cumm., Brit. & Baxt. Mycologia 61: 932. 1969.

Spermogonia in adaxial leaf surface. Aecia too old for accurate measurements, few in groups, with peridium; spores about 20-25 x 18-22 µm, wall about 1.5 µm thick, verrucose. Uredinia mostly on abaxial surface, cinnamon brown; spores (26-)28-32(-36) x 22-24(-26) µm, obovoid or ellipsoid, wall 1-1.5 µm thick, cinnamon brown, echinulate except around pores, spines spaced 3-4(-5) µm, pores 2, equatorial in slightly flattened sides, with conspicuous caps. Telia amphigenous, exposed, blackish brown, pulverulent; spores (40-)43-50(-53) x (25-)29-32(-34) µm, elongately obovoid or ellipsoid, wall 1.5-2(-3) µm thick, dark chestnut brown at sides, (4-)5-6(-7) µm and slightly paler apically, smooth, pore apical in each cell; pedicels colorless, to 70 µm long.

Hosts and distribution: *Eupatorium*(?) sp.: Sinaloa, Nayarit and Jalisco, Mexico.

Type: highway 40 east of Mazatlán, Sin., Hennen No. 67-526 (PUR 61886).

All host material has been sterile but is believed to be a *Eupatorium*.

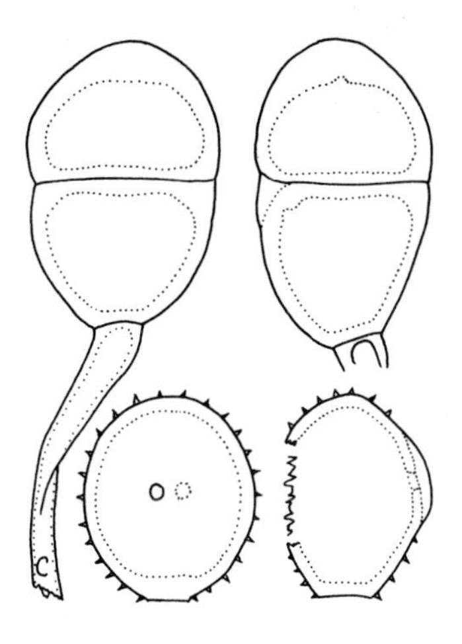

47. *PUCCINIA FRANSERIAE* H. Syd. & P. Syd. Ann. Mycol. 1:326. 1903.
Puccinia caborcensis J. Parm. Can J. Bot. 47:1396. 1969.

Spermogonia on adaxial leaf surface. Aecia amphigenous, uredinoid, around the spermogonia, chocolate brown; spores (23-)25-32(-38) x (18-)20-28(-30) μm, globoid or ellipsoid, wall 1.5-2(-2.5) μm thick, nearly chestnut brown, echinulate, spines spaced (2-)3-4 μm, pores 2, equatorial in flattened sides. Uredinia similar to the aecia but not with spermogonia; spores as the aeciospores. Telia amphigenous and on small stem galls, exposed, compact, blackish brown; spores (32-)37-52(-59) x (23-)25-30(-33) μm, ellipsoid or elongately obovoid, wall (1-)1.5-2(-3) μm thick at sides, chestnut brown or slightly paler, (4-)5-7(-9) μm at apex which is nearly concolorous, smooth, pore apical in each cell; pedicel colorless or pale yellowish, variable in length, longer from stem galls, to 175 μm long.

Hosts and distribution: *Ambrosia* (*Franseria*) spp., *Hymenoclea pentalepis* Rydb.: Utah to California south to Sonora and Baja California, Mexico. Perhaps also in Ecuador and Peru.

Type: on *Franseria ambrosioides* (= error for *Ambrosia deltoidea* (Torr.) Payne), Tucson Mts., Arizona, Griffiths (S; isotypes Griffiths West Amer. F. 257).

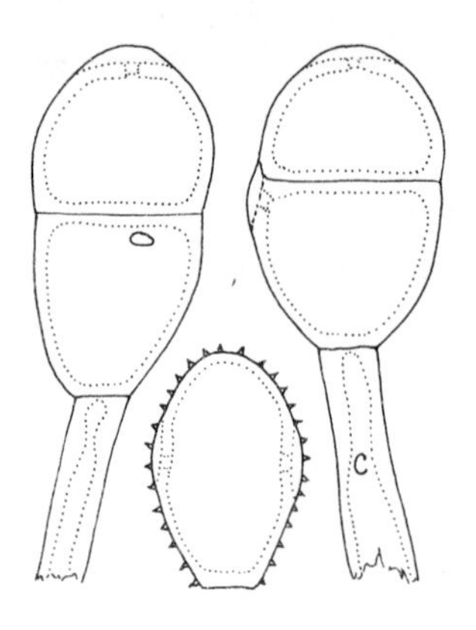

48a. *PUCCINIA CALEAE* Arth. Bot. Gaz. 40:201. 1905 var. *CALEAE*.

Spermogonia on adaxial leaf surface. Aecia on abaxial surface in small groups, peridium cylindrical, lacerate; spores 23-29(-33) x 19-24 µm, globoid or nearly so, wall colorless, about 1 µm thick, verrucose with flat warts 1 µm diam. Uredinia amphigenous, cinnamon brown; spores (23-)26-32(-37) x (18-)20-25 µm, obovoid or ellipsoid, wall (1-)1.5-2 µm thick, dark cinnamon brown, uniformly echinulate with spines spaced 2-4(-4.5) µm, pores 2, equatorial in slightly flattened sides, with caps. Telia amphigenous or on adaxial surface, exposed, becoming pulverulent, blackish brown; spores (36-)40-50(-54) x (24-)26-30(-34) µm, ellipsoid tending obovoid, wall (1.5-)2-3(-3.5) µm at sides, dark chestnut brown (3-)4-6(-7.5) µm over pores as a slightly paler, low umbo, pore apical in upper cell, next to septum in lower cell, smooth; pedicels colorless except yellowish near hilum, 100 µm or less long.

Hosts and distribution: *Calea urticifolia* (Mill.) DC. Jalisco, Mexico south to Costa Rica; also in Brazil.

Type: Sayula, Jal., Mexico, Holway No. 5126 (PUR 33940; isotypes Barth. N. Amer. Ured. 930).

48b. *PUCCINIA CALEAE* var. *CUERNAVACAE* J. Parm. Can. J. Bot. 45:2273. 1967.

The variety differs primarily in that the germ pores of the urediniospores are in smooth areas and the apical wall of the urediniospores usually is slightly thicker than the side wall. It is doubtful that the punctate wall of the teliospore, mentioned by Parmelee, is distinctive.

Hosts and distribution: ?*Agiabampoa congesta* Rose, *Calea hypoleuca* B. L. Rob. & Greenm., *C. zacatechichi* Schlecht. and vars.: Jalisco, Mexico south to Costa Rica.

Type: on *Calea zacatechichi* var. *rugosa* B. L. Rob & Greenm., Cuernavaca, Mor., Mexico, Holway No. 5301 (PUR 33956).

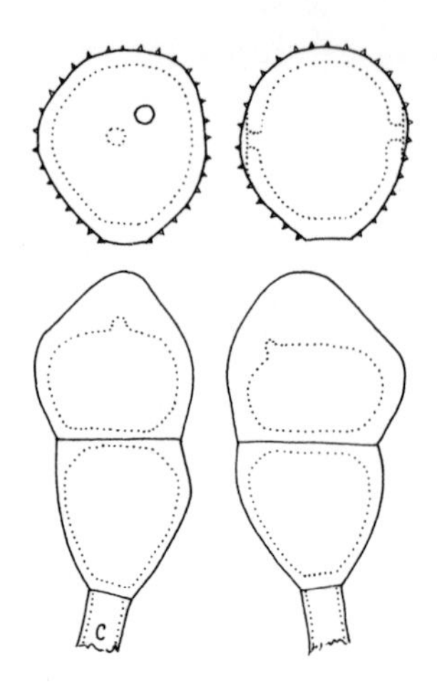

49. *PUCCINIA NUDA* Ellis & Ever. J. Mycol. 3:57. 1887.

Spermogonia and aecia unknown. Uredinia amphigenous, cinnamon brown; spores (23-)25-30(-33) x (22-)25-28(-29) µm, mostly globoid, wall (1.5-)2-3 µm thick, golden brown, uniformly echinulate, pores 2(3), equatorial or nearly so in not or only slightly flattened sides, with slight or no caps. Telia amphigenous, exposed, blackish brown, compact; spores (34-)40-50(-56) x (18-)22-28(-30) µm, mostly ellipsoid or obovoid, wall (1.5-)2-3 µm thick at sides of upper cell, thinner in lower cell, (5-)6.5-10(-11) µm thick at apex, from deep golden brown to chestnut brown with the apical thickening becoming paler but not as a defined umbo, smooth, pore apical in each cell; pedicels colorless, to 100 µm long.

Hosts and distribution: species of *Calcadenia, Hemizonia, Lagophylla* and *Madia*: on the Pacific Coast of the U.S.; also in South America.

Type: on *Arnica foliosa* (error for *Madia* sp.), Falcon Valley, Washington, Suksdorf No. 200 (NY: isotype PUR 42589).

50. *PUCCINIA HELIANTHI* Schw. Schr. Nat. Ges. Leipzig 1:73. 1822.
Puccinia helianthorum Schw. Trans. Amer. Phil. Soc. II. 4:296. 1832.
Puccinia xanthifoliae Ellis & Ever. J. Mycol. 6:120. 1891.

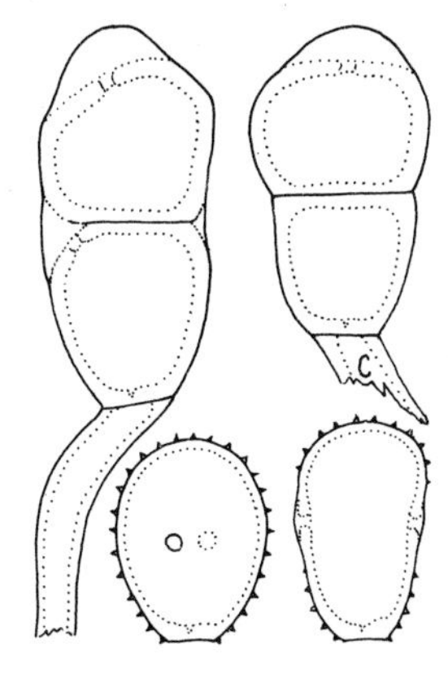

Spermogonia mostly on adaxial leaf surface. Aecia on abaxial surface in groups, peridium short; spores (16-) 20-25(-30)x (13-)16-21(-23) µm, ellipsoid or globoid, wall (0.5-)1-1.5 (-2.5) µm thick, colorless, minutely verrucose. Uredinia mostly on abaxial surface, cinnamon brown; spores (23-)26-33(-38) x (14-)18-28(-32) µm, broadly ellipsoid or obovoid with pores face view, oblong ellipsoid or narrowly obovoid with pores lateral, wall 1-1.5(-2) µm thick, cinnamon brown, echinulate except around pores, pores 2, equatorial, in flattened sides. Telia amphigenous or mostly on abaxial surface, erumpent, compact, blackish brown; spores (33-)38-60(-70) x (18-)21-30(-33) µm, oblong ellipsoid or elongately obovoid, wall (1-)1.5-2(-2.5) µm thick at sides, clear chestnut or golden brown, (5-)7-10(-12) µm at apex and at pore of lower cell, the umbos pale, smooth, pore of each cell apical; pedicels colorless, to 170 µm long but commonly less than 100 µm.

Hosts and distribution: species of *Helianthus*, *Heliopsis helianthoides* (L.) Sweet, *Iva xanthifolia* (Fres.) Nutt.: circumglobal.

Neotype: *Puccinia helianthorum* Schw., on *Helianthus tuberosus* L., Bethlehem, Pennsylvania, Syn. Fung. Amer. Bor. No. 2923 (PH). Neotype designated by Parmelee (25).

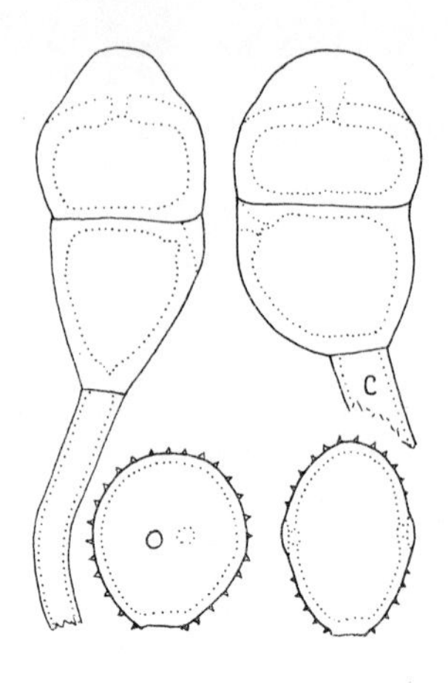

51. *PUCCINIA COGNATA* P. Syd. & H. Syd. Monogr. Ured. 1:172. 1902.

Spermogonia on adaxial leaf surface. Aecia on abaxial surface in groups, with lacerate peridium; spores 22-27 x 16-23 µm, ellipsoid or globoid, wall 1-1.5(-2) µm thick, colorless, coarsely verrucose. Uredinia mostly on abaxial surface, cinnamon brown; spores (21-)24-28(-30) x (18-)19-25 (-27) µm, broadly ellipsoid or obovoid, wall (1-)1.5-2 µm thick, cinnamon brown or somewhat paler, echinulate, spines spaced 1.5-2.5(-3) µm, pores 2, equatorial in smooth areas of slightly flattened sides. Telia mostly on abaxial surface, exposed, blackish brown, pulvinate, compact; spores (32-)38-55(-64) x (20-)23-30(-35) µm, ellipsoid, oblong ellipsoid or elongately obovoid, wall (1-)2-3(-3.5) µm thick at sides, (6-)8-11(-13) µm at apex, clear chestnut brown or the apical thickening paler but scarcely a defined umbo, pore apical in each cell, smooth; pedicel yellowish, to 150 µm long but usually less than 100 µm.

Hosts and distribution: *Verbesina* spp.: Florida to Arizona and south to Guatelmala.

Type: on *Verbesina virginica* L., Austin, Texas, Long No. 50 (S; isotype PUR 34556).

Also see *Puccinia invelata* H. S. Jack. and varieties.

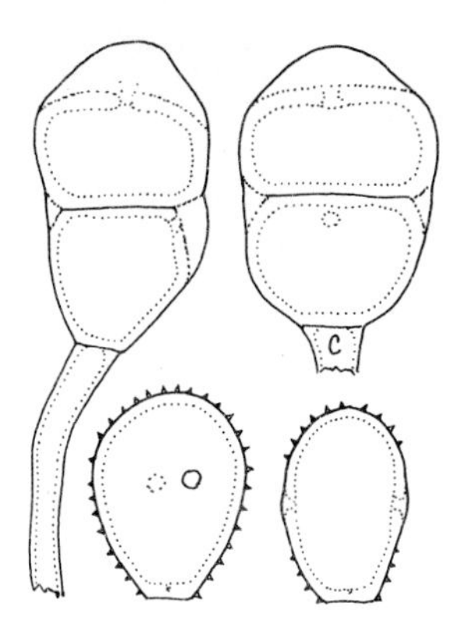

52a. *PUCCINIA INVELATA* H. S. Jack. in Arthur, Bull. Torrey Bot. Club 46:119. 1919 var. *INVELATA*.
Puccinia parthenices H. S. Jack. Mycologia 14:108. 1922.
Puccinia cognata var. *fraseri* J. Parm. Can. J. Bot. 45: 2288. 1967.

Spermogonia on adaxial leaf surface. Aecia on abaxial surface in groups or singly, peridium short, lacerate; spores 19-30 x 16-24 µm, globoid or ellipsoid, wall 1-1.5 (-2) µm thick, colorless, coarsely verrucose. Uredinia mostly on abaxial surface, dark cinnamon brown; spores (19-) 23-28(-30) x (16-)18-22(-24) µm, broadly ellipsoid or obovoid, wall (1-)1.5-2(-2.5) µm thick, about cinnamon brown, echinulate except around pores, pores 2, equatorial in slightly flattened sides. Telia mostly on abaxial surface, exposed, compact, blackish brown; spores (32-)35-50(-53) x (19-)22-28(-30) µm, ellipsoid or obovoid, often broadly so, wall (1-)1.5-2.5 µm thick at sides, (5-)6-8(-10) µm over pores, clear chestnut brown except a paler umbo over pores, smooth, pore apical in upper cell, next to septum in lower cell; pedicel colorless, to 130 µm long but usually about 100 µm or less.

Hosts and distribution: *Parthenice mollis* Gray, *Verbesina* spp.: southern Arizona to southern Mexico.

Type: on *Verbesina montanoifolia* B. L. Rob. & Greenm., Patzcuaro, Mich., Mexico, Holway No. 3606 (PUR 33568).

52b. *PUCCINIA INVELATA* var. *ECHINULATA* (J. Parm.) Cumm.

Mycotaxon 5:405. 1977.
Puccinia cognata Syd. var. *echinulata* J. Parm. Can. J. Bot. 45:2287. 1967.

Spermogonia and aecia unknown; urediniospores as in var. *invelata* except without smooth areas around the pores; teliospores (32-)34-45(-50) x (19-)22-28(-32) µm, mostly ellipsoid or elongately obovoid, wall 1-1.5(-2.5) µm thick at sides, (4-)5-7(-9) µm at apex, chestnut brown except the paler, differentiated umbo; pedicels to 125 µm long but often broken shorter.

Hosts and distribution: *Verbesina* spp.: from central Mexico to Panama.

Type: on *Verbesina turbacensis* H.B.K., vicinity of Zunil, Guatemala, Standley No. 83189 (PUR 49983).

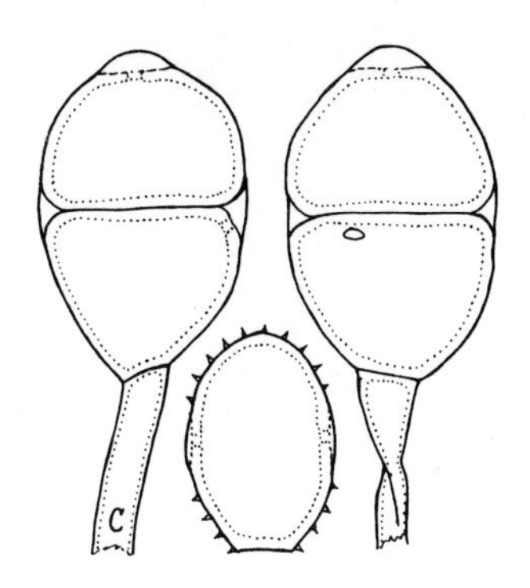

53. *PUCCINIA EUPATORII* Diet. Hedwigia 36:32. 1897.

Spermogonia amphigenous. Aecia mostly on abaxial leaf surface, uredinoid, grouped around the spermogonia, chestnut brown; spores (30-)33-40(-43) x (26-)28-33 µm, mostly broadly ellipsoid or broadly obovoid, wall (2-)2.5-3 µm thick, nearly chestnut brown, echinulate except an area around pores, pores 2, equatorial in slightly flattened sides. Uredinia mostly on abaxial surface, scattered; spores similar to aeciospores except 28-34(-38) x 25-30(-32) µm and wall (1.5-)2(-2.5) µm thick. Telia mostly on abaxial surface, exposed, blackish brown, rather pulverulent; spores (38-)40-52(-56) x (28-)30-34 µm, mostly broadly obovoid or ellipsoid, wall 2(-2.5) µm thick at sides, smooth, clear chestnut brown except the 4-7 µm thick, pale umbos over pores, pore apical in upper cell, at septum in lower cell; pedicels colorless, to 65 µm long.

Hosts and distribution: *Eupatorium sagittatum* Gray: central Sonora to Nayarit, Mexico; also in South America.

Lectotype: on *Eupatorium macrocephalum* Less., Serra Geral, Brazil, ULe No. 1687 (S).

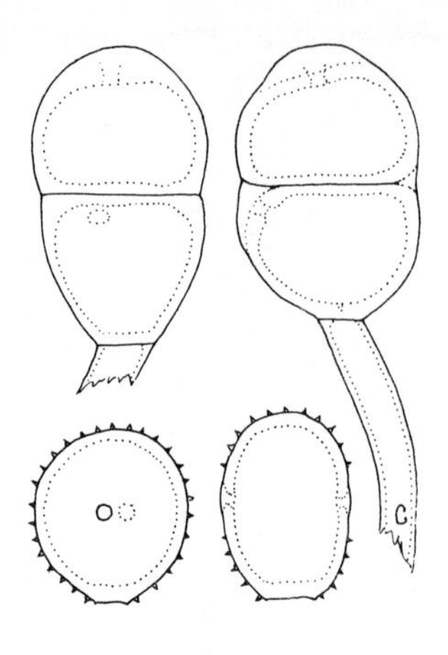

54. *PUCCINIA ABRAMSII* J. Parm. Can. J. Bot. 45:2291. 1967.

Spermogonia amphigenous. Aecia uredinoid, confluent around the spermogonia, amphigenous, orange brown; spores 25-35 x 21-29 µm, globoid, broadly ellipsoid or oblong ellipsoid, wall 1.5-2 µm thick, about cinnamon brown, echinulate, pores 2, equatorial, in flattened smooth sides. Uredinia uncertain, perhaps not produced. Teliospores in the aecia 40-48 x 29-32 µm, broadly ellipsoid or obovoid, wall 1.5-2.5(-3) µm thick at sides, mostly 6-7.5 µm at apex, uniformly clear chestnut brown, or paler externally, smooth, pore apical in each cell; pedicel colorless, usually broken at 40 µm or less.

Type: on *Geraea viscida* (Gray) Blake, Campo, San Diego County, Calif., Abrams 3633 (PUR 37627); not otherwise known.

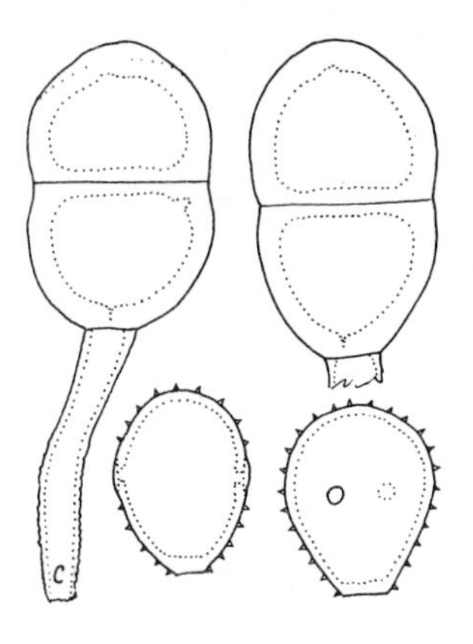

55. *PUCCINIA VAGA* H. S. Jack. Mycologia 14:112. 1922.

Spermogonia on adaxial leaf surface, few. Aecia on abaxial surface singly or in groups, peridium yellowish to buff; spores 21-28(-32) x (16-)18-22(-24) µm, globoid or broadly ellipsoid, wall 1 µm thick, colorless, finely verrucose. Uredinia amphigenous, pale cinnamon brown; spores (20-)23-27(-29) x (17-)19-23 µm, broadly ellipsoid or obovoid, wall 1(-1.5) µm thick, golden or pale cinnamon brown, echinulate except around pores, pores 2, equatorial in slightly flattened sides, with inconspicuous caps. Telia amphigenous and on stems, exposed, blackish brown, tending to be pulverulent; spores (35-)38-46(-52) x 24-28(-30) µm, mostly ellipsoid, wall 2-3 µm thick at sides, (3.5-)4.5-6 (-6.5) µm at apex, chestnut brown or slightly paler at apex but not as a defined umbo, smooth, pore of each cell apical; pedicels colorless, rugose basally, to 115 µm long.

Hosts and distribution: *Verbesina oreophila* Woot. & Standl., *V. sphaerocephala* Gray, *V.* sp.: southwestern Texas and southwestern Mexico.

Type: on *Verbesina* sp., Cuernavaca, Mor., Mexico, Holway (PUR 42610).

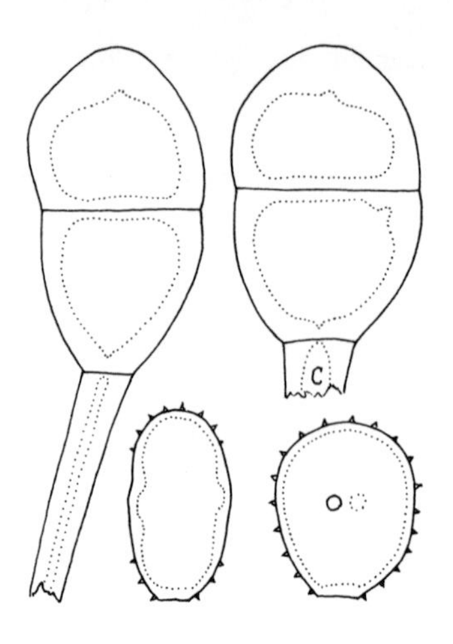

56. *PUCCINIA SPLENDENS* Vize, Grevillea 7:11. 1878.

Spermogonia amphigenous. Aecia amphigenous, cupulate to short cylindrical, the peridium erose, white; spores (22-)24-32 x (19-)21-25(-27) µm, angularly ellipsoid or more or less globoid, wall 1.5-2 µm thick, colorless or yellowish. Uredinia mostly on abaxial leaf surface, usually only on specimens having aecia, or sometimes on small witches' brooms, dark cinnamon brown; spores (24-)26-33(-35) µm long, (22-)24-27(-29) µm wide with pores face view, (15-)16-18 (-20) µm wide with pores lateral, wall (1-)1.5(-2) µm thick, echinulate, spines spaced (2-)3-4 µm, cinnamon brown, pores 2, equatorial in flattened sides. Telia on fusiform stem galls to at least 5 cm long and 2 cm diam, small and discrete on leaves, blackish brown; spores (40-)44-65(-68) x (26-)29-35(-40) µm, mostly ellipsoid, wall (2-)2.5-4(-5) µm thick at sides, (5-)6-10(-12) µm at apex, chestnut brown, smooth, pedicels colorless, often exceeding 200 µm long.

Hosts and distribution: *Hymenoclea salsola* Torr. & Gray, *H. monogyra* Torr. & Gray: the southwestern United States and adjacent Mexico.

Type: on "onion or rabbit bush" (= *Hymenoclea* sp.), California, Harkness No. 678 (K; isotype PUR 33748).

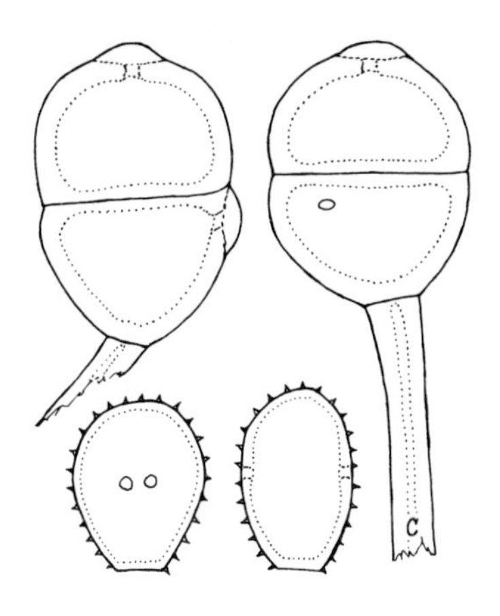

57. *PUCCINIA SOLIDIPES* H. S. Jack. & Holw. in Arthur, Amer. J. Bot. 5:527. 1918.

Spermogonia on adaxial leaf surface. Aecia on abaxial surface, yellowish, in groups of 2 to 6, in poor condition; spores about 23-29 x 18-24 µm, broadly ellipsoid or globoid, wall 1.5-2.5 µm thick including verrucae, coarsely rugose verrucose with irregularly cuboidal warts. Uredinia amphigenous, cinnamon brown; spores (24-)27-32(-34) x 24-28(-32) µm, broadly ellipsoid or obovoid with pores face view mostly 19-22 µm wide with pores lateral, wall 1.5-2 µm thick, cinnamon brown, echinulate, pores 2, approximately equatorial in flattened sides. Telia amphigenous, exposed, blackish brown, pulverulent; spores (40-)42-53(-56) x (33-)36-39 µm, broadly ellipsoid, wall 3-4 µm thick at sides, chestnut brown, 5-8 µm over pores as pale umbos, smooth, pore apical in each cell; pedicels colorless, usually roughened basally, up to 150 µm long but often less.

Hosts and distribution: *Piptothrix areolare* (DC.) King & H. Robins., *P. palmeri* Gray: western Chihuahua and southern Sonora south to Nayarit and Michoacán, Mexico and in Guatemala.

Type: on *Eupatorium tubiflorum* (= *P. areolare*), Volcán de Agua, Guatemala, Holway No. 557 (PUR 37451).

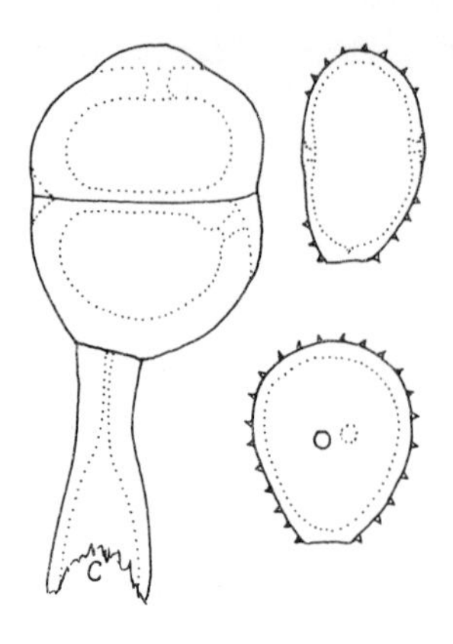

58. *PUCCINIA TURGIDIPES* H. S. Jack. Mycologia 14:110. 1922.

Spermogonia few, on abaxial leaf surface on slightly hypertrophied areas along the veins. Aecia aecidioid; spores (20-)23-27(-30) x (16-)18-22 µm, oblong ellipsoid, broadly ellipsoid or globoid, wall 1.5-2 µm thick, finely verrucose, colorless. Uredinia mostly on adaxial surface, chocolate brown; spores (24-)26-31(-33) x (15-)17-25(-28) µm, broadly ellipsoid, broadly obovoid or globoid with pores face view, wall 2-2.5 µm thick, dark cinnamon brown, echinulate, pores 2, equatorial in smooth flattened sides. Telia amphigenous, exposed, blackish brown, becoming pulverulent; spores (36-) 38-44(-48) x (29-)30-35 µm, broadly ellipsoid, wall 4-5 µm thick at sides 6-8(-9) µm over pores, dark chestnut brown except the paler umbos over pores, smooth, pore of each cell apical; pedicels colorless, to 100 µm long, swelling basally to 12-24 µm wide.

Hosts and distribution: *Viguiera deltoidea* Gray: southern Arizona to southern California and Baja California.

Type: Estrella Mts., Maricopa County, Arizona, Gooding No. 48 (PUR 42550).

This fungus has been collected rarely but is common in the Organ Pipe Cactus National Monument, Arizona, where the aecia were found. In this area, telia occasionally occur also on fusiform stem galls.

59. *PUCCINIA XIMENESIAE* Long, Bull. Torrey Bot. Club 29: 114. 1902.

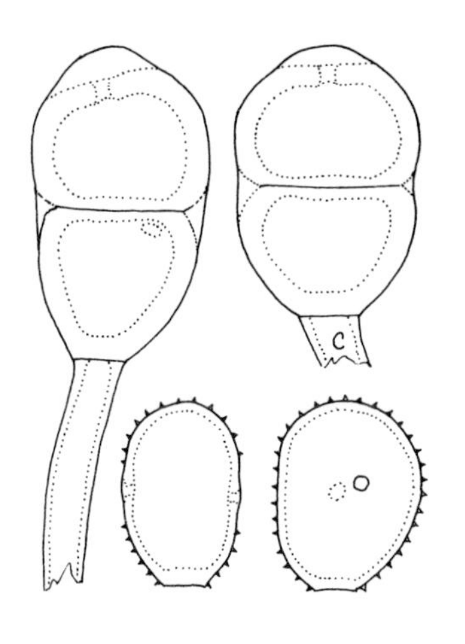

Spermogonia amphigenous, few. Aecia amphigenous in groups, peridium whitish, fragile; spores (20-)26-34(-37) x (15-) 18-23 μm, from globoid to oblong ellipsoid, wall about 1 μm thick, finely verrucose, colorless. Uredinia amphigenous, cinnamon brown; spores (23-)25-32(-36) x (18-)20-24(-27) μm, ellipsoid or obovoid, wall 1.5-2 μm thick, echinulate except around pores, cinnamon brown or slightly darker, pores 2, in slightly flattened sides. Telia amphigenous, exposed, blackish brown, more or less pulverulent; spores (30-) 35-46(-52) x (24-)26-32(-35) μm, ellipsoid or broadly ellipsoid, wall (2-)2.5-4(-5) μm thick at sides, dark chestnut brown, or thinner and golden brown in some spores, (5.5-)6-8(-10) μm at pores with pale brown umbos, defined in pale spores, less so in dark colored spores, smooth, pore apical in upper cell, next to septum in lower cell, pedicels nearly colorless, to 135 μm long on dark spores, more fragile on pale spores.

Hosts and distribution: *Verbesina* spp.: in the United States and Mexico along the boundary from Texas to Baja California and in Durango, Mexico.

Neotype: on *Verbesina encelioides* (Cav.) Benth. & Hook., Austin, Texas, 14 Nov. 1899, Long No. 65 (PUR 34614).

A neotype is established because I have found no specimen that matches the published record: "Nov. 16, 1899, No. 65, by W. L. Bray." A specimen in BPI, where the Long Herbarium is located, bears the number 65 and the date Nov. 14, 1899 but no collector's name and the script is not that of Long. Parmelee (25) accepted PUR 34614 as an isotype, which it cannot be, but it does fortify its designation as neotype. Perhaps Long assigned his No. 65 to Bray's collection, but this is not certain.

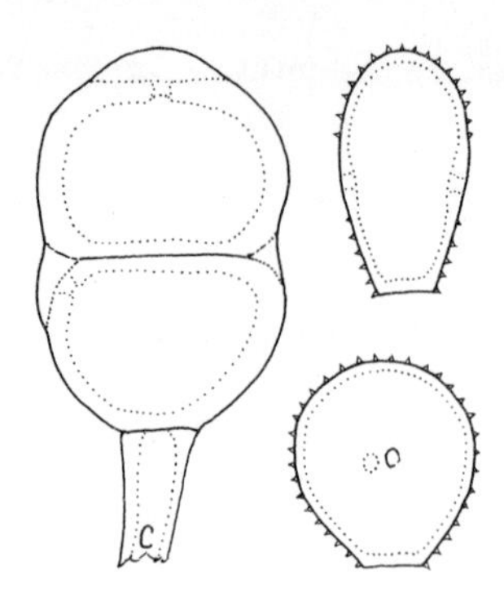

60. *PUCCINIA TAGETICOLA* Diet. & Holw. in Holway, Bot. Gaz. 24:26. 1897.

Spermogonia and aecia unknown. Uredinia amphigenous and on stems, pale cinnamon brown; spores (24-)26-30(-34) x (16-)19-24(-26) µm, ellipsoid or obovoid, wall 1.5(-2) µm thick, about cinnamon brown, echinulate except around the pores, pores 2, equatorial in flattened sides. Telia amphigenous and on stems, exposed, blackish brown, pulvinate; spores (40-)42-50(-58) x (26-)28-35(-40) µm, mostly broadly ellipsoid but variable in some collections, wall (2-)2.5-4 (-4.5) µm thick at sides, (5.5-)7-9(-10) µm over pores as pale defined umbos, pore apical in each cell, chestnut brown, smooth; pedicel colorless, to 200 µm long.

Hosts and distribution: *Tagetes* spp.: central Mexico to Costa Rica; also in South America and Puerto Rico.

Type: on *Tagetes tenuifolia* Cav. (= *T. patula* L.), Guadalajara, Jal., Mexico, Holway (S; isotype PUR 42644).

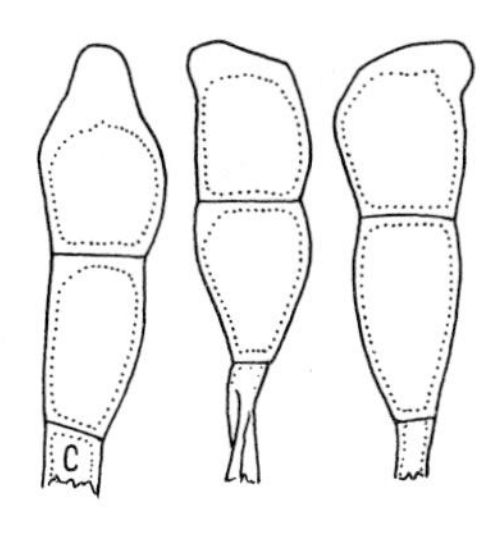

61. *PUCCINIA TENUIS* Burr. Bot. Gaz. 9:188. 1884.

Spermogonia few on adaxial leaf surface or absent. Aecia mostly on abaxial surface in groups, peridium short and fragile; spores 15-17 x 12-16 µm, mostly globoid, wall 1-1.5 µm thick, colorless, verrucose. Uredinia lacking. Telia mostly on abaxial surface, associated with aecia or separate, covered by the epidermis, blackish, loculate with brownish stromatic paraphyses; spores (28-)32-43(-47) x (11-)12-15(-18) µm, mostly narrowly oblong or oblong ellipsoid, sometimes fusiform, wall 1-1.5 µm thick at sides, 5-8 (-10) µm at apex, yellowish or golden brown, smooth; pedicels colorless, 10-25 µm long.

Hosts and distribution: *Eupatorium rugosum* Hout.: Quebec west to Minnesota and Nebraska and southward to Florida.

Type: on *Eupatorium ageratoides* (= *E. rugosum*), Bloomington, Illinois, Seymour (ILL 24724; isotype PUR 61219).

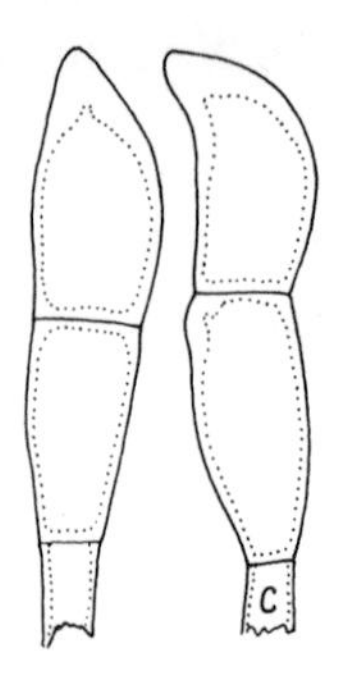

62. *PUCCINIA BATESIANA* Arth. Bull. Torrey Bot. Club 28:661. 1901.

Spermogonia on adaxial surface. Aecia mostly on abaxial surface in small groups, the peridium short, whitish; spores (15-)17-22(-24) x 14-20 µm, mostly broadly ellipsoid or globoid, wall 1 µm thick, colorless, finely verrucose. Uredinia lacking. Telia in small, compact groups on the abaxial surface, covered by the epidermis, blackish brown, loculate with brown, stromatic paraphyses; spores (35-)40-60(-64) x (10-)12-17(-19) µm, mostly more or less cylindrical, often curved, the apex obtuse, rounded or usually narrowed, wall 0.5-1(-1.5) µm thick at sides, (3-)5-8(-10) µm at apex, nearly uniformly chestnut brown or deep golden brown, smooth, pores apical but obscure; pedicels yellowish to pale brownish, to 25 µm long.

Hosts and distribution: *Heliopsis helianthoides* (L.) Sweet var. *helianthoides* and var. *scabra* (Dunal) Fern.: Nebraska to Maryland.

Type: on *H. scabra* Dunal, Long Pine, Nebr., Bates (PUR 36442; probable isotypes Bartholomew N. Amer. F. 125; Griffiths W. Amer. F. 322).

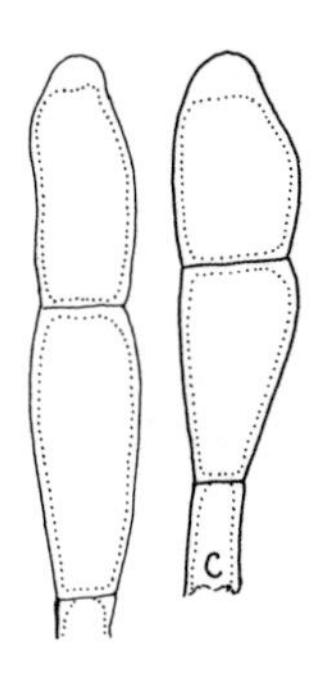

63. *PUCCINIA DESMANTHODII* Diet. & Holw. in Holway, Bot. Gaz. 31:334. 1901.

Spermogonia not seen, probably not produced. Aecia amphigenous in small groups in necrotic areas, peridium short, erose, white; spores 17-22(-24) x 14-19 µm, mostly nearly globoid, wall 1-1.5 µm thick, colorless, finely verrucose rugose, pore plugs present. Uredinia lacking. Telia on abaxial leaf surface in close groups, covered by the epidermis, blackish brown, loculate with brown stromatic paraphyses; spores (35-)40-55(-60) x (10-)12-18(-20; -26) µm, mostly nearly cylindrical, the apex obtuse, rounded or acute, wall 0.5-1(-1.5) µm thick at sides, 3.5-7(-9) µm at apex, uniformly clear chestnut brown, smooth; pedicels brown, to 25 µm long.

Hosts and distribution: *Desmanthodium fruticosum* Greenm., *D. ovatum* Benth.: Nayarit to Oaxaca, Mexico.

Type: on *Desmanthodium ovatum*, Oaxaca, Oax., Holway No. 3365 (S; isotypes PUR 36481; Bartholomew N. Amer. Ured. 1540).

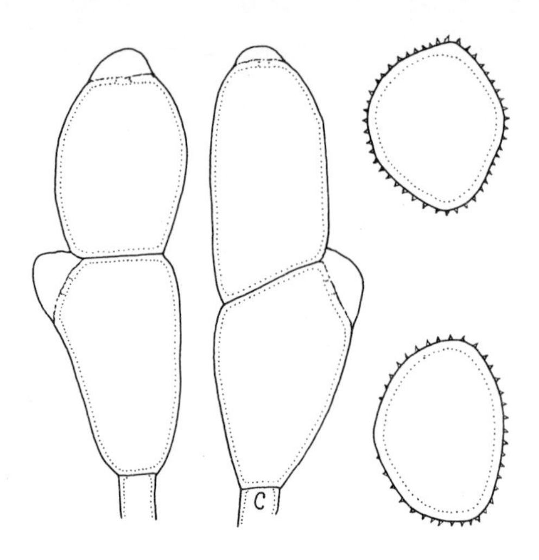

64. *PUCCINIA INTERJECTA* H. S. Jack. Mycologia 24:148. 1932.
Allodus ancizari sensu Arth. & Orton, N. Amer. Flora 7: 476. 1921, not *Puccinia ancizari* Mayor 1913.

Spermogonia amphigenous in groups. Aecia amphigenous or mostly on adaxial leaf surface, without peridium, opening by a pore in the host epidermis, yellowish; spores (30-)33-38(-42) x (21-)23-28(-30) µm, mostly ellipsoid or broadly ellipsoid, wall 1.5-2(-2.5) µm thick, colorless, echinulate. Uredinia lacking. Telia on abaxial leaf surface, exposed, golden brown, compact; spores (56-)62-78(-86) x (21-)23-27 (-30) µm, elongately ellipsoid, wall 1 µm thick at sides, yellowish, (3-)4.5-6.5(-8) µm over pores as colorless umbos, smooth, pore apical in each cell; pedicel colorless, to 60 µm long, usually broken shorter; germinating without dormancy.

Hosts and distribution: *Baccharis* spp.: San Luis Potosí, Mexico to Guatemala.

Type: on *Baccharis lancifolia* Less., Cerro Quemado, Quezaltenango, Guatemala, Holway No. 103 (PUR 36483).

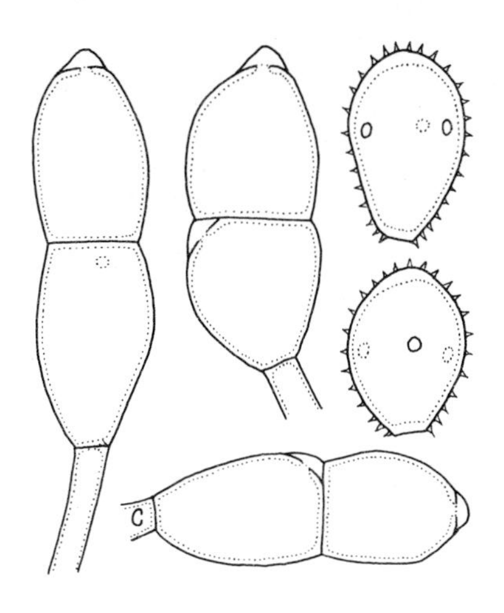

65. *PUCCINIA VALLARTENSIS* Hennen & Cumm. Rept. Tottori Mycol. Inst. 10:175. 1973.

Spermogonia mostly on adaxial leaf surface. Aecia on abaxial surface, deep seated, without peridium but opening by a pore, yellow when fresh; spores (24-)27-34(-37) x (17-) 20-24 µm, broadly ellipsoid or broadly obovoid, wall 1(-1.5) µm thick or 1.5-3(-4) µm at apex, colorless, echinulate with spines spaced about 3 µm, pores 3, equatorial, obscure. Uredinia lacking? Telia on abaxial surface around the aecia, pulvinate, cinnamon brown becoming gray from germination; spores (42-)50-66(-70) x (16-)19-24(-25) µm, narrowly ellipsoid or more or less cylindrical, wall 0.5 µm thick at sides and pale golden brown, umbonate over pore at apex and septum, the umbo 2.5-4 µm thick and nearly colorless, smooth; pedicel colorless, to 70 µm long but often shorter; teliospores germinate without dormancy.

Hosts and distribution: *Baccharis trinervis* (Lam.) Pers., Jalisco and Nayarit, Mexico.

Type: south of Puerto Vallarta, Jal., Cummins No. 71-525 (PUR 64083).

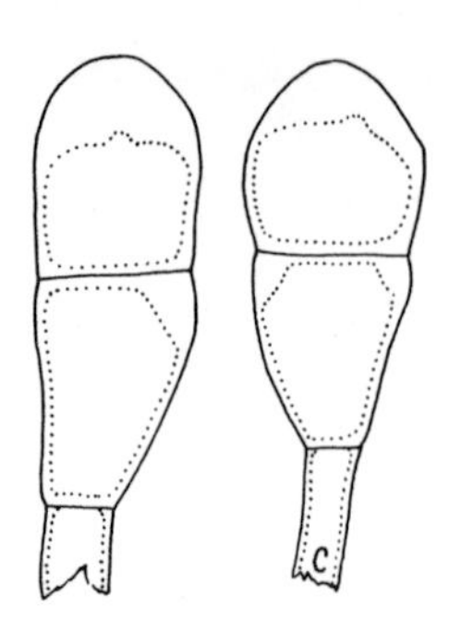

66. *PUCCINIA INVESTITA* Schw. Trans. Amer. Phil. Soc. II. 4: 296. 1832.

Spermogonia on adaxial leaf surface. Aecia mostly on abaxial surface in loose groups or singly, peridium cylindrical becoming lacerate, whitish; spores (19-)20-26(-29) x (16-)18-22(-24) µm, globoid or nearly so, wall 1-1.5 µm thick, colorless, finely verrucose. Uredinia lacking (possibly aecidioid?). Telia on abaxial surface and on stems, exposed, blackish brown, compact; spores (36-)40-53(-58) x (14-)18-23(-25) µm, more or less oblong or elongately obovoid, wall 1-1.5(-2) µm thick at sides, (5-)8-10(-13) µm at apex, commonly also thickened in the angles of the lower cell at septum, uniformly chestnut brown, smooth, pore apical in each cell; pedicels colorless or yellowish, to about 60 µm long.

Hosts and distribution: *Gnaphalium* spp.: Ontario and Vermont south to Mexico and Central America; also in South America.

Neotype: on *G. obtusifolium* L., Shelter Island, Long Island, N.Y., 20 Oct. 1905, Farlow (PUR 36470; isotypes Reliq. Farl. No. 258).

Arthur and Bisby (3) report that there is no original specimen, only an empty packet in PH.

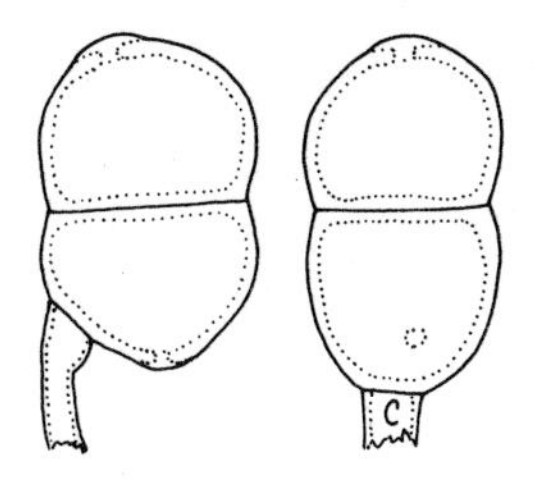

67. *PUCCINIA INTERMIXTA* Peck, Bot. Gaz. 4:218. 1879.

Spermogonia abundant, systemic, amphigenous. Aecia mostly on abaxial leaf surface, systemic, peridium hemisphaeric, rupturing irregularly, yellowish; spores 19-30 (-35) x 16-25 µm, variable in size and shape, mostly from ellipsoid to globoid, wall 1-1.5 µm thick, nearly colorless, finely verrucose. Uredinia lacking. Telia amphigenous, exposed, most often systemic (as collected) but sometimes localized, pulverulent, chocolate brown; spores (30-)32-42 (-48) x (18-)20-26(-30) µm but usually with much smaller spores intermixed, mostly ellipsoid or broadly ellipsoid, wall 1-1.5 µm thick except 2-3 µm at the pores by illdefined low umbos, uniformly clear chestnut or deep golden brown, smooth, pore in upper cell apical, in lower cell in lower 1/4; pedicels colorless, always broken near hilum in mature spores.

Hosts and distribution: *Iva axillaris* Pursh: Manitoba to Alberta south to New Mexico and California.

Type: Green River, Wyoming, 1879, Jones (NYS; isotype PUR 36354).

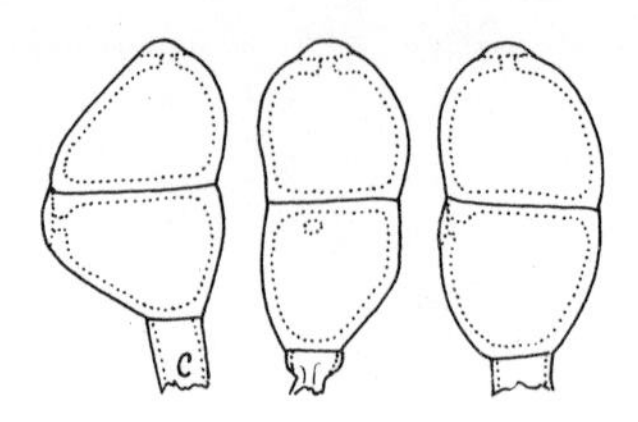

68\. *PUCCINIA SENECIONIS* Lib. Pl. Crypt. Ard. Exs. No. 92, Cent. I. 1830.
Puccinia subcircinata Ellis & Ever. J. Mycol. 3:56. 1887.

Spermogonia not seen. Aecia mostly amphigenous, peridium cup like, the margin erose; spores 17-21 x 16-19 µm, more or less globoid, wall 1-1.5 µm thick, colorless, finely verrucose. Uredinia aecidioid, like the aecia but occurring singly; spores like the aeciospores. Telia mostly amphigenous, surrounding the aecia or uredinia or in separate groups, exposed, dark chestnut brown, pulverulent; spores (21-)23-33(-36) x (14-)16-21 µm, mostly more or less ellipsoid, wall 1-1.5(-2) µm thick at sides, chestnut brown, 2-3.5(-4) µm thick over pores with low, colorless umbos, smooth, pore usually apical in each cell; pedicels colorless, always broken near hilum.

Hosts and distribution: *Senecio* spp.: North Dakota and Alberta to New Mexico and California.

Type: on *S. saracenicus* L., Ardennes Mts., France (isotypes Pl. Crypt. Ard. Exs. No. 92, Cent. I; BPI).

69. *PUCCINIA VIRGAE-AUREA* (DC.) Lib. Pl. Crypt. Ard. 393. 1837.
Xyloma virgae-aurea DC. in Lam. & DC. Syn. Pl. p. 63. 1806.

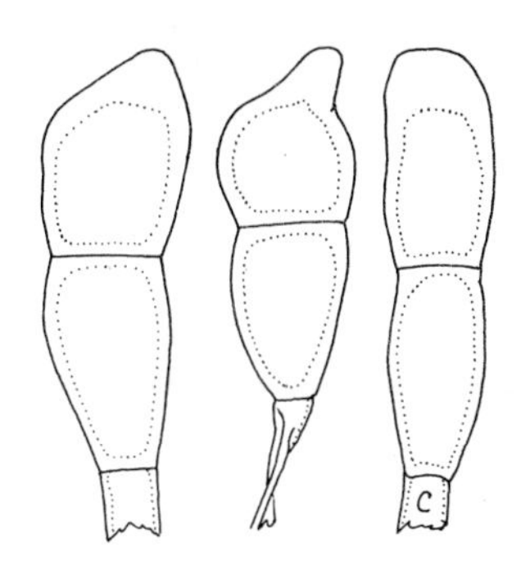

Telia on abaxial leaf surface, covered by the epidermis, closely grouped but the sori discrete, bounded by compacted hyphae, grayish black; spores variable (33-)40-55(-62) x (11-)15-20(-22) µm, mostly elongately obovoid, oblong ellipsoid or even cylindrical or fusiform, wall 1.5-2.5(-3) µm thick at sides, (4-)5-8(-11) µm at apex, golden brown, smooth, pore in each cell apical but obscure; pedicels brownish, to 30 µm long.

Hosts and distribution: *Solidago* spp.: the Great Lakes region east to Nova Scotia; also in Europe and Asia.

Type: on *Solidago virgaurea* L., France (G). Not seen.

70. *PUCCINIA GALLULA* Hennen & Cumm. Rept. Tottori Mycol. Inst. 10:137-138. 1973.

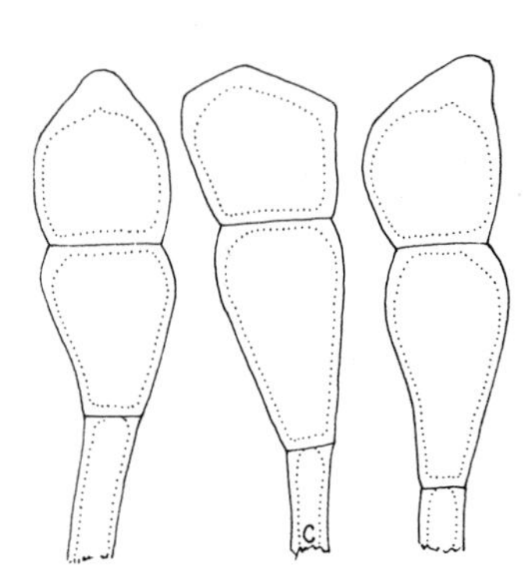

Telia encircling stems on slightly hypertrophied areas as much as 4 cm long, blackish, with abundant brown stromatic paraphyses that delimit locules containing spores; spores variable in size and shape, (37-)44-58(-66) x (13-)17-21(-25) µm, narrowly ellipsoid, oblong or narrowly obovoid, wall 1-1.5(-2) µm thick at sides, (3-)5-7(-9) µm apically, chestnut brown, smooth; pedicel brownish, to 45 µm long but usually shorter.

Hosts and distribution: *Porophyllum scoparium* Gray, southwestern Texas and in Chihuahua and Coahuila, Mexico.

Type: Big Bend National Park, Texas, Cummins No. 70-37 (PUR 63751).

71. *PUCCINIA STROMATIFERA* Hennen, Leon-Gall. & Cumm. Southw. Nat. 16:374. 1972.

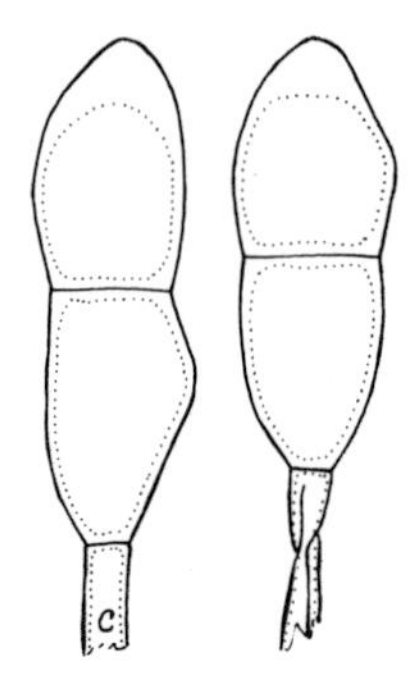

Telia amphigenous, in close groups, blackish, covered by the epidermis, loculate with brown stromatic paraphyses; spores (40-)45-55(-60) x (13-)16-20 µm, oblong ellipsoid, nearly cylindrical or sometimes fusiform, apex from rounded to acute, wall 1-1.5(-2) µm thick at sides, 5-10 µm thick at apex, clear chestnut brown or the apex paler, smooth; pore in each cell apical, obscure; pedicels pale brown, to 25 µm long, often shorter.

Hosts and distribution: *Perezia nudicaulis* Gray and *P. palmeri* Wats.: Chiapis and Baja California Sur, Mexico.

Type: on *P. nudicaulis*, east of La Trinitaria on road to Layo de Montebello, Chis., Breedlove and Raven No. 8343 (PUR 62570).

72. *PUCCINIA FRASERI* Arth. Bull Torrey Bot. Club 42:591. 1915.

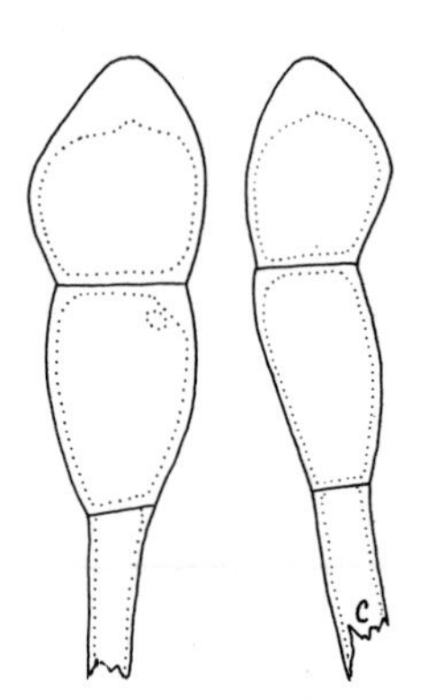

Telia mostly on abaxial leaf surface, systemic in extensive chlorotic areas, densely grouped, exposed, compact, pale yellowish when old, bright yellow when fresh; spores (30-)36-46(-50) x (13-)16-20 µm, mostly elongately obovoid, less often narrowly ellipsoid, wall 1-1.5 µm thick at sides, (4-)6-9(-10) µm at apex, colorless or faintly yellowish, smooth, pore apical in each cell, obscure; pedicels colorless, to 70 µm long.

Hosts and distribution: *Hieracium* spp.: Nova Scotia and Quebec to Pennsylvania and Montana.

Type: on *Hieracium scabrum* Michx., Pictou, Nova Scotia, Fraser, (PUR 42346).

73. *PUCCINIA EXCURSIONIS* Savile, Mycologia 57:476. 1965.

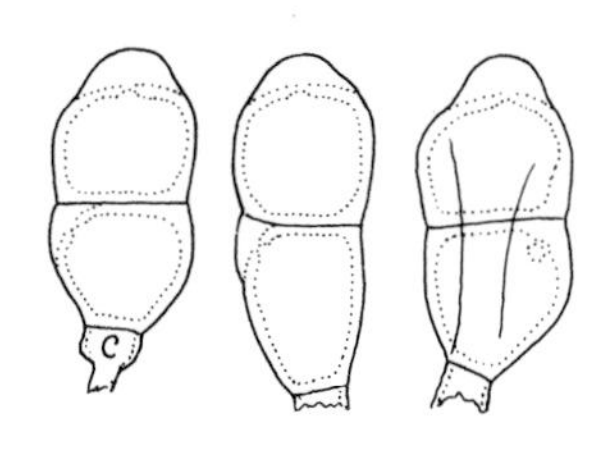

Telia amphigenous, exposed, confluent as large pulverulent sori, chocolate brown; spores (22-)26-38(-42) x (11-)13-18(-20) µm, mostly ellipsoid, oblong ellipsoid, or elongately obovoid, wall (1-)1.5(-2) µm thick at sides, (4-)4.5-6(-6.5) µm thick at apex with a broad, yellowish umbo over the pore, a less conspicuous umbo over lower pore, clear chestnut or golden brown, smooth but often with refractive ridges; pedicels colorless, broken near hilum.

Type: on *Erigeron peregrinus* (Pursh) Greene ssp. *callianthemus* (Greene) Cronq., Fourth of July Camp, Boulder County, Colorado, Savile 4886 (DAOM; isotype PUR 59359). There is another specimen from Gunnison County, Colo. Both collections are from approximately 11,000 ft.

74. *PUCCINIA CONGLOMERATA* (Strauss) Roehl. Deutschl. Fl. Ed. 2. 3(3):130. 1813.
Uredo conglomerata Strauss, Ann. Wetter. Ges. 2:100. 1810; telia present.

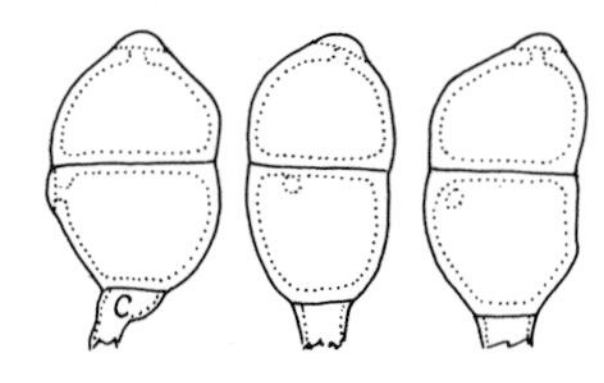

Telia mostly on abaxial leaf surface in groups to 1 cm diam, exposed, chestnut brown, pulverulent; spores (22-)24-34(-38) x (13-)15-19 (-21) µm, mostly ellipsoid, wall 1-1.5(-2) µm thick at sides, chestnut brown, (2.5-)3-3.5(-4) µm thick over pores with low colorless umbos, smooth or rarely with a few ridges, pore apical or nearly so in each cell; pedicels colorless, broken near hilum.

Hosts and distribution: *Petasites frigidus* (L.) Fries, *P. sagittatus* (Banks) Gray: New York and Minnesota to Alaska; also in Europe.

Type: on *Homogyne alpina* (L.) Cass. (as *Tussilago alpina*), mountains of Bavaria (type deposited?).

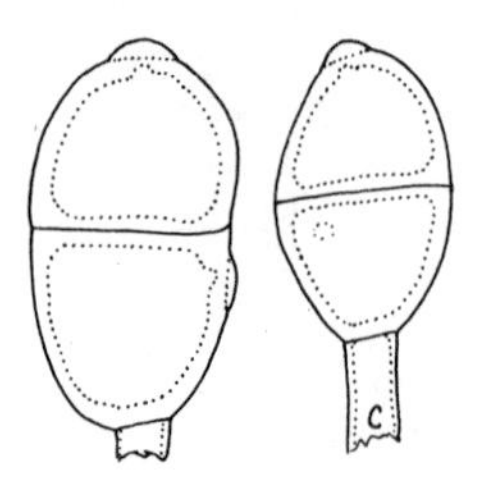

75. *PUCCINIA GLOMERATA* Grev. Fl. Edin. P. 433. 1824.
Puccinia expansa Link in Willdenow, Species Plantarum Ed. 4, VI, 2. p. 75. 1825.

Spermogonia and rudimentary aecia reported in Europe but rare. Uredinia lacking. Telia abaxial or sometimes on both leaf surfaces and petioles, exposed, chestnut brown, pulverulent; spores dimorphic, (20-)23-31(-37) x 14-19(-21) µm or (25-)29-41(-45) x (17-)18-27(-29) µm, the larger more abundant, mostly more or less ellipsoid, wall 1.5-2 µm thick at sides, golden brown to chestnut brown, (2.5-)3-5(-6) µm over pores with small, colorless umbos, smooth; pore in each cell usually apical or nearly so; pedicels colorless, always broken near hilum.

Hosts and distribution: *Senecio* spp.: Montana and Washington to Utah and California; also in the Old World.

Lectotype: on *Senecio jacobaea* L., Scotland, locality and date not recorded (E).

The original publication cites "... spring and summer. Caroline Park and on the coast of Fife, not rare." A lectotype designation is used because the specimen in (E) cannot be associated with either Caroline Park or Fife but obviously is authentic because the packet bears the notation "mihi".

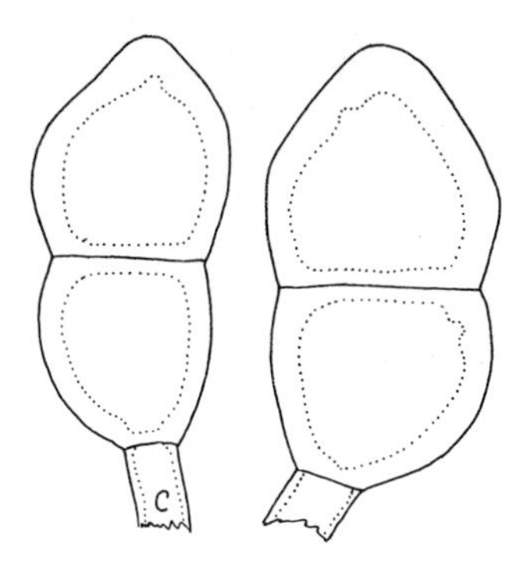

76. *PUCCINIA MARIANAE* P. Syd. & H. Syd. Hedwigia 40(Beibl.): 127. 1901.

Spermogonia, aecia and uredinia unknown, probably not formed. Telia amphigenous, loosely grouped, exposed, compact, blackish brown; spores (34-)42-62(-68) x (18-)22-30 (-33) μm, variable in size and shape, from narrowly ellipsoid or nearly fusiform to broadly ellipsoid, wall (2.5-)3-4(-4.5) μm thick at sides, (4-)5-8(-10) μm at apex, clear chestnut or deep golden brown, smooth, pore apical in upper cell, next to septum in lower cell, inconspicuous, without umbos; pedicels persistent, colorless, to 125 μm long; 1 celled spores frequent.

Hosts and distribution: *Sideranthus megacephalus* (Nash) Small: southern Florida.

Type: on *Chrysopsis mariana* (= *Sideranthus megacephalus*), Sanibel Island, Florida, Tracy, Plants of the Gulf States No. 7240 (S: isotype PUR 41995).

77. *PUCCINIA GRINDELIAE* Peck, Bot. Gaz. 4:127. 1879.
Puccinia xylorhizae Arth. Bull. Torrey Bot. Club 28: 662. 1901.
Gymnoconia riddellii D. Griff. Bull. Torrey Bot. Club 29:296. 1902.
(See N. Amer. Flora for other synonyms.)

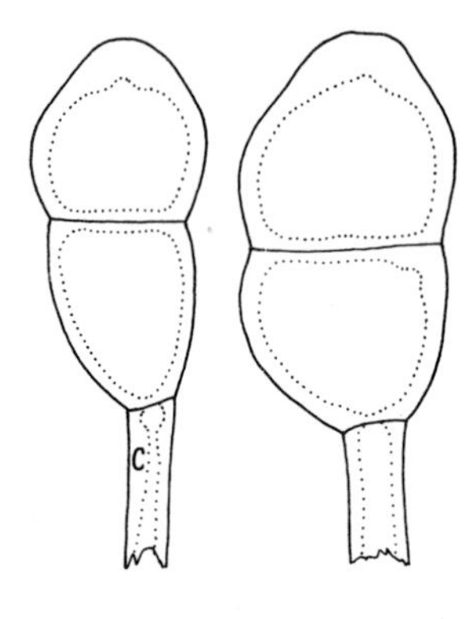

Spermogonia present on some hosts. Aecia usually lacking but occasionally present with spermogonia early in season; spores 24-36 x 20-26 μm, wall (2-)2.5-3.5(-4) μm thick, verrucose, yellowish. Uredinia wanting. Telia amphigenous and on stems, exposed, in groups of various sizes or rarely solitary, compact, blackish brown; spores (34-)40-58(-64) x (18-)20-26(-28) μm, mostly elongately obovoid, sometimes oblong ellipsoid, wall 1.5-2.5(-3.5) μm thick at sides, (4-)7-10(-14) μm at apex, mostly chestnut brown or long narrow spores often golden, smooth; pore in each cell apical but inconspicuous; pedicels essentially colorless, persistent, to 200 μm long; 1, 3, and 4 celled spores occur in some collections.

Hosts and distribution: species of *Aster, Baileya, Chrysoma (= Ericameria), Chrysopsis (= Heterotheca), Chrysothamnus, Ericameria, Erigeron, Grindelia, Gutierrezia, (= Xanthocephalum), Haplopappus, Heterotheca, Hymenopappus, Hymenoxys, Isocoma, Lygodesmia, Machaeranthera, Oliveae, Prionopsis, Psilostrophe, Solidago, Tetradymia, Tetraneuris (= Hymenoxys), Xanthocephalum, Xylorhiza (= Aster)*: Wisconsin to Alberta and south to central Mexico.

Type: on *Grindelia squarrosa* (Pursh) Dunal, Colorado (without locality), Brandegee (NYS).

The size of the spores varies considerably between and within collections but the spores typically are long and always have long pedicels. The occurrence of aecia or of peridial cells and aeciospores with the teliospores is of uncertain significance but it is doubtful that a stable opsis form exists. The descriptions of *Puccinia xylorhizae* and *Gymnoconia riddellii* include the aeciospores. In Arizona, spermogonia, aecia and a few urediniospores are common in early season infections (*G. riddellii*) of *Psilostrophe* and *Baileya*.

78. *PUCCINIA SPEGAZZINII* De-Toni in Sacc. Syll. Fung. 7:704. 1888.
Puccinia australis Speg. An. Soc. Cient. Argent. 10:8. 1880, not Koernicke 1876.

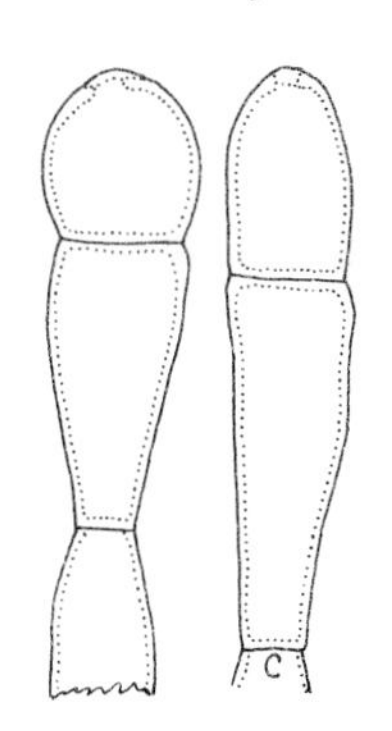

Telia on abaxial leaf surface, exposed, in close groups, pale cinnamon brown, becoming gray from germination, compact; spores (35-) 40-60(-70) x (10-)13-18(-20) µm, narrowly ellipsoid or cylindrical, wall 0.5-1 µm thick at sides, 2-3(-4) µm thick at apex, yellowish to nearly colorless, smooth, pore of each cell apical; pedicels about as wide as the spore but collapsing, colorless, to 70 µm long.

Hosts and distribution: *Mikania* spp.: southern United States southward to Panama; also in South America.

Type: on *Mikania scandens* (L.) Willd. var. *periplocifolia* (Hook. & Arn.) Baker, La Boca del Riachuelo, Argentina, 1880, Schnyder (LPS).

79. *PUCCINIA SCHISTOCARPHAE* H. S. Jack. & Holw. in Arthur, Amer. J. Bot. 5:534. 1918.

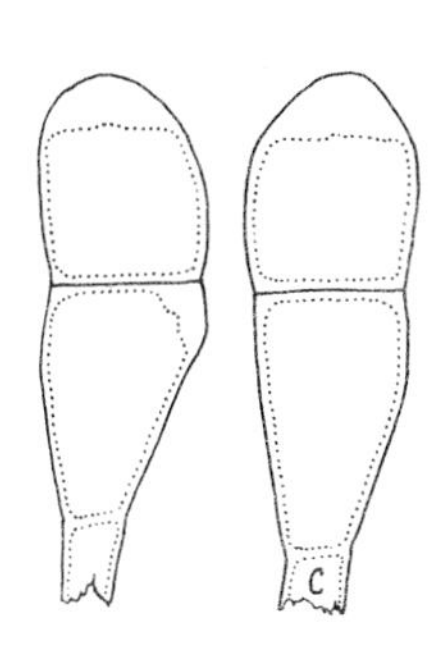

Telia on abaxial leaf surface, exposed, in small close groups, becoming confluent, compact and almost waxy in appearance, hard when dry, pale yellowish brown; spores (33-)40-52(-64) x (15-)17-20(-22) µm, oblong ellipsoid or elongately obovoid, wall 1 µm thick at sides, 5-9.5 µm thick at apex, the thickening abrupt, smooth, essentially colorless, pore apical in each cell; pedicels colorless, mostly less than 25 µm long.

Hosts and distribution: *Schistocarpha* spp.: Guatemala.

Type: on *S. platyphylla* Greenm., San Rafael, Holway No. 42 (PUR 42238).

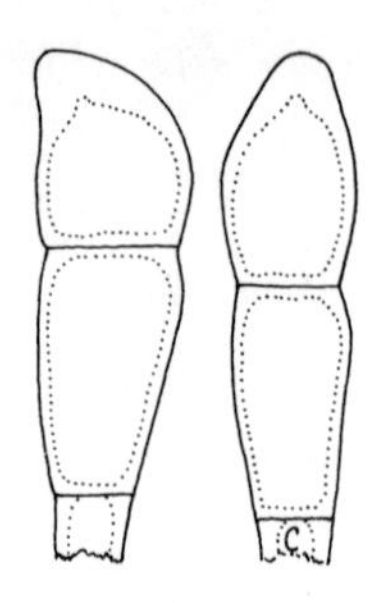

80. *PUCCINIA XANTHII* Schw. Schr. Nat. Ges. Leipzig 1:73. 1822.

Telia mostly on abaxial leaf surface, exposed, in close groups of various sizes, the sori often confluent, not loculate but usually with some stromatic paraphyses, the sporogenous layer dark brown, sori chocolate brown becoming gray from germination, compact; spores (30-)36-60(-70) x 13-19 µm, variable in size and shape, narrowly ellipsoid, narrowly obovoid or nearly cylindrical, wall 1-1.5 µm thick at sides, (4-)5-8(-11) µm at apex, clear chestnut brown or golden brown, the apex somewhat paler, smooth, pore apical in each cell; pedicel pale golden, to 50 µm long but usually shorter.

Hosts and distribution: on species of *Ambrosia* and *Xanthium*: circumglobal.

Type: on *Xanthium* sp., Sal. & Beth (= Salem, North Carolina and Bethlehem, Pennsylvania), no collector or date (PH).

81. *PUCCINIA DYSSODIAE* Cumm. Mycotaxon 5:403. 1977.

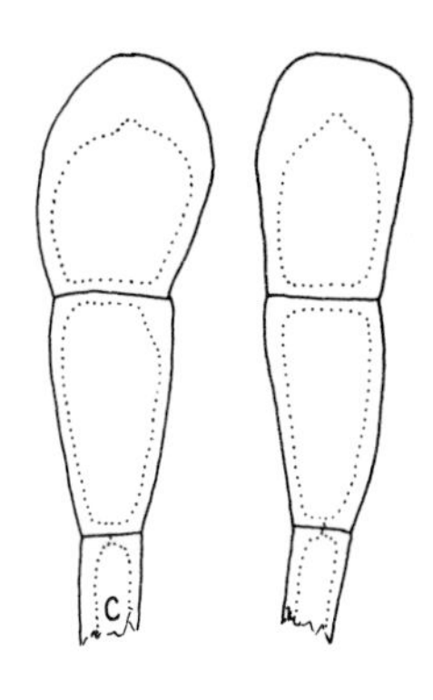

Telia single or usually grouped along the leaves and stems, exposed, not loculate but occasional groups of stromatic paraphyses occur, the sporogenous layers dark brown, compact, blackish brown or gray from germination; spores (39-)44-60(-65) x (17-)18-23(-25) µm, variable but mostly elongately clavate or oblong ellipsoid, the apex mostly broadly rounded or obtuse, side wall of lower cell (1.5-)2-2.5(-3) µm thick, of upper cell usually 2.5-4 µm thickening toward apex, apical wall (5-)7-9(-11) µm, chestnut brown; pedicel 35-40 µm.

Hosts and distribution: *Dyssodia greggii* (Gray) B. L. Rob., *D. pentachaeta* (DC.) B. L. Rob.: southeastern Arizona and central Nuevo Leon, Mexico; two collections known.

Type: on *D. pentachaeta*, Portal, Arizona, Cummins No. 61-173 (PUR 59039).

82. *PUCCINIA PRAEMORSA* Diet. & Holw. in Holway, Bot. Gaz. 31:332. 1901.

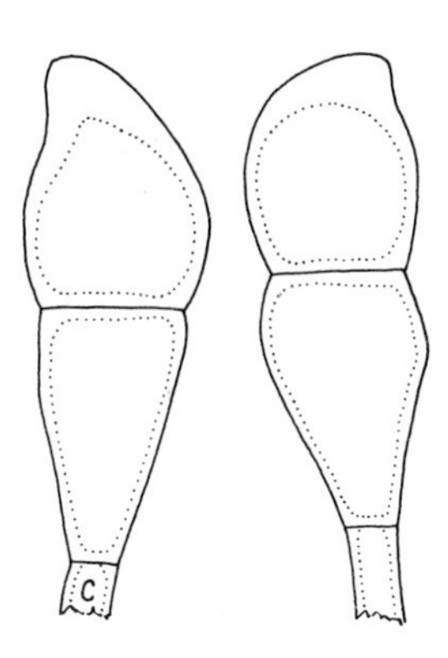

Telia on abaxial leaf surface, exposed, in circular groups, compact, chocolate brown; spores (42-)50-70 (-78) x (16-)20-27(-32) µm, oblong, clavate or sometimes fusiform, wall (1.5-)2-3(-4) µm thick at sides, (3-)4-8(-12) µm at apex, chestnut brown apically, cinnamon brown or yellowish basally, smooth, pore apical in each cell, indistinct; pedicels yellow or yellowish brown, to 30 µm long.

Hosts and Distribution: *Brickellia veronicaefolia* Gray: Mexico from central Nuevo Leon to Oaxaca.

Type: Oaxaca, Oax., Holway No. 3686 (S; isotype PUR 41402).

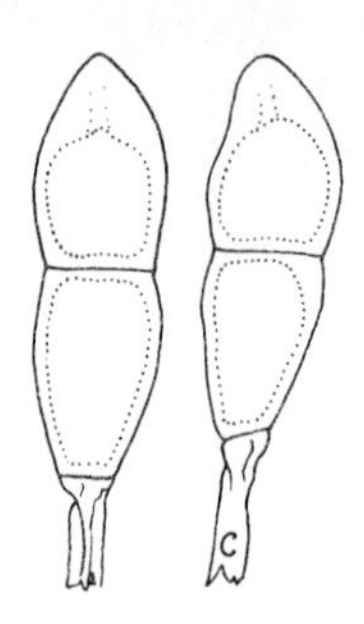

83. *PUCCINIA SILPHII* Schw. Trans. Amer. Phil. Soc. II. 4: 296. 1832.

Telia mostly on abaxial leaf surface and stems, exposed, closely grouped and often confluent, compact, germinating form about cinnamon brown, dormant form blackish brown; spores (30-)35-50(-57) x (8-)13-18(-20) µm, variable but mostly elongately ellipsoid, apex from obtusely rounded to attenuate, wall 1-1.5(-2.5) µm thick at sides, (4-)6-9 (-12) µm at apex, uniformly golden brown to clear chestnut brown, smooth; pedicels mostly pale golden, to 65 µm long but usually less than 50 µm.

Hosts and distribution: *Silphium* spp.: the midwestern and eastern U.S. and in southwestern Ontario.

Type: on *S. trifoliatum* L.: North Carolina, Denke (PH).

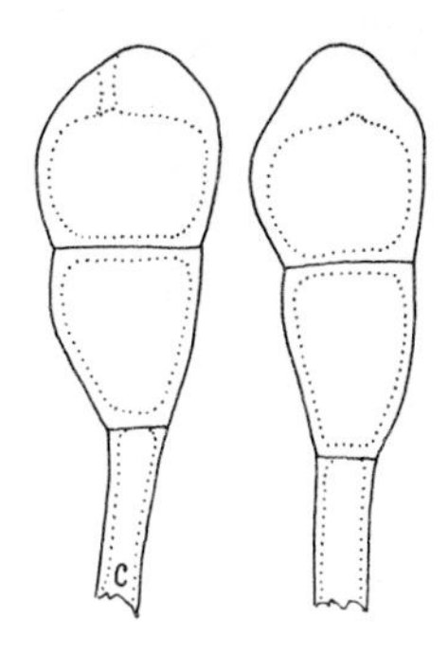

84. *PUCCINIA CNICI-OLERACEI* Pers. ex Desm. Cat. Pl. Omis. p. 24. 1823.
Puccinia asteris Duby, Bot. Gall. p. 888. 1830.
Puccinia maculosa Schw. Trans. Amer. Phil. Soc. II. 4: 295. 1832, not Roehling 1813.
Puccinia millefolii Fckl. Jahrb. Nass. Ver. Nat. 23-24: 55. 1869.
Puccinia ptarmicae Karst. Bidr. Kaenned. Finl. Nat. Folk. 31:41. 1879.
Puccinia columbiensis Ellis & Ever. Proc. Acad. Nat. Sci. Phila. 1893:153. 1893.
Puccinia rudbeckiae Barth. in Arthur, N. Amer. Flora 7: 580. 1922.

Telia mostly on the abaxial leaf surface, exposed, mostly in tight groups, blackish brown or becoming gray from germination, compact; spores (32-)37-50(-55) x (13-)15-20 (-23) µm, mostly elongately obovoid or more or less oblong, wall (1-)1.5(-2.5) µm thick at sides, (4-)6-10(-16) µm thick at apex, smooth, pore apical in each cell; pedicel colorless or yellowish, to about 50 µm long.

Hosts and distribution: species of *Achillea, Agoseris, Artemisia, Aster, Erigeron, Hieracium, Krigia, Prenanthes,* and *Rudbeckia*: Alaska and Canada southward to Costa Rica; also in South America, Europe and Asia.

Type: on *Cnicus oleraceus* L. (= *Cirsium oleraceum*), northern France (deposited?).

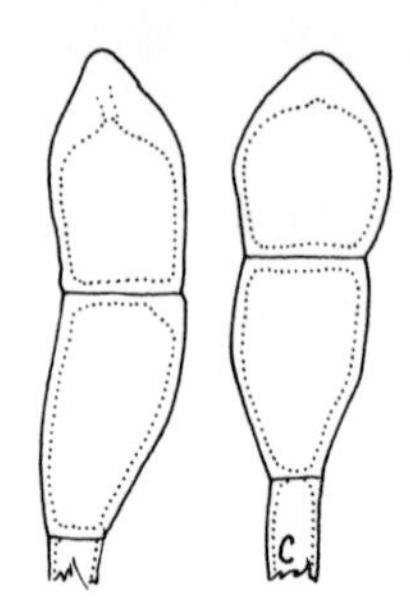

85. *PUCCINIA MELAMPODII* Diet. & Holw. in Holway, Bot. Gaz. 24:32. 1897.
Puccinia emiliae P. Henn. Hedwigia 37:278. 1898.
Puccinia paupercula Arth. Bot. Gaz. 40:206. 1905.
Puccinia ordinata H. S. Jack. & Holw. in Arthur, Amer. J. Bot. 5:530. 1918.
Puccinia flaveriae H. S. Jack. Mycologia 14:117. 1922.
Puccinia riparia Mains, Papers Mich. Acad. Sci. Arts Letters 22:156-157. 1937, not Holway, 1904.
Puccinia ripulae Mains Bull. Torrey Bot. Club 66:620. 1939.
(For additional synonyms see the North American Flora)

Telia mostly on abaxial leaf surface, exposed, in close groups, often confluent, compact, cinnamon brown becoming gray from germination; spores (31-)35-58(-64) x (13-)16-19 (-22) µm, mostly elongately ellipsoid or elongately obovoid, the apex various but usually narrowly rounded, wall (0.5-)1 (-1.5) µm thick at sides, (3-)4-7(-9) µm at apex, pale golden to clear chestnut brown, smooth, pore apical in each cell; pedicels from colorless to brownish, to 55 µm long but commonly shorter.

Hosts and distribution: species of *Baccharis*, *Calea*, *Calendula*, *Eleutheranthera*, *Emilia*, *Flaveria*, *Lagascea*, *Loxothysanus*, *Melampodium*, *Parthenium*, *Pectis*, *Plagiolophus*, *Pseudelephantopus*, *Spiranthes*, *Synedrella*, *Tetranthus*, *Tridax*, *Verbesina*, and *Zinnia*: the central United States southward into South America.

Type: on *Melampodium divaricatum* (Rich.) DC., Cuernavaca, Mor., Mexico, Holway (S; isotype PUR 42095).

86. *PUCCINIA RECEDENS* P. Syd. & H. Syd. Monogr. Ured. 1:146. 1902.

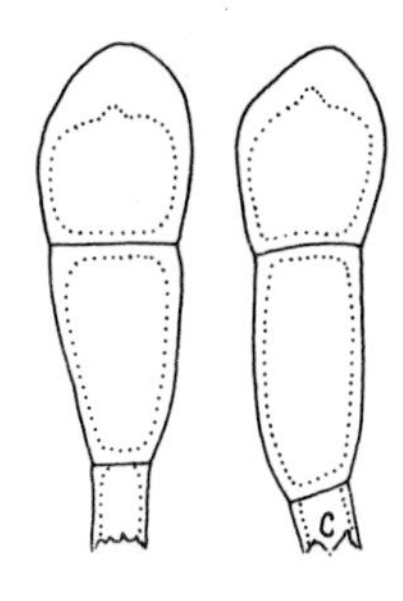

Telia mostly on abaxial leaf surface, exposed, in tight groups and often more or less confluent, compact, about chestnut brown; spores (32-)36-48(-51) x (12-)14-18(-21) µm, mostly oblong or more or less elongately obovoid, wall (1-)1.5-2(-2.5) µm thick at sides, (5-)6-8(-10) µm at apex, wall golden or light cinnamon brown, smooth, pore in each cell apical, obscure; pedicels colorless or nearly so, to about 35 µm long but often broken shorter.

Hosts and distribution: *Senecio* spp.: across the northern half of the U.S. and into Canada.

Type: on *Senecio aureus* L., Ann Arbor, Michigan, Holway (S; isotype PUR 42318).

87. *PUCCINIA TOLIMENSIS* Mayor Mem. Soc. Neuch. Sci. Nat. 5: 516. 1913.

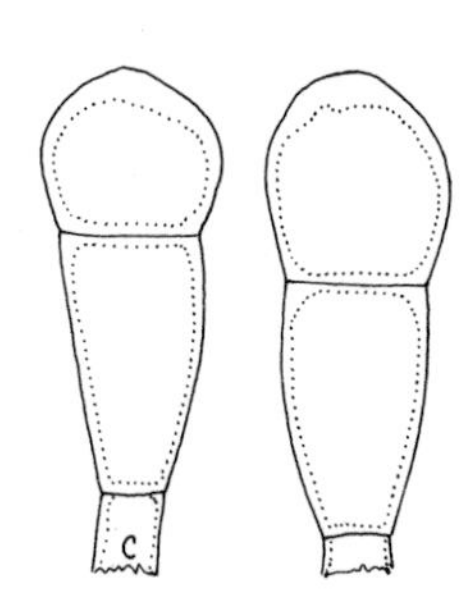

Telia on abaxial surface, exposed, in close groups, about cinnamon brown becoming gray from germination, compact; spores (32-)40-47(-55) x (17-)19-22 (-25) µm, mostly narrowly obovoid, wall 1-1.5 µm thick at sides, 3-7(-9) µm apically, uniformly golden brown, smooth, pore of upper cell apical, of lower cell at septum, obscure; pedicels yellow or pale brownish, to 20 µm long.

Hosts and distribution: *Eupatorium* spp.: United States (New York only), Guatemala and in South America.

Type: on *Eupatorium* sp., Soledad, Dept. Tolima, Colombia, Mayor (NEU; isotype PUR F8056).

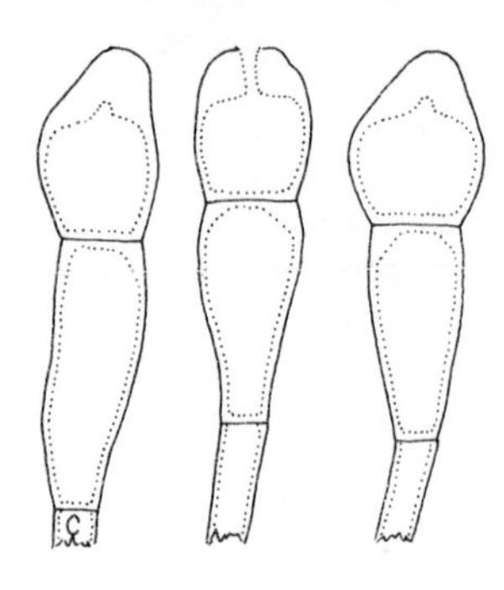

88. *PUCCINIA SEMOTA* H. S. Jack. & Holw. in Arthur, Amer. J. Bot. 5:531. 1918.

Telia on abaxial leaf surface, exposed, in tight, confluent groups, chestnut brown, compact; spores variable, (33-)40-58(-62) x (11-)13-17(-19) µm, elongately clavate or more or less cylindrical, wall 0.5-1 µm thick at sides, (5.5-)6.5-10(-11) µm at apex, about golden brown apically, paler below, smooth; pore apical in each cell; pedicels colorless, to 20 µm long but often shorter. At least some spores germinate without dormancy.

Hosts and distribution: *Hymenostephium cordatum* (Hook. & Arn.) Blake: Guatemala.

Type: on *Gymnolomia subflexuosa* Benth. (= *H. cordatum*), Solola, Holway No. 146 (PUR 42007).

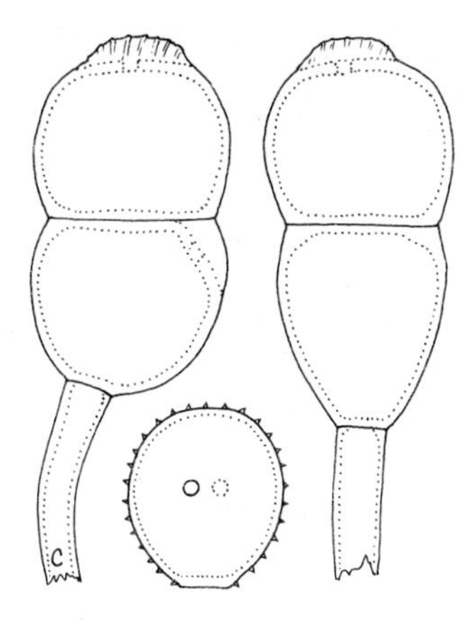

89. *PUCCINIA ARCHIBACCHARIDIS* Hennen, Leon-Gall. & Cumm. Southw. Nat. 16:358-359. 1972.

Spermogonia unknown. Aecia on abaxial leaf surface, peridium cylindrical, whitish; spores 24-29(-34) x 21-24 µm, more or less globoid, wall 1 µm thick, colorless, verrucose with cubical warts about 2 µm high spaced 1.5-2 µm on centers. Urediniospores in the telia 22-27 x 22-27 µm, globoid or broadly obovoid, wall 1(-1.5) µm thick, cinnamon brown, echinulate, pores 2, equatorial. Telia on abaxial surface, exposed, dark brown, compact; spores (42-)49-60(-66) x (23-) 25-32(-34) µm, oblong ellipsoid or broadly ellipsoid, wall 1-2(-2.5) µm thick at sides, 5-8 µm thick over pores as pale umbos, smooth except indistinctly striate on apical umbo, pore apical in each cell; pedicel colorless, to 110 µm long but usually broken short.

Hosts and distribution: *Archibaccharis hieracioides* (Blake) Blake: southern Mexico.

Type: Dist. Temascaltepec, Mex., Hinton No. 3271 (PUR 61550).

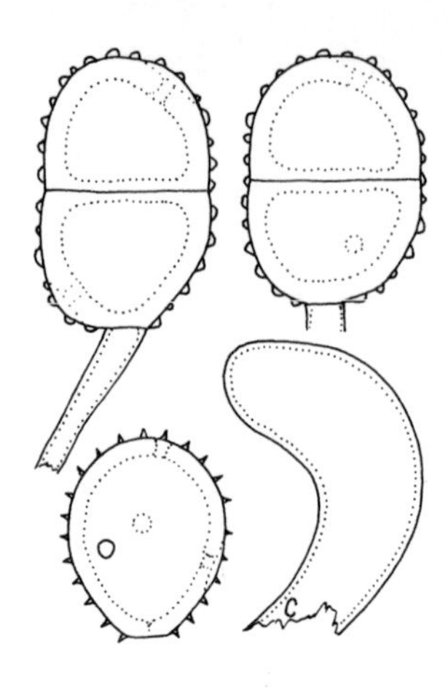

90. *PUCCINIA RATA* H. S. Jack. & Holw. in Jackson, Bot. Gaz. 65:303. 1918.

Spermogonia and aecia unknown. Uredinia mostly on abaxial surface, pale cinnamon brown, with peripheral paraphyses to 125 μm long and 25 μm wide, more or less clavate or cylindrical, wall 0.5-1 μm thick, colorless or nearly so; spores (23-)25-28(-32) x (21-)24-28 μm, mostly obovoid or broadly ellipsoid, wall 2-3 μm thick, pale golden brown, echinulate, pores 4-6, scattered, commonly 3 equatorial and 1 apical. Telia on abaxial surface, exposed, cinnamon brown, paraphyses as in the uredinia or lacking; spores (32-)35-42(-44) x (25-)26-29(-30) μm, broadly ellipsoid, wall (3-)3.5-4.5 μm thick at sides, 5-6.5 μm over pores but not as defined umbos, clear chestnut or golden brown, verrucose with verrucae 1-1.5 μm high, 1-3 μm diam spaced 2-5 μm on centers, pore in upper cell apical, of lower cell mostly midway to hilum; pedicels colorless, fragile and broken within 35 μm of the hilum.

Hosts and distribution: *Vernonia leiocarpa* DC., southern Mexico to Guatemala and Honduras.

Type: Guatemala City, Holway No. 490 (PUR 33790).

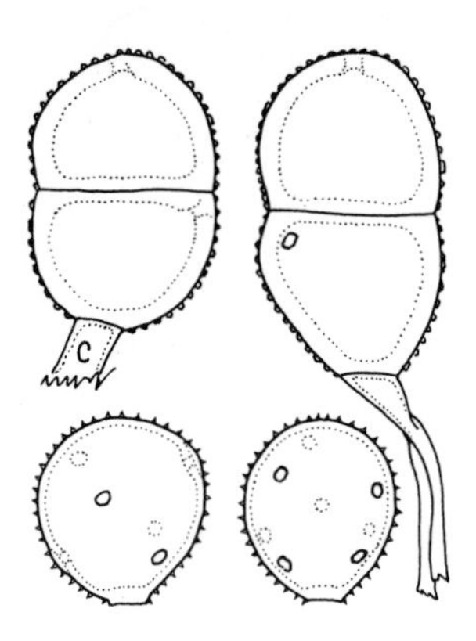

91. *PUCCINIA VIATICA* Hennen & Cumm. Rept. Tottori Mycol. Inst. 10:177. 1973.

Spermogonia and aecia unknown. Uredinia mostly on abaxial leaf surface, yellowish brown; spores (23-)24-28 x (20-)22-25 µm, broadly ellipsoid, broadly obovoid or globoid, wall 1-1.5 µm thick, pale golden brown, echinulate with spines spaced 1-1.5(-2) µm, pores obscure, scattered, about 6. Telia mostly on abaxial surface, blackish brown, pulverulent; spores (30-)36-42(-44) x 23-29(-31) µm, broadly ellipsoid, wall (2-)2.5-3(-4) µm thick, chestnut brown, finely rugose with warts of various shapes in labyrinthiform or pseudoreticulate patterns, pore apical and at septum, obscure, pedicels colorless, length of spore or shorter.

Hosts and distribution: on *Porophyllum punctatum* Blake, Jalisco, Mexico.

Type: south of Puerto Vallarta along hgw. 200, Cummins No. 71-511 (PUR 64768); not otherwise known.

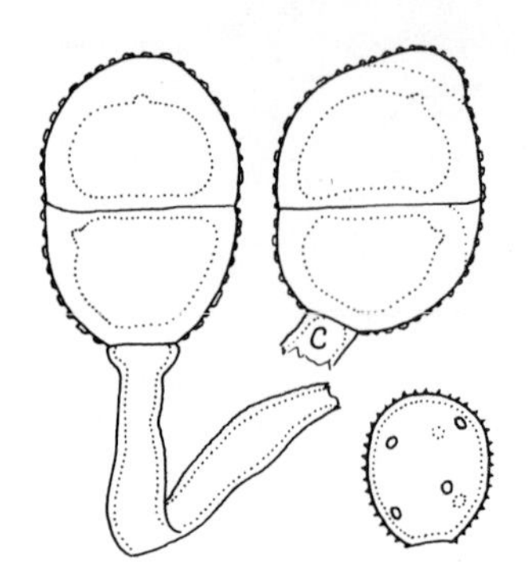

92. *PUCCINIA JALISCANA* Arth. Bot. Gaz. 40:202. 1905.

Spermogonia and aecia unknown. Uredinia on abaxial leaf surface, yellowish; spores (17-)20-23(-25) x (14-)17-20(-23) µm, mostly broadly ellipsoid, wall 1(-1.5) µm thick, pale yellowish, echinulate, pores 6-8, scattered or commonly 6 and bizonate. Telia amphigenous, exposed, blackish brown, more or less pulverulent; spores (32-)36-44(-48) x (24-)26-33 µm, mostly broadly ellipsoid, wall 2.5-3.5 µm thick at sides, 4-6(-7) µm over the pores but not as differentiated umbos, dark chestnut brown, verrucose with irregular, plate like warts 1-4 µm diam, pore apical in each cell; pedicel colorless except apically, to about 90 µm long.

Hosts and distribution: *Porophyllum holwayanum* Greenm., *P. nutans* B. L. Rob. & Greenm.: Jalisco and Mexico states, Mexico.

Type: on *Porophyllum holwayanum*, Sayula, Jal., Holway No. 5130 (PUR 42649).

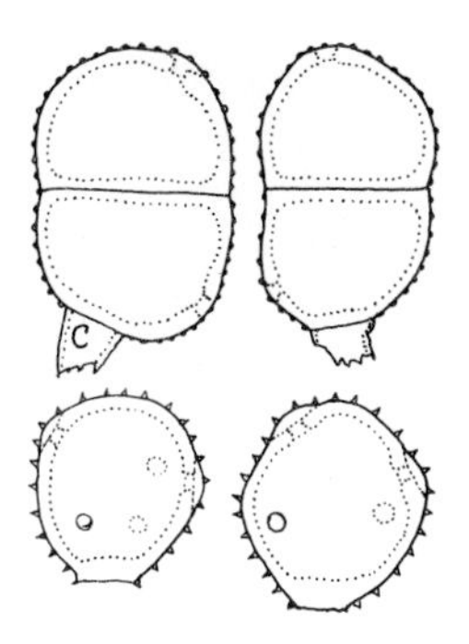

93. *PUCCINIA MINUSSENSIS* Thuem. Bull. Soc. Imp. Nat. Moscow 53:214. 1878.

Spermogonia amphigenous. Aecia mostly on abaxial leaf surface and on stems, systemic, without peridia, opening by a pore in the host epidermis; spores (19-)21-28(-30) x (16-) 17-22(-25) µm, ellipsoid or nearly globoid, wall (1.5-)2(-3) µm thick, colorless or pale yellowish, verrucose, commonly in striolate patterns. Uredinia amphigenous, cinnamon brown; spores (19-)23-26(-28) x (17-)20-25 µm, broadly ellipsoid or globoid, wall 1.5-2 µm thick, golden or cinnamon brown, uniformly echinulate, pores scattered, (3)4-6, with hyaline caps. Telia amphigenous, exposed, pulverulent, blackish brown; spores often variable, (26-)30-38(-42) x (15-)19-25 (-27) µm, mostly ellipsoid, wall uniformly (1-)1.5-2(-2.5) µm thick, chestnut brown, verrucose with small verrucae spaced (1.5-)2-2.5(-3) µm, pore of upper cell mostly apical, pore of lower cell depressed 1/2-3/4, with no or only slight caps; pedicels colorless, always broken near hilum; 1 celled spores occasionally predominate.

Hosts and distribution: *Lactuca canadensis* L. but principally *L. pulchella* (Pursh) DC.: the northern half of the United States and adjacent Canada west of the Great Lakes, and recorded in New York; also in Europe and Asia.

Type: on *Mulgedium sibiricum* L. (= *Lactuca sibirica*) near Minussinsk, U.S.S.R. (isotypes Thuemen Myc. Univ. No. 1430).

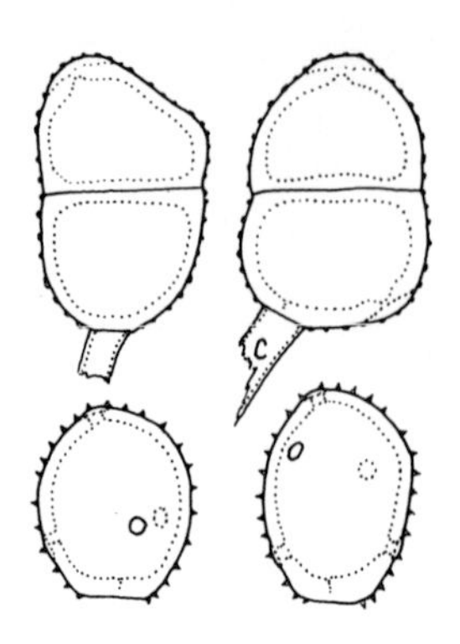

94. *PUCCINIA ARTHURELLA* Trott. in Sacc. Syll. Fung. 23:694. 1925.
Puccinia proximella Arth. Bull. Torrey Bot. Club 47:471. 1920. Not Sydow 1912.

Spermogonia and aecia unknown. Uredinia mostly on abaxial leaf surface, cinnamon brown; spores (19-)21-26(-28) x (16-)18-22(-25) μm, ellipsoid or broadly so, often variable and misshapen, wall 1.5-2 μm thick, golden to cinnamon brown, uniformly echinulate, pores 4-6, variously distributed, often 3 near the base and 1 apical, with small caps. Telia on abaxial surface, early exposed, chestnut brown, pulverulent; spores (26-)30-36(-42) x (16-)18-22(-25) μm, broadly ellipsoid or oblong ellipsoid, wall (1.5-)2-2.5 μm thick at sides, chestnut brown, slightly thicker over pores with pale, low umbos, verrucose with small verrucae mostly spaced 1.5-2(-3) μm, irregular in arrangement, sometimes in lines, pore of upper cell apical or nearly so, of lower cell midway or below; pedicels colorless, always broken near hilum.

Hosts and distribution: *Lactuca intybacea* Jacq.: eastern Mexico; also in islands of the Caribbean.

Type: on *L. intybacea*, Sabana Grande, Puerto Rico, Stevens No. 318 (PUR 34887).

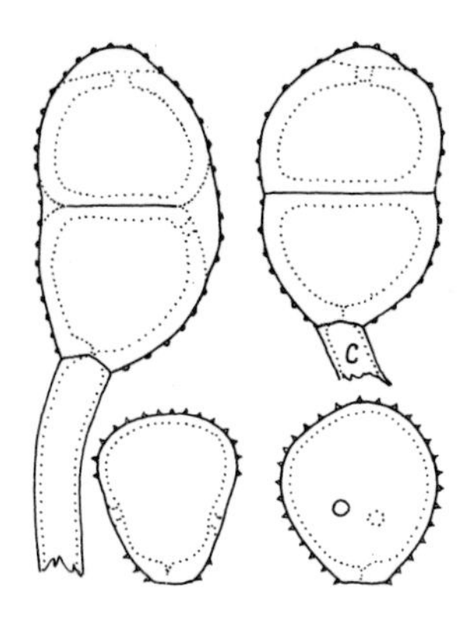

95. *PUCCINIA PRAETERMISSA* J. Parm. Can. J. Bot. 45:2310. 1967.

Spermogonia and aecia unknown. Uredinia amphigenous, dark cinnamon brown; spores (19-)21-26(-28) x (17-)20-23(-25) µm, broadly ellipsoid or obovoid with pores face view, triangularly obovoid with pores lateral, wall 1-1.5(-2) µm thick, about cinnamon brown, echinulate except around pores, pores 2, slightly subequatorial in flattened sides. Telia amphigenous, exposed, pulverulent, blackish brown; spores (32-)35-45(-53) x (20-)23-30(-35) µm, ellipsoid or broadly ellipsoid, wall (2-)2.5-3(-3.5) µm thick at sides, (4.5-)6-8 µm over pores, dark chestnut brown, verrucose with small, rounded cones spaced 2-4 µm, pore of each cell apical, each with a poorly defined umbo; pedicel colorless, rugose basally, to 80 µm long.

Hosts and distribution: *Lagascea* spp.: Sinaloa and Jalisco, Mexico.

Type: on *Lagascea decipiens* Hemsl., near El Caballo, km 1997 N of Guaymas, Son., Cummins No. 62-18 (PUR 61246).

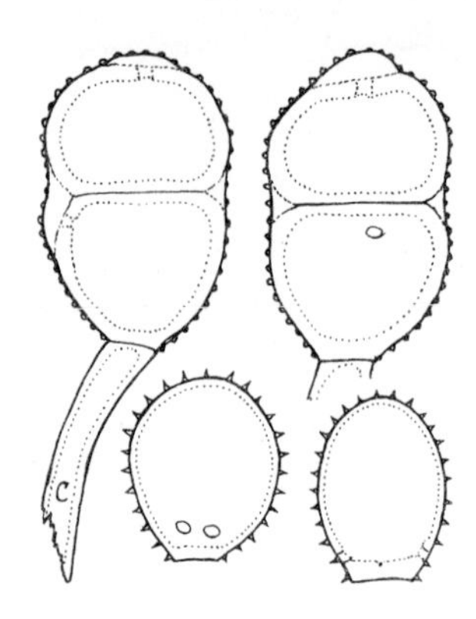

96. *PUCCINIA HODGSONIANA* Kern in Arthur, Amer. J. Bot. 5: 526. 1918.

Spermogonia and aecia unknown. Uredinia mostly on abaxial leaf surface, cinnamon brown; spores (25-)27-32(-35) x (21-)23-26(-27) µm, broadly ellipsoid or mostly obovoid with pores face view, oblong ellipsoid with pores lateral, wall 1-1.5(-2) µm thick except 3 µm at hilum, echinulate, about cinnamon brown, pores 2 adjacent to hilum, rarely a third one near base or near apex. Telia mostly on abaxial surface, exposed, more or less pulverulent, blackish brown; spores (40-)42-48(-53) x (25-)28-31(-33) µm, ellipsoid or oblong ellipsoid, wall 2.5-3.5(-4) µm thick at sides, 5-9 µm over pores, verrucose, chestnut brown but with a paler umbo over each pore, pore apical in upper cell, near septum in lower cell; pedicels colorless, to 150 µm long, often rugose basally.

Hosts and distribution: *Eupatorium* spp.: Nuevo Leon, Mexico to Nicaragua.

Type: on *Eupatorium phoenicolepis* B. L. Rob. var. *guatemalensis* B. L. Rob., Volcán Acatenango, Guatemala, Kellerman No. 6087 (PUR 37440).

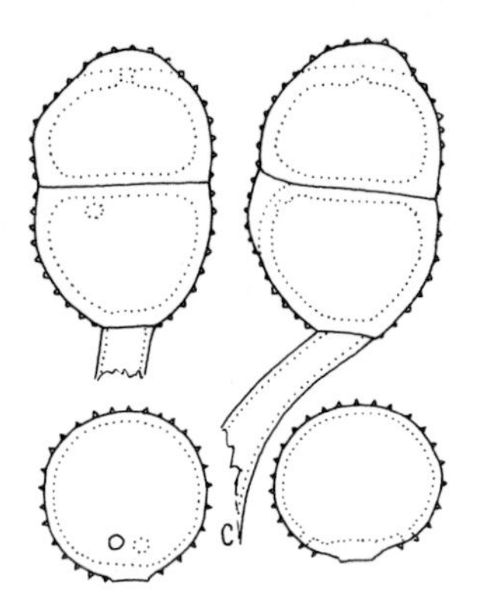

97. *PUCCINIA BASIPORULA* H. S. Jack. & Holw. in Arthur, Amer. J. Bot. 5:528. 1918.

Spermogonia and aecia unknown. Uredinia mostly on abaxial leaf surface, pale cinnamon brown; spores (20-)22-25 (-26) x 22-25(-27) µm slightly depressed globoid or globoid, wall 1-1.5 µm thick, golden or cinnamon brown, echinulate with spines spaced 1.5-2.5 µm, pores 2, near hilum. Telia mostly on abaxial surface, exposed, more or less pulverulent, blackish brown; spores (32-)34-39(-42) x (23-)25-28 (-30) µm, broadly ellipsoid, wall (1.5-)2-3 µm thick at sides (4-)5-7 µm over pores, chestnut brown except the pale umbo over each pore, verrucose with small cones spaced (1.5-)2-3.5 µm, pore apical in upper cell, near septum in lower cell; pedicel colorless, to 100 µm long.

Hosts and distribution: *Eupatorium mairetianum* DC. var. *adenopoda* B. L. Rob., *E. phoenicolepis* B. L. Rob., *E. rafaelense* Coulter: Guatemala.

Type: on *E. mairetianum* var. *adenopoda*, Quezaltenango, Guatemala, Holway No. 98 (PUR 37438).

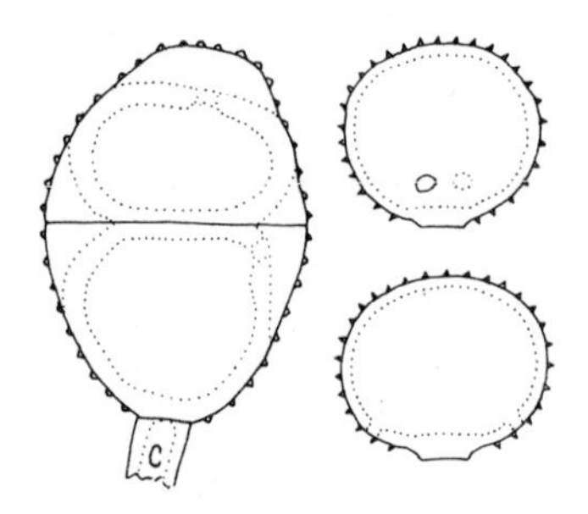

98. *PUCCINIA OBESISEPTATA* Cumm., Brit. & Baxt. Mycologia 61: 939. 1969.

Spermogonia and peridiate aecia present but too old for accurate description; spores about 20-26 µm long, probably broadly ellipsoid, wall about 2 µm thick, colorless, verrucose. Uredinia on abaxial leaf surface, cinnamon brown; spores 20-24 µm high, 22-25 µm wide, mostly slightly depressed globoid or globoid, wall 1-1.5 µm thick, cinnamon brown or darker apically, uniformly echinulate with spines spaced (1.5-)2-2.5 µm, pores 2, next to the hilum. Telia on abaxial leaf surface, exposed, rather pulverulent, blackish brown; spores (40-)43-48(-50) x (30-)33-40(-43) µm, broadly ellipsoid, wall (2-)2.5-3.5(-4) µm thick in the sides of each cell, chestnut brown, (5-)6-10(-12) µm over each pore and as an equatorial belt, the umbo and belt differentiated from the inner wall, verrucose with low conical warts spaced 2.5-4 µm, pore apical in each cell; pedicel colorless, to 150 µm long, rugose basally.

Type: on *Eupatorium deltoideum Jacq.*, Oaxaca, Oax., Mexico, Holway (PUR 37393; isotypes Barth. N. Amer. Ured. 57 as *Puccinia rosea* (Diet. & Holw.) Arth.). Not otherwise known.

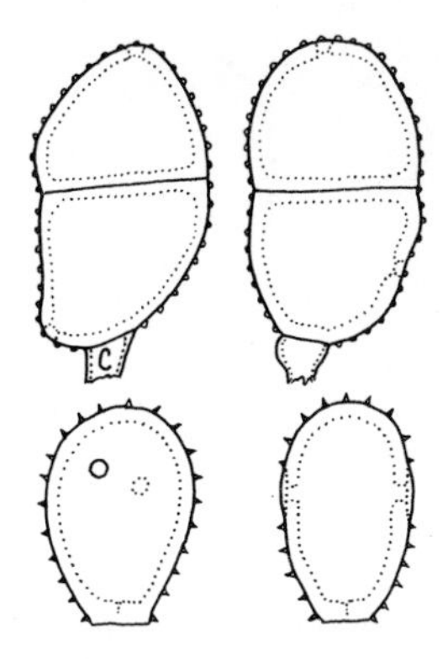

99. *PUCCINIA MIRIFICA* Diet. & Holw. in Dietel, Erythea 3:79. 1895.

Spermogonia mostly on abaxial leaf surface, systemic. Aecia mostly on abaxial surface, systemic, uredinoid, chocolate brown; spores (22-)24-30(-35) x (14-)17-23(-27) µm, mostly obovoid with pores face view, wall (1.5-)2-2.5 µm thick, nearly chestnut brown, echinulate except around the pores, pores 2, equatorial or above, in slightly to much flattened sides. Uredinia in groups, not systemic, amphigenous; spores like the aeciospores. Telia mostly on abaxial surface, exposed, chocolate brown, pulverulent; spores mostly 30-42 x 19-27 µm but much larger, much smaller and 1 celled spores are common, mostly ellipsoid, wall 2-3 µm thick, chestnut brown, verrucose with rounded or cone shaped warts mostly spaced 2-2.5 µm, pore mostly apical in upper cell, in lower half of lower cell, without umbos; pedicels colorless, broken near hilum.

Hosts and distribution: *Borrichia arborescens* (L.) DC., *B. frutescens* (L.) DC.: South Carolina and Florida to Texas and northern Mexico; also on the islands of the Caribbean.

Type: on *Borrichia frutescens*, Corpus Christi, Texas, Heller (S; isotype PUR 37537).

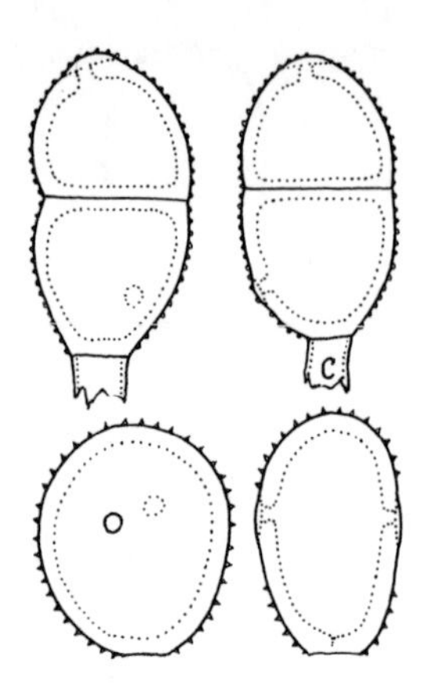

100. *PUCCINIA BALASAMORHIZAE* Peck Bull. Torrey Bot. Club. 11:49. 1884.

Spermogonia amphigenous, associated with veins. Aecia amphigenous, uredinoid, along veins, chocolate brown; spores like the urediniospores. Uredinia amphigenous, scattered, chocolate brown, spores (24-)27-36(-38) x (20-)23-31(-33) µm, mostly globoid or broadly ellipsoid, wall (1.5-)2-3 µm thick, cinnamon or near chestnut brown, echinulate, pores 2, equatorial or above, in smooth, somewhat flattened sides. Telia amphigenous, exposed, chocolate brown, pulverulent; spores (30-)33-44(-50) x (17-)20-26(-30) µm, ellipsoid or oblong ellipsoid, wall 1.5-2.5(-3) µm thick, chestnut brown, slightly thicker and paler over pores as small umbos, verrucose with small verrucae spaced 0.5-1.5(-2) µm, sometimes more or less in longitudinal lines, pore of upper cell apical, of lower cell in lower half; pedicels colorless, always broken near hilum.

Hosts and distribution: species of *Balsamorhiza* and *Wyethia*: British Columbia to Wyoming, Colorado, Arizona and California.

Lectotype: on *B. macrophylla* Nutt., Salt Lake City, Utah, M. E. Jones (NYS).

101a. *PUCCINIA HIERACII* (Roehl.) Mart. Prodr. Flora Mosq. Ed. 2, p. 227. 1817 var. *HIERACII*.
Puccinia flosculosorum var. *hieracii* Roehl. Deutschl. Flora, Ed. 2. III. 3:131. 1813.
Puccinia arnicalis Peck, Bot. Gaz. 6:227. 1881.
Puccinia chondrillina Bub. & Syd. in Sydow Monogr. Ured. 1:44. 1902.
Puccinia eriophylli H. S. Jack. Brooklyn Bot. Gard. Mem. 1:246. 1918.

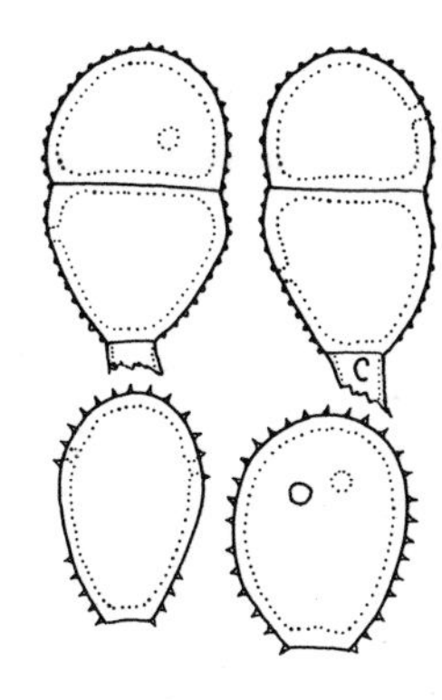

Spermogonia amphigenous. Aecia around the spermogonia, uredinoid, amphigenous, dark cinnamon brown; spores (21-)24-30 (-35) x (17-)19-25(-29) µm, broadly ellipsoid or obovoid with pores face view, wall 1.5-2(-2.5) µm thick, cinnamon brown, echinulate except below each pore, pores 2(3), from superequatorial to near apex, in flattened sides, with slight or no caps. Uredinia and spores similar to the aecia and spores. Telia amphigenous, exposed, blackish brown, pulverulent; spores (26-)30-40(-45) x (17-)20-26(-29) µm, ellipsoid or oblong ellipsoid, wall uniformly (1-)1.5-2(-3) µm thick, chestnut brown, verrucose with small verrucae spaced about 2-2.5 µm, pore of upper cell apical or depressed, of lower cell usually depressed 1/2 or more, with slight or no caps; pedicel colorless, always broken near the hilum.

Hosts and distribution: species of *Agoseris, Apargia, Arnica, Chondrilla, Cichorium, Crepis, Eriophyllum, Hieracium, Krigia, Lactuca, Microseris, Pyrrhopappus, Scorzonella, Stephanomeria, Taraxacum:* circumglobal.

Type: on *Hieracium* sp., Germany (not seen; extant?).

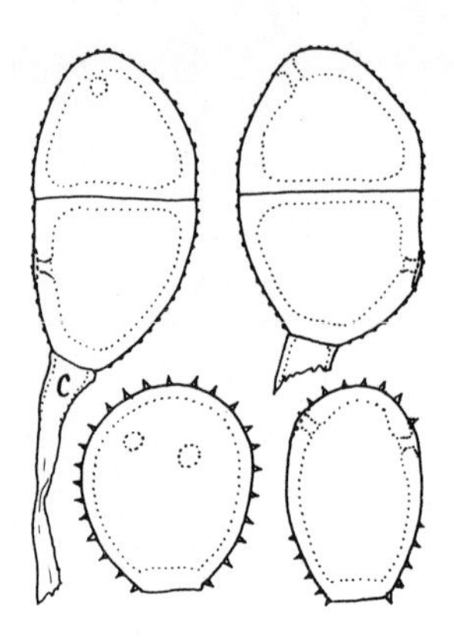

101b. *PUCCINIA HIERACII* var. *HARKNESSII* (Vize) Cumm. Mycotaxon 5:404. 1977.
Puccinia harknessii Vize, Grevillea 7:11. 1878.

Spermogonia in groups in hypertrophied areas of stems. Aecia around the spermogonia, uredinioid, dark cinnamon brown, spores like the urediniospores. Uredinia amphigenous, dark cinnamon or chocolate brown; spores (22-)23-26(-30) x (16-)19-24(-26) µm, broadly obovoid or globoid with pores face view, wall 1.5-2 µm thick, cinnamon brown or darker, echinulate except below the pores, pores 2 (rarely 3), near apex. Telia amphigenous, exposed, pulverulent, chocolate brown; spores (30-)33-40(-46) x (18-)21-25(-28) µm, mostly ellipsoid, wall (1.5-)2-2.5(-3) µm thick or slightly thicker around the pores, chestnut brown, punctately or sometimes obviously verrucose, often in lines, with verrucae spaced 1.5-2(-2.5) µm, chestnut brown, pore of upper cell usually depressed 1/4-1/3 toward septum, of lower cell depressed 1/3 -2/3 toward hilum with slight umbos; pedicels colorless, to 110 µm long, more persistent than those of var. *hieracii*.

Hosts and distribution: *Rafinesquia neomexicana* Gray and species of *Lygodesmia* and *Malacothrix*: Montana to Texas and westward in the United States.

Neotype: on *Lygodesmia spinosa* Nutt., Sierra Nevada, 7000 ft, without date or locality, Harkness No. 742 (PUR 34727; NYS).

Harkness and Moore (13), under *Puccinia*, list "Harknessii n. sp. Vize, on *Zigodesmus spinosa*. Mt. Rosa, Nevada 7000 ft." The label on the neotype agrees with this locality.

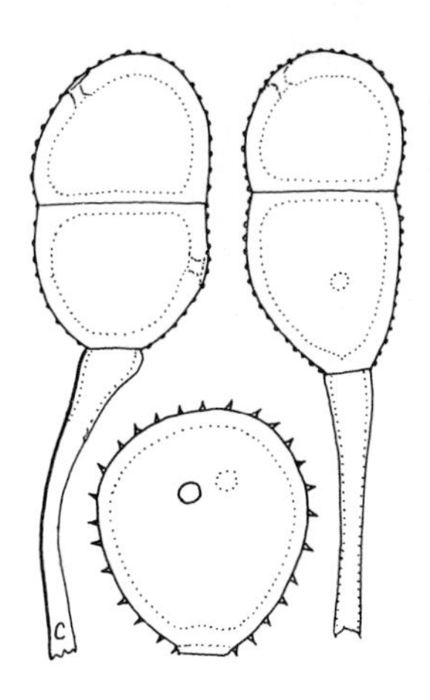

101c. *PUCCINIA HIERACII* var. *STEPHANOMERIAE* (P. Syd. & H. Syd.) Cumm. Mycotaxon 5:404. 1977.
Puccinia stephanomeriae P. Syd. & H. Syd. Monog. Ured. 1:157. 1902.
Puccinia harknessii major Arth. Manual Rusts U.S. and Canada. p. 353. 1934.

Spermogonia and aecia unknown. Urediniospores in telia (28-)30-36(-40) x (22-)25-30(-34) µm, globoid or broadly obovoid with pores face view, wall (1.5-)2-2.5 µm thick, cinnamon brown or slightly darker, echinulate except below the pores, pores 2, above the equator in flattened sides. Telia amphigenous and on stems, exposed, pulvinate, blackish brown; spores (33-)35-46(-50) x (21-)23-28(-30) µm, ellipsoid or oblong ellipsoid, wall 2-3 µm thick or to 3.5 µm around the pores, punctately verrucose with verrucae spaced (1.5-)2-2.5(-3) µm, pore of upper cell usually depressed 1/4 -1/3, pore of lower cell depressed 1/3-1/2, with slight or no caps; pedicel colorless, semipersistent.

Hosts and distribution: *Stephanomeria cichoriacea* Gray, *S. virgata* Benth. ssp. *pleurocarpa* (Greene) Gottl.: California.

Type: on *S. cichoriacea*, Pasadena, McClatchie (S).

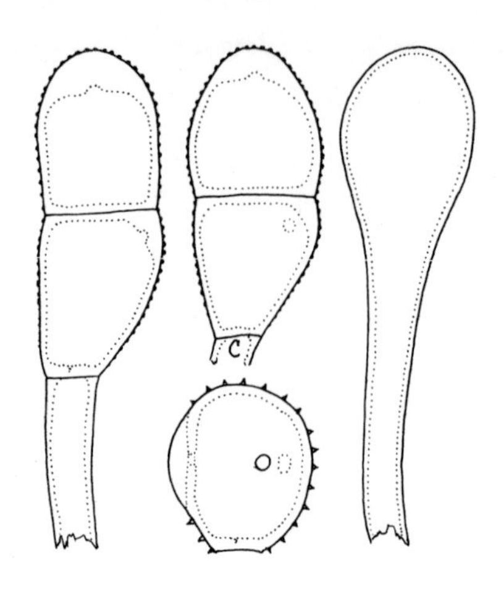

102. *PUCCINIA LUDOVICIANAE* Fahr. Ann. Mycol. 39:181. 1941.

Spermogonia and aecia unknown. Uredinia on abaxial leaf surface, about cinnamon brown, with colorless to brownish, thin walled paraphyses to at least 120 μm long, more or less cylindrical to capitate and to 35 μm diam; spores (25-)28-35(-40) x (22-)25-28(-30) μm, obovoid or broadly ellipsoid, wall (1.5-)2-2.5 μm thick, golden to cinnamon brown, echinulate except over pores, pores 3, equatorial, each with a colorless cap. Telia on abaxial surface, exposed, usually with paraphyses as the uredinia, compact, blackish brown; spores (40-)46-58(-65) x (19-)22-28(-30) μm, oblong ellipsoid or elongately obovoid, wall 1.5-2(-3) μm thick at sides, (5-)6-8(-9) μm at apex, clear chestnut brown, the apical thickening slightly paler, punctately verrucose especially apically, often appearing striate at apex, pore of each cell apical; pedicels colorless, to 115 μm long.

Hosts and distribution: *Artemisia* spp., especially *A. ludoviciana* Nutt. and relatives: western Canada south to at least Durango, Mexico.

Lectotype: on *A. ludoviciana*, Silesia, Montana, Bartholomew (PUR 37779; isotypes Barth. N. Amer. Ured. 1022 as *Puccinia absinthii*). A type was not designated by Fahrendorf but he cited this collection.

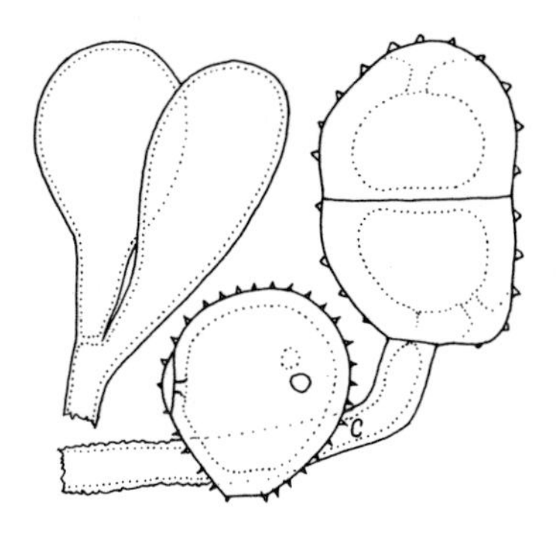

103a. *PUCCINIA EGREGIA* Arth. Bot. Gaz. 40:204. 1905. var. *EGREGIA*.

Spermogonia and aecia unknown. Uredinia amphigenous, yellowish to cinnamon brown, with peripheral, colorless, thin walled, saccate or cylindrically capitate paraphyses, 12-25 µm wide and to 65 µm long; spores (25-)28-32(-33) x (20-)23-28(-30) µm, globoid or obovoid, wall 2-3(-3.5) µm thick, yellowish or pale cinnamon brown, uniformly echinulate, pores 3 or 4, equatorial, with low caps. Telia amphigenous or mostly on abaxial surface, exposed, chocolate brown, pulverulent; spores (35-)40-45(-50) x (26-)28-32(-34) µm, broadly ellipsoid, wall (3.5-)4-5 µm thick, 5-8(-10) µm over pores but not as defined umbos, uniformly clear chestnut brown, echinulately verrucose with cones spaced 4-6(-8) µm, upper pore apical, pore in lower cell near hilum; pedicel colorless, to about 85 µm long, rugose basally.

Hosts and distribution: *Vernonia* spp.: Mexico from southern Sonora southward.

Type: on *Vernonia uniflora* Sch. Bip. (= *V. salicifolia* (DC.) Sch. Bip., Oaxaca, Oax., Seler No. 1739 (PUR 37336).

All hosts are of the section *Eremosis*.

The following variety differs in having urediniospores with typically 2 pores and darker pigmentation and smaller teliospores.

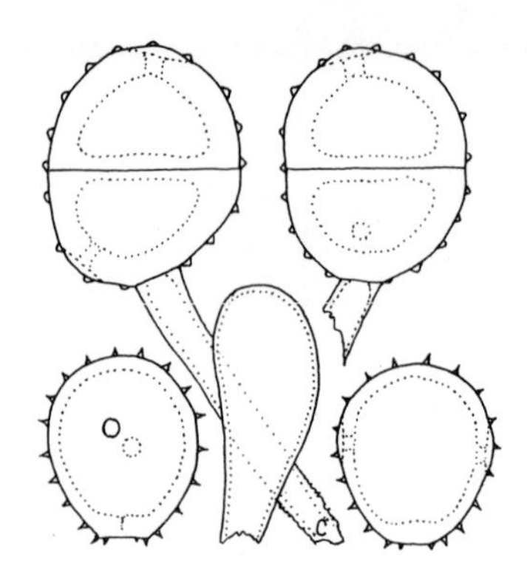

103b. *PUCCINIA EGREGIA* var. *CUMMINSIANA* Urban, Acta Univ. Carolinae Biol. 1971:62. 1973.

Uredinia on abaxial surface of leaves, with colorless peripheral paraphyses; spores (23-)25-28(-30) x (21-)23-26 (-27) µm, mostly globoid, wall (1.5-)2(-2.5) µm thick, cinnamon brown, uniformly echinulate, pores 2, rarely 3, equatorial, with slight caps. Telia mostly on abaxial surface; spores (30-)33-37(-39) x (25-)28-30(-31) µm, broadly ellipsoid, wall 4-5 µm thick or slightly thicker over pores, chestnut brown or slightly paler over pores but not as defined umbos, echinulately verrucose with cones spaced mostly 4-6 µm, pore of upper cell apical, of lower cell near hilum; pedicel nearly colorless, to about 55 µm long, rugose basally.

Hosts and distribution: *Vernonia palmeri* Rose, vicinity of Alamos, Son., Mexico.

Type: trail to Sierra Alamos, Cummins No. 63-699 (PUR 60440).

104. *PUCCINIA SEMIINSCULPTA* Arth. Bot. Gaz. 40:204. 1905.

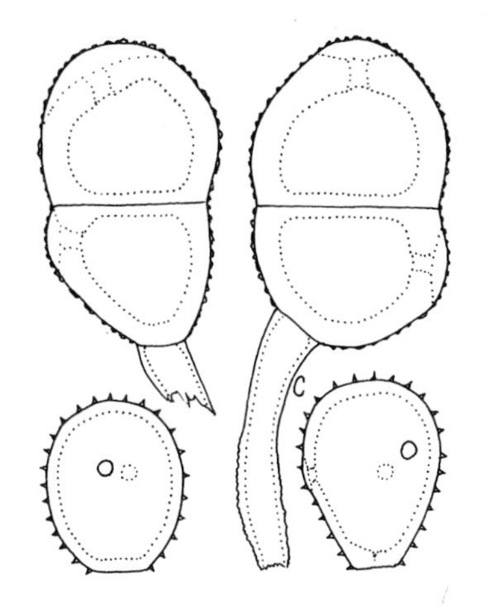

Spermogonia on adaxial leaf surface. Aecia on adaxial surface in slightly hypertrophied spots, uredinoid, pale yellowish brown; spores (22-) 24-30(-32) x (18-)21-24(-26) µm, broadly ellipsoid or obovoid, wall 1.5-2(-3) µm thick, pale golden to nearly colorless, uniformly echinulate, pores 2 or 3(4), equatorial. Uredinia amphigenous, scattered, pale yellowish brown; spores similar to the aeciospores. Telia mostly on adaxial surface, exposed, mostly not compact; spores of two types, the germinating kind (38-)42-48 x (20-)22-30(-33) µm, from narrowly to broadly obovoid, golden brown, the pore apical in each cell under an umbo, resting kind (40-)44-56 (-60) x (28-)30-40 µm, broadly ellipsoid, wall (3-)5-7(-8) µm at sides, 6-10(-11) µm over pores, dark chestnut brown or slightly paler over pores but not as a defined umbo, wall in both types from rugose with short anastomosing ridges to rugosely reticulate, pore of upper cell apical, of lower cell next to septum or midway to hilum; pedicels colorless, to 85 µm long but usually shorter, rugose basally in intact pedicels.

Hosts and distribution: *Vernonia* spp.: Durango and Sinaloa southward to Oaxaca; also in South America.

Type: on *Vernonia alamanii* DC., Amecameca, Mex., Mexico, Holway No. 3754 (PUR 37311).

Urban (30) recognized var. *hennenii* Urban, based on slight differences in the wall thickness and pigmentation of the urediniospores and on narrower teliospores.

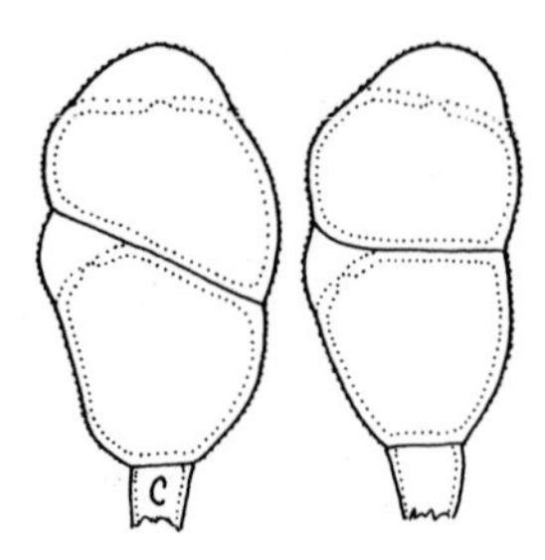

105. *PUCCINIA EGRESSA* Arth. Bull. Torrey Bot. Club 46:108. 1919.
Puccinia egregia Arth. Bull. Torrey Bot. Club 38:370. 1911, not Arth. 1905.

Spermogonia on abaxial leaf surface, commonly along veins. Aecia mostly on abaxial surface and on stems, systemic, altering the habit of the plant, peridium conspicuous, yellowish, lacerated, spores (19-)22-28(-30) x (16-)18-22 µm, mostly broadly ellipsoid or globoid, wall including warts 2-3 µm thick, verrucose with flat, irregular verrucae about 1-1.5 µm wide, discrete or pseudoreticulately joined, colorless. Urediniospores as originally reported about 23-27 µm diam, wall 1.5 µm thick, finely echinulate, cinnamon brown, pores indistinct, apparently 2, equatorial. Telia on abaxial surface among the aecia, exposed, dark cinnamon brown, compact; spores (35-)38-45(-48) x 21-26(-28) µm, broadly ellipsoid, wall (0.5-)1(-1.5) µm thick and pale chestnut or golden brown at sides, (4-)5-7(-9) µm over pores as defined umbos, punctately verrucose with verrucae spaced about 1-2(-2.5) µm, appearing striate on apical umbo, pore apical in each cell; pedicels colorless, always broken short.

Type: on *Archibaccharis oaxacanum* (Greenm.) Blake (as *Baccharis oaxacanum*), Mt. Oaxaca, Oax., Mexico, Pringle (PUR 33798). Known from one other collection by Pringle in the same locality.

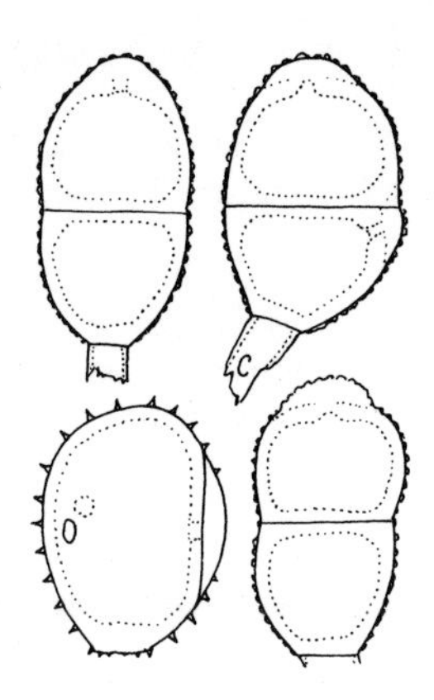

106. *PUCCINIA ARTEMISIAE-NORVEGICAE* Tranz. & Woron. Publ. Riabouchinsky Exped., Bot. 2:563. 1914.

Spermogonia and aecia unknown. Uredinia on abaxial leaf surface, cinnamon brown; spores (27-)30-35(-37) x (18-) 21-25(-27) µm, broadly ellipsoid or obovoid, wall 2 µm thick, pale cinnamon or golden brown, pores 3, equatorial, in smooth areas, with colorless caps about 3 µm thick and 15 µm diam. Telia amphigenous and on stems, exposed, blackish brown, pulverulent; spores (30-)33-38(-40;45) x (17-)19-24(-26) µm, ellipsoid or oblong ellipsoid, wall (1.5-)2-2.5 (-3) µm thick at sides, (3-)4-6(-7) µm over pores by distinct but concolorous umbos, chestnut brown, rugose or pseudoreticulate with small verrucae and ridges that are discrete or fused in various patterns, pore of each cell apical; pedicels colorless, always broken near hilum.

Hosts and distribution: *Artemisia arctica* Less.: Alaska; also in the U.S.S.R.

Type: on *Artemisia norvegica* (= *A. arctica*), crater of Volcano Uzon, Kamchatka, 1909, Komarov (LE; not seen).

107a. *PUCCINIA TANACETI* DC. Flore Fr. 2:222. 1805 var. *TANACETI*.
Puccinia absinthii DC. Flore Fr. 5:56. 1815.
Puccinia chrysanthemi Roze Bull. Soc. Mycol. Fr. 16: 92. 1900.

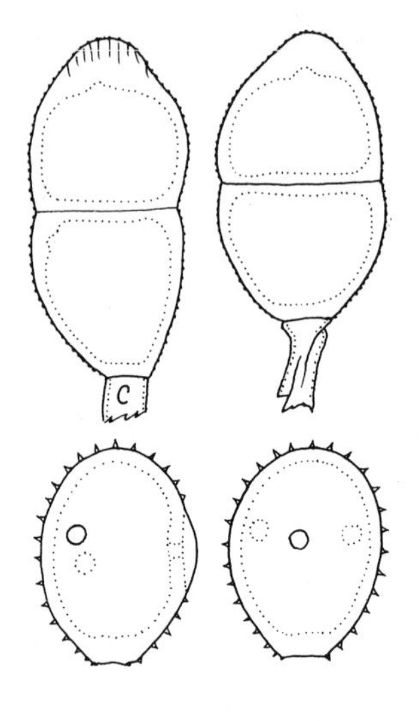

Spermogonia and aecia unknown. Uredinia on the abaxial leaf surface, commonly in circles, cinnamon brown; spores (25-)28-32(-35) x (20-)22-26(-28) µm, mostly broadly ellipsoid or obovoid, wall 1.5-2(-2.5) µm thick, echinulate except over pores, cinnamon brown, pores 3, equatorial, with conspicuous caps. Telia on abaxial surface, exposed, commonly in circles, compact, blackish brown; spores (36-)40-56(-60) x (19-)22-28(-30) µm, mostly ellipsoid, wall 1.5-2 (-2.5) µm thick at sides, (4-)6-8 (-10) µm at apex, nearly uniformly clear chestnut brown, punctate with verrucae spaced (.5-)1-1.5(-2) µm, appearing striate at apex, pore apical in each cell; pedicel colorless, to 110 m long.

Hosts and distribution: *Chrysanthemum morifolium* (Ramat.) Hemsl.: circumglobal on cultivated "mums".

Type: on *Chrysanthemum vulgare* (L.) Bernh., France, Delaroche & Leman (G).

The following variety differs mainly in having narrower urediniospores with thinner walls.

107b. *PUCCINIA TANACETI* DC. var. *DRACUNCULINA* (Fahr.) Cumm. Mycotaxon 5:406. 1977.
Puccinia dracunculina Fahr. Ann. Mycol. 39:181. 1941.

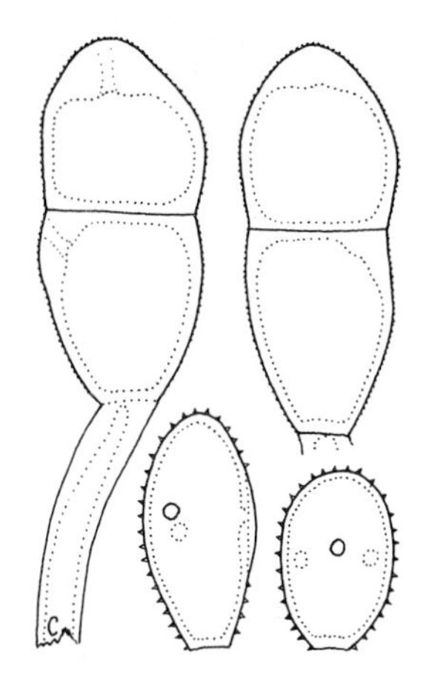

Spermogonia on adaxial leaf surface. Aecia mostly on abaxial surface opposite the spermogonia, uredinoid; spores as the urediniospores. Uredinia mostly on abaxial surface, cinnamon brown; spores (23-)26-33(-35) x (15-)17-20 µm, mostly ellipsoid, wall 1-1.5 µm thick, pale cinnamon or golden brown, echinulate, pores 3, equatorial in smooth, flattened sides, with low caps. Telia mostly on abaxial surface and on stems, exposed, pulverulent, blackish brown; spores (36-)42-55(-60) x (18-)22-28 (-30) µm, ellipsoid or obovoid, wall (1-)1.5-2(2.5) µm thick at sides, 4-6(-8) µm at apex, chestnut brown or slightly paler over pores but not as defined umbos, punctately verrucose with verrucae spaced 0.5-1(-1.5) µm or these irregularly united in part, pore of upper cell apical, of lower cell at septum; pedicel colorless, to 115 µm long.

Hosts and distribution: *Artemisia dracunculus* L. (*A. glaucus* Pall., *A. dracunculoides* Pursh): the western half of the United States and in northern Baja California; also in Europe.

Lectotype: on *A. dracunculoides*, Boulder, Colorado, Bartholomew and Bethel (PUR 37727; isotypes Barth. F. Columb. 2753).

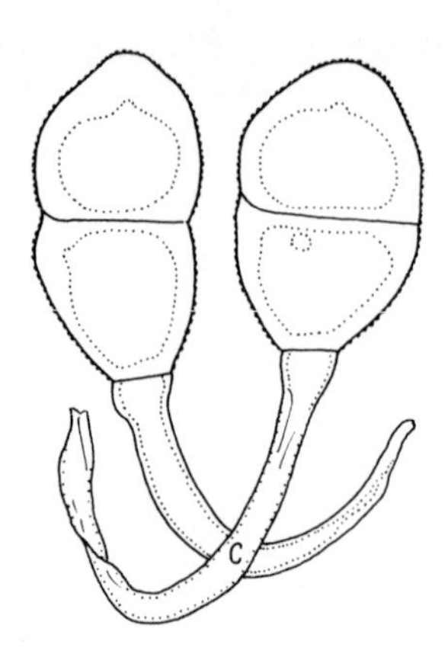

108a. *PUCCINIA LONGIPES* Lagerh. Tromsoe Mus. Aarsh. 17:64. 1895 var. *LONGIPES.*
Puccinia bullata Schw. Schr. Nat. Ges. Leipzig 1:74. 1822 not Link 1815.
Puccinia vernoniae Schw. Trans. Amer. Phil. Soc. II. 4:296. 1832 (based on uredinia).

Spermogonia on adaxial leaf surface. Aecia mostly on adaxial surface, uredinoid, cinnamon brown; spores as the urediniospores. Uredinia amphigenous or mostly on abaxial surface, cinnamon brown; spores (24-)27-32(-37) x (22-)24-28(-30) µm, broadly ellipsoid or obovoid, wall (2-)2.5-3 (-3.5) µm thick, cinnamon or golden brown, uniformly echinulate, pores mostly 3, equatorial, with caps. Telia mostly on abaxial or forming large sori on stems, exposed, blackish brown; spores (35-)40-52(-60) x (20-)22-28(-32) µm, oblong ellipsoid, broadly ellipsoid or obovoid, wall (2-)2.5-3.5 (-4) µm thick at sides, (4-)6-9(-10) µm over pores, clear chestnut brown, paler over pores but scarcely as defined umbos, punctately verrucose or reticulately rugose, pore of each cell apical; pedicels colorless, to 160 µm long but often shorter.

Hosts and distribution: *Vernonia* spp.: the United States east of the Mississippi River.

Type: "on stems of various plants, e.g., Ambrosia, Chenopodium" (listed as *Vernonia novoboracensis* in 1832), Salem (North Carolina) and Bethlehem (Pennsylvania); holotype in pH is labelled "*Puccinia bullata* LVS Salem & Bethl in Caulibus variis"; isotype PUR 37299).

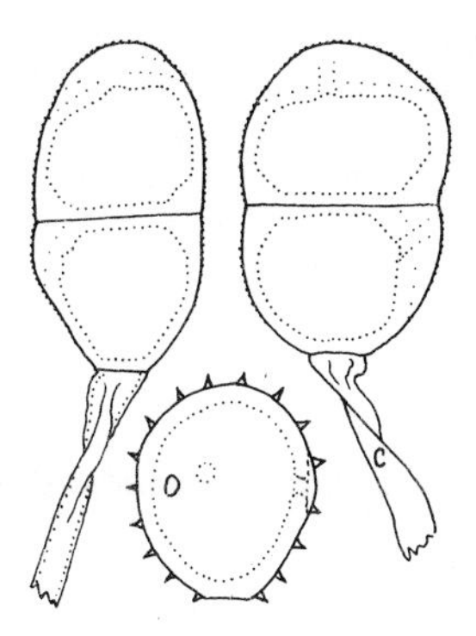

108b. *PUCCINIA LONGIPES* var. *BREVIPES* (Diet.) Urban, Acta Univ. Carolinae Biol. 1971:70. 1973.
Puccinia vernoniae Diet. var. *brevipes* Diet. J. Mycol. 7:43. 1891.

Spermogonia and aecia unknown. Uredinia amphigenous or mostly on abaxial surface, cinnamon brown; spores (20-)22-27 (-30) x (18-)20-24(-26) µm, broadly ellipsoid or obovoid, wall (1.5-)2-2.5(-3) µm thick, cinnamon brown or golden, uniformly echinulate, pores (2)3, equatorial, with caps. Telia mostly on abaxial surface or forming large sori on stems, exposed, blackish brown; spores (32-)36-44(-48) x (18-)21-28(-30) µm, oblong ellipsoid, broadly ellipsoid or obovoid, wall (1.5-)2-2.5(-3) µm thick at sides, 4-7(-10) µm over pores, clear chestnut brown, paler over pores but scarcely as differentiated umbos, punctately verrucose or reticulately rugose, pore apical in each cell; pedicel colorless, to 90 µm long or longer in the sori on stems but often broken shorter.

Hosts and distributions: *Vernonia* spp.: the United States east of the Rocky Mountains and mostly west of the Mississippi River, south to Costa Rica.

Lectotype: on *Vernonia baldwinii* Torr., Perryville, Missouri, Demetrio (S; designated by Urban).

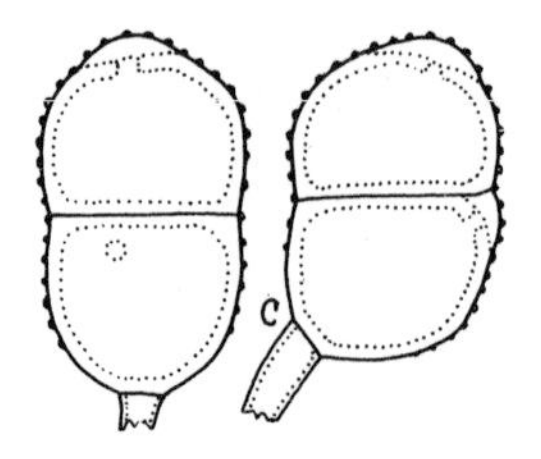

109. *PUCCINIA OBLATA* Mains Contrib. Univ. Mich. Herb. 1:12. 1939.

Spermogonia on adaxial leaf surface. Aecia amphigenous, with whitish cylindrical peridium; spores 26-29 x 20-25 µm, globoid or ellipsoid, wall 1-2 µm thick, colorless, verrucose. Uredinia mostly on abaxial surface, cinnamon brown; spores (16-)19-24 x 22-26 µm, oblate sphaeroid, wall 1.5-2 µm thick, pale cinnamon brown, uniformly echinulate, pores 2 or 3, equatorial or slightly below. Telia mostly on abaxial surface, exposed, chocolate brown, pulverulent; spores (27-)32-37(-40) x (20-)22-25(-27) µm, ellipsoid, wall 1-1.5 µm thick at sides, cinnamon brown or clear chestnut brown, (3-)4-6(-7) µm over pores by pale umbos, verrucose with low warts or rounded cones, becoming smooth basally, pore in upper cell apical or nearly so, of lower cell at septum; pedicels colorless and broken near hilum.

Hosts and distribution: *Notoptera brevipes* (B. L. Rob.) Blake, *N. scabridula* Blake: Guatemala and British Honduras.

Type: on *Otopappus curviflorus* (R. Br.) Hemsl. (= possible error for *N. brevipes*), Los Amates, Guatemala, Deam No. 89A (MICH).

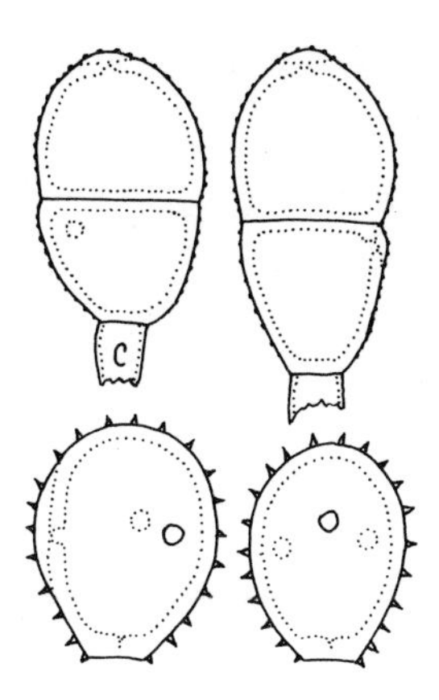

110. *PUCCINIA CNICI* Mart. Prodr. Fl. Mosq. Ed. 2. p. 227. 1817.

Spermogonia mostly on adaxial leaf surface. Aecia amphigenous or mostly on abaxial surface, without peridium but bounded by coarse hyphae; spores (23-)25-33(-38) x 21-30 (-32) µm, globoid or broadly ellipsoid, wall 2-3 µm thick, conspicuously verrucose with round but sometimes irregular verrucae or these variously united, colorless. Uredinia amphigenous, chocolate brown; spores (25-)28-38(-42) x (22-) 24-28(-32) µm, broadly ellipsoid or obovoid, wall 1.5-2 (-2.5) µm thick, dark cinnamon brown, uniformly echinulate, pores 3 (rarely 2 or 4) equatorial, with caps. Telia amphigenous or mostly on abaxial surface, exposed, chocolate brown, pulverulent; spores (31-)35-47(-52) x (19-)21-26(-30) µm, ellipsoid, wall at sides 1.5(-2.5) µm thick, clear chestnut brown, slightly thicker over pores by small umbos, punctately verrucose with verrucae spaced 1.5-3 µm, appearing almost smooth in silhouette, pore apical or nearly so in each cell; pedicels colorless, usually broken near hilum.

Hosts and distribution: *Cirsium vulgare* (Savi) Tenore (*C. lanceolatum* Auth.), rarely other?: Canada and the northern half of the United States; also in Europe.

Type: on *Cnicus lanceolatus* (= *Cirsium vulgare*), near Moscow, USSR (extant?).

Savile (29) lists with (?) *Cirsium brevistylum and C. callilepis*.

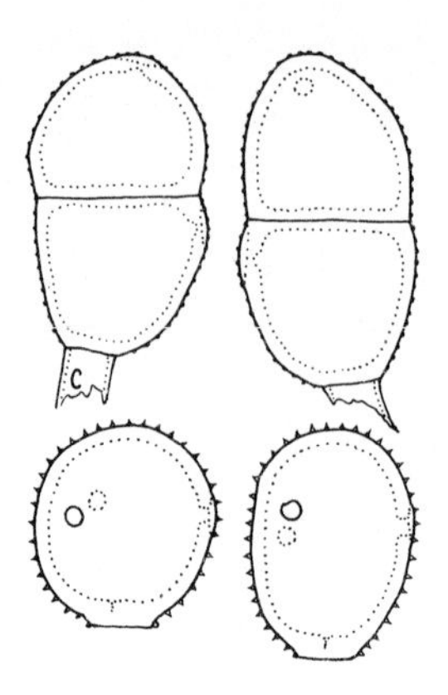

111. *PUCCINIA INCLUSA* P. Syd. & H. Syd. Monogr. Ured. 1:56. 1902.

Spermogonia mostly on adaxial leaf surface. Aecia mostly on adaxial surface around the spermogonia, dark cinnamon brown, uredinoid; spores like the urediniospores. Uredinia amphigenous, dark cinnamon brown; spores (21-)23-29 (-31) x (21-)22-27(-29) µm, globoid or broadly ellipsoid, wall 1.5-2(-2.5) µm,thick, cinnamon to dark cinnamon brown, uniformly echinulate, pores 3 (rarely 2 or 4), equatorial, with slight caps. Telia amphigenous or mostly on abaxial surface, exposed, blackish brown, pulverulent; spores (30-) 35-46(-50) x (20-)22-27(-30) µm, mostly ellipsoid, wall 1.5-2(-2.5) µm thick at sides, slightly thicker over pores by illdefined umbos, punctately verrucose with verrucae spaced 1.5-2(-3) µm, chestnut brown, pore of upper cell apical or slightly depressed, of lower cell at or near septum; pedicels colorless, to 100 µm long but commonly broken short.

Hosts and distribution: *Cirsium* spp.: the western half of the United States and Canada and in central Mexico.

Type: on *Cirsium undulatum* (Nutt.) Spreng., Rooks County, Kans., Bartholomew (S; isotypes Barth. F. Columb. 1571).

Savile (29) established vars. *flodmanii, boreohesperia, brevifolii* and *mexicana* on the basis of minor variations of the echinulation and in the sizes of teliospores. *P. inclusa* is scarcely more than a variety of *P. californica.*

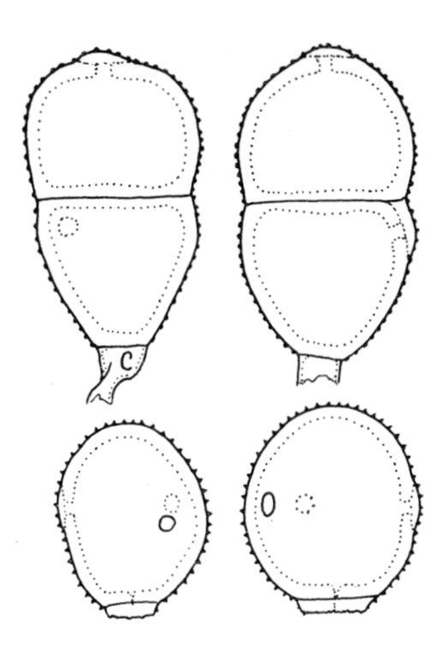

112. *PUCCINIA CALIFORNICA* Diet. & Holw. in Dietel, Bot. Gaz. 18:254. 1893.

Spermogonia and aecia unknown. Uredinia amphigenous, dark cinnamon brown; spores (25-)27-32(-36) x (20-)23-27(-30) µm, globoid or broadly ellipsoid, wall (1.5-)2-2.5(-3) µm thick, dark cinnamon or chestnut brown, uniformly echinulate, pores 3 (rarely 2 or 4), equatorial, with small caps. Telia amphigenous, exposed, blackish brown, pulverulent; spores (33-)37-53(-59) x (21-)25-30(-33) µm, mostly ellipsoid, wall (1.5-)2-2.5(-3) µm thick at sides, slightly thicker over pores by illdefined umbos, chestnut brown, punctately verrucose with verrucae spaced about 2 µm, pore of each cell apical or slightly depressed; pedicels colorless, more than 100 µm long but usually broken short.

Hosts and distribution: *Cirsium* spp.: California east to Utah and Arizona.

Type: on *Cirsium breweri* (Gray) Jeps. (but perhaps is *C. tioganum* (Congd.) Petr.), Kings River Canyon, California, Holway (S; isotype MIN 317382).

Savile (29) established var. *deserticola* on *C. arizonicum* (Gray) Petr. because of urediniospores having thicker and darker walls.

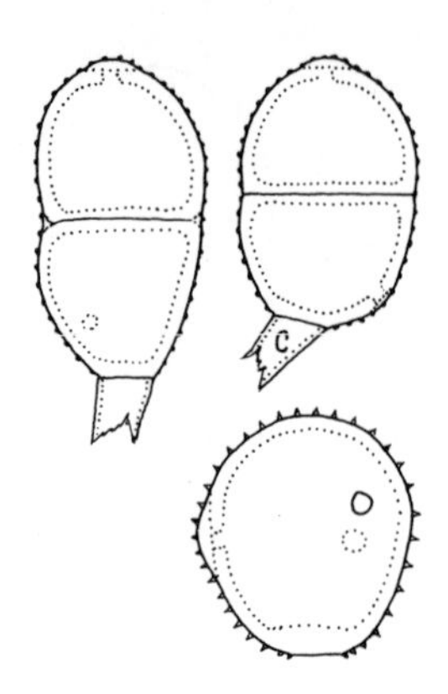

113. *PUCCINIA PUNCTIFORMIS* (Strauss) Roehl. Deutschl. Fl. Ed. 2. III, 3:132. 1813.
Uredo punctiformis Strauss Ann. Wett. Ges. 2:103. 1810. Telia present.

Spermogonia mostly on abaxial leaf surface, systemic. Aecia mostly on abaxial surface, systemic, uredinoid; spores like the urediniospores. Uredinia localized, cinnamon brown; spores (24-)26-30(-33) x (21-)23-28 µm, mostly globoid, wall 1.5-2 µm thick, cinnamon brown, uniformly echinulate, pores 3 (rarely 2 or 4), equatorial, with slight to conspicuous caps. Telia systemic when with spermogonia and aecia, later localized, exposed, chocolate brown, pulverulent; spores (28-)32-38(-42) x (18-)20-24(-26) µm, ellipsoid or broadly so, wall 1.5-2 µm thick, deep golden or clear chestnut brown, punctately verrucose with verrucae spaced about 1.5-2 µm, pore of upper cell apical or somewhat depressed, of lower cell midway to hilum or below, each with a low umbo; pedicels colorless, always broken near hilum.

Hosts and distribution: *Cirsium arvense* (L.) Scop.: widespread; to be expected wherever the host occurs.

Type: on *Cnicus arvensis* (L.) Sm. (= *Cirsium arvense*), Germany (not seen).

114a. *PUCCINIA CALCITRAPAE* DC. Flore Fr. 2:221. 1805 var. *BARDANAE* (Wallr.) Cumm. Mycotaxon 5:402. 1977.
Puccinia inquinans bardanae Wallr. Flora Crypt. Germ. 2:219. 1833.
Puccinia bardanae (Wallr.) Corda, Icones Fung. 4:17. 1840.

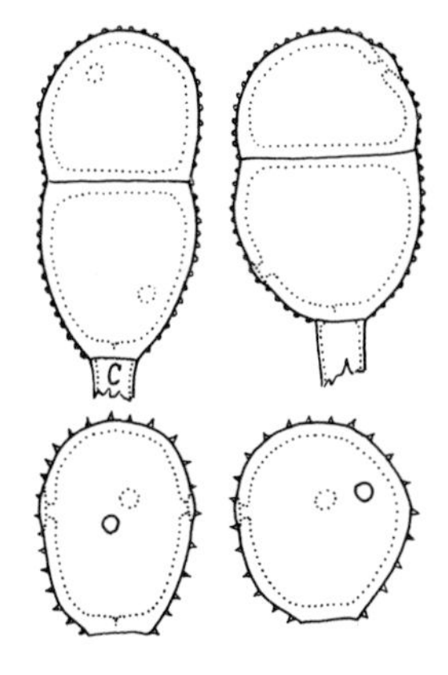

Spermogonia on adaxial leaf surface. Aecia uredinoid, on adaxial surface in a ring around the spermogonia, cinnamon brown; spores similar to the urediniospores. Uredinia mostly on abaxial surface, scattered, cinnamon brown; spores (24-)26-30(-35) x (20-)23-28(-31) µm, nearly globoid, broadly ellipsoid or broadly obovoid, wall 1.5-2(-2.5) µm thick, cinnamon brown or golden, uniformly echinulate, pores 3 (rarely 2 or 4), equatorial, with caps. Telia mostly on abaxial surface, exposed, blackish brown, pulverulent; spores (30-)32-44(-50) x (20-)22-28(-30) µm, mostly ellipsoid, wall (1.5-)2-2.5 µm thick, slightly thicker over pores by low umbos, punctately verrucose with verrucae spaced (1.5-)2-2.5(-3) µm, chestnut brown, pore of upper cell apical or depressed 1/3, of lower cell midway or below; pedicel colorless, always broken near hilum.

Hosts and distribution: *Arctium minus* (Hill) Bernh.: southern Canada to North Carolina and Utah; also in Europe.

Type: on *Arctium bardana* Willd. (= *A. lapa* L.), Germany (not seen).

This differs from var. *centaureae* because the urediniospores are echinulate to the hilum. Whether this is a distinction of sufficient significance to warrant recognizing a variety is open to question.

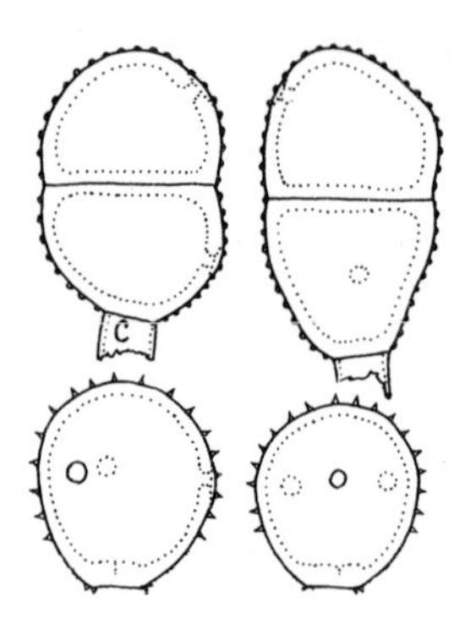

114b. *PUCCINIA CALCITRAPAE* var. *CENTAUREAE* (DC.) Cumm. Mycotaxon 5:402. 1977.
Puccinia centaureae DC. Flore Fr. 5:59. 1815.
Puccinia carthami Corda Icones Fung. 4:15. 1840.
Puccinia cirsii Lasch in Rabh. F. Eur. No. 89. 1859, not Kirchner 1856.
Puccinia laschii Lagerh. Tromsoe Mus. Aarsh. 17:63. 1895.
Puccinia irrequisita H. S. Jack. in Arthur, Bull. Torrey Bot. Club 48:32. 1921.

Spermogonia amphigenous or mainly on the adaxial leaf surface. Aecia amphigenous or on abaxial surface, uredinoid, with spermogonia, otherwise as the uredinia. Uredinia amphigenous, cinnamon brown; spores (22-)25-28(-29) x (20-)22-25(-27) µm, broadly obovoid or broadly ellipsoid, wall (1.5-)2-3 µm thick except about twice as thick at hilum, cinnamon or golden brown, echinulate except a smooth zone above the hilum, pores 3, rarely 2 or 4, equatorial, with slight or no caps. Telia amphigenous, exposed, blackish brown, pulverulent; spores (28-)33-42(-45) x (18-)20-24(-26) µm, mostly ellipsoid, wall verrucose with discrete small verrucae spaced 1.5-2(-2.5) µm, pore of upper cell apical or usually depressed 1/4-1/3, pore of lower cell depressed 1/4-1/2, with slight or no caps; pedicel colorless, to 55 µm but usually broken near hilum.

Hosts and distribution: *Carthamus tinctorius* L., *Centaurea americana* Nutt., *C. nigra* L., species of *Cirsium*: Canada, Mexico and the United States, especially the western parts; also in Europe, Asia and New Zealand.

Type: on *Centaurea*, environs de Mende, Prost (G).

The variety *calcitrapae*, unknown in North America, differs mostly because the verrucae are prominent over the pores but inconspicuous and often lacking elsewhere.

The rust fungi combined here have the principal characteristics in common i.e., uredinoid aecia (unknown in *P. irrequisita*), teliospores with small verrucae, depressed pores with slight or no umbonate caps, fragile pedicels, and similar dimensions and urediniospores with 3 equatorial pores with slight or no caps, a smooth zone above the hilum, which is conspicuously thicker than the sidewall, and similar dimensions. Savile (29) recognized the above listed entities as species and added *P. laschii* var. *xerophylla* Savile.

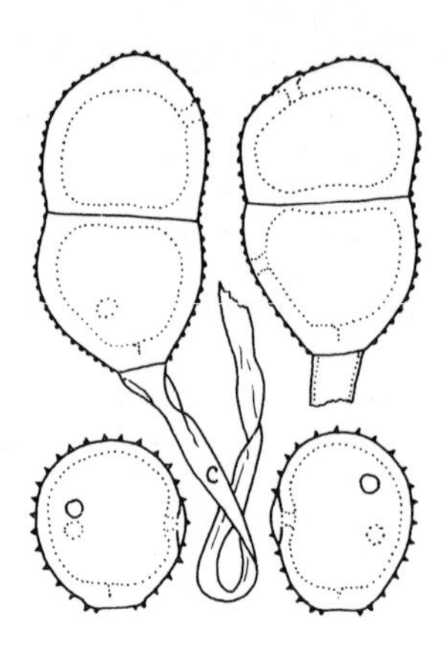

115. *PUCCINIA ACROPTILI* P. Syd. & H. Syd. Monogr. Ured. 1:4. 1902.

Spermogonia and aecia unknown. Uredinia amphigenous, about cinnamon brown; spores (21-)23-26(-29) x (18-)19-23 (-24) µm, broadly obovoid or globoid, wall 1.5-2(-2.5) µm thick, thicker basally, finely and uniformly echinulate, cinnamon brown, pores 3 (very rarely 2), equatorial, with obvious caps. Telia amphigenous, exposed, blackish brown, moderately pulverulent; spores (30-)35-40(-50) x (18-)22-27 (-30) µm, mostly ellipsoid, wall (1.5-)2-2.5(-3) µm thick at sides, usually 3.5-5 µm at apex, chestnut brown, finely verrucose or punctately verrucose, with verrucae spaced 1-1.5(-2) µm, uniformly distributed or sometimes tending to be in longitudinal lines, pore of upper cell depressed 1/4-1/2, pore of lower cell depressed 1/3-3/4, with no or only inconspicuous caps; pedicels colorless, usually persistent, to 100 µm.

Hosts and distribution: *Centaurea repens* L.: British Columbia to California and southern Arizona.

Lectotype: on *Acroptilon picris* (= *Centaurea repens*), Kurdistan, Persia, Haussknecht (S). Lectotype designated here.

116. *PUCCINIA CREPIDIS-MONTANAE* Magn. in Fischer, Beitr. Kryptogfl. Schweiz 2(2):212. 1904.

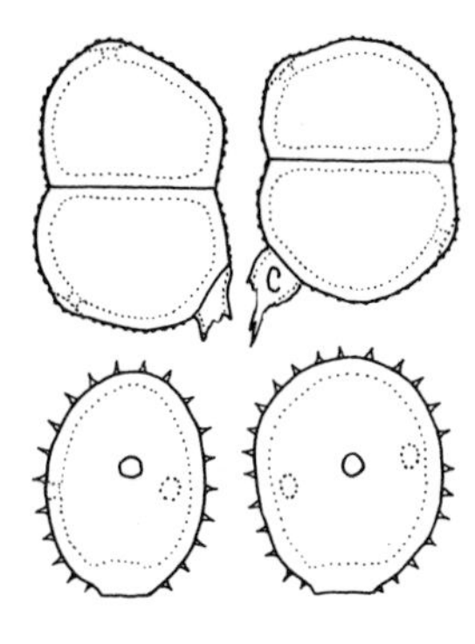

Spermogonia amphigenous. Aecia amphigenous or mostly on abaxial leaf surface, in groups, peridium whitish, short, spores 21-26 x 16-19 µm, broadly ellipsoid or oblong ellipsoid, wall 1 µm thick, pale yellowish, verrucose. Uredinia mostly on abaxial surface, yellowish brown; spores (22-)25-28(-30) x (18-)20-24(-26) µm, ellipsoid or broadly so, wall (1.5-) 2(-2.5) µm thick, pale cinnamon brown, uniformly echinulate, pores 3 (4) equatorial or nearly so, with slight or no caps. Telia amphigenous or mostly on abaxial surface, exposed, blackish brown, pulverulent; spores (27-) 31-40(-42) x (19-)21-26(-29) µm, ellipsoid or oblong ellipsoid, variable, wall uniformly 1.5-2(-2.5) µm, chestnut brown, minutely punctately verrucose with verrucae spaced 1.5-2 µm, sometimes striately arranged, pore of upper cell apical or commonly depressed, of lower cell midway or below, rarely at septum, with slight colorless caps; pedicels colorless, always broken near hilum.

Hosts and distribution: *Crepis* spp.: Montana and Washington southward to Colorado; also in Europe.

Lectotype: on *Crepis montana* Reichb., Kashalde, Furstenalp, Graubunden, 2000 m, 30 July 1901, Volkart (HBG).

Fischer, in "Bemerkungen", states, "Obige Beschreibung ist nach den von Volkark gesammelten und von Magnus bestimmten Exemplaren entworfen". Despite this statement and because two other specimens, one with telia, were listed by Fischer, I designate the above specimen as lectotype.

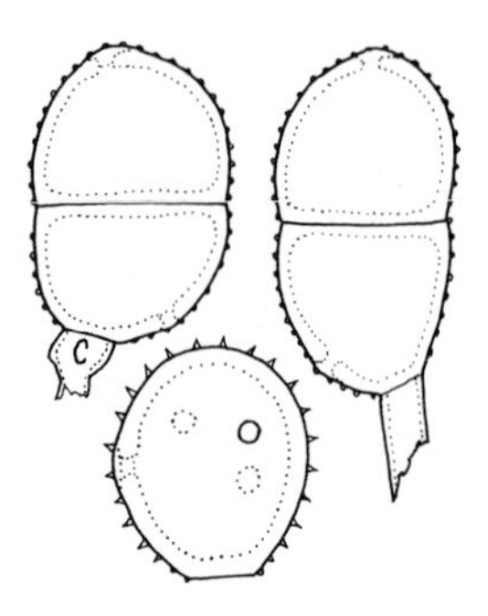

117. *PUCCINIA ORBICULA* Peck & G. W. Clint. in Peck, Ann. Rep. N.Y. State Mus. 30:53. 1879.

Spermogonia amphigenous, aecia on abaxial leaf surface, opening by a pore in the epidermis, with rudimentary or no peridium; spores catenulate, (20-)23-30(-33) x (16-)18-22 (-24) µm globoid or broadly ellipsoid, wall 1-1.5(-2) µm thick, colorless, minutely verrucose. Uredinia mostly on abaxial surface, pale cinnamon brown; spores (23-)25-29(-31) x (21-)23-25(-27) µm, globoid or broadly obovoid, wall (1.5-)2-2.5(-3) µm thick, yellowish to golden brown, uniformly echinulate, pores (2)3 or 4(5), usually scattered, with slight caps. Telia amphigenous, exposed, blackish brown, pulverulent; spores (30-)33-42(-46) x (18-)21-25(-28) µm, ellipsoid or oblong ellipsoid, wall 1.5-2(-2.5) µm thick, chestnut brown, finely verrucose with verrucae spaced mostly 2-2.5 µm, pore of upper cell mostly apical, of lower cell depressed 2/3 toward the hilum, with slight or no caps; pedicels colorless, always broken near hilum.

Hosts and distribution: on *Prenanthes* spp.: Newfoundland to Saskatchewan and southeastward to Virginia.

Type: on *Nabalus* (= *Prenanthes*) sp., Buffalo, New York (NYS).

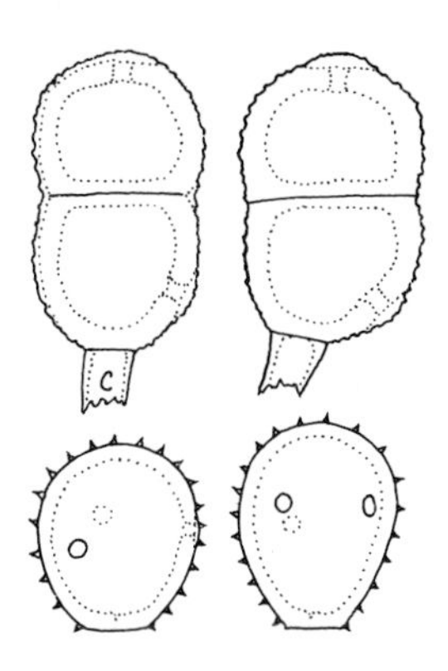

118. *PUCCINIA INAEQUATA* H. S. Jack. & Holw. in Jackson, Bot. Gaz. 65:309. 1918.

Spermogonia on adaxial leaf surface. Aecia mostly on adaxial surface around the spermogonia, uredinoid, cinnamon brown; spores like the urediniospores. Uredinia amphigenous, scattered, cinnamon brown; spores (21-)23-26(-28) x (17-)18-21(-23) µm, obovoid or broadly ellipsoid, wall 1.5-2.5(-3) µm thick, pale golden or nearly colorless, uniformly echinulate, pores 2 or mostly 3, approximately equatorial, with slight caps. Telia amphigenous, exposed, becoming pulverulent, blackish brown; spores (29-)33-38(-41) x (20-)22-25(-27) µm, broadly ellipsoid or oblong ellipsoid, wall 2.5-3 µm thick at sides, 4-5 µm at apex as a low, scarcely defined umbo, rugose or pseudoreticulate, tending to be bilaminate, pale chestnut brown, pore of upper cell apical, of lower cell in lower half; pedicels colorless, to 50 µm long but usually broken short.

Hosts and distribution: *Vernonia* spp.: Guatemala to Panama; also in South America.

Type: on *Vernonia patens* H.B.K., Escuintla, Guatemala, Holway No. 502 (PUR 37307).

The type cited by Jackson has telia, although his listing indicates only 0, I, II.

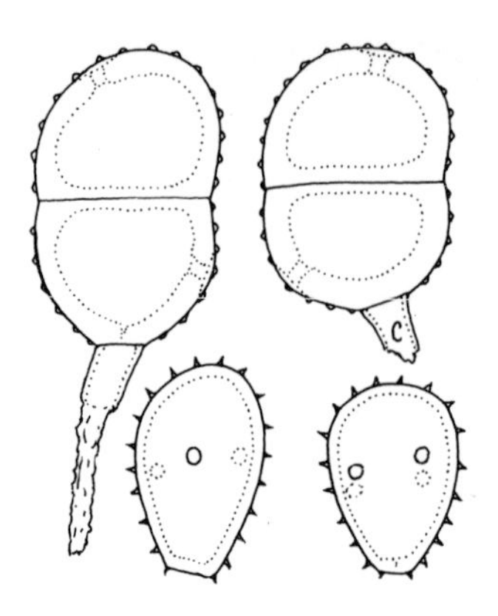

119. *PUCCINIA PRAEALTA* H. S. Jack. & Holw. in Jackson, Bot. Gaz. 65:306. 1918.

Spermogonia and aecia unknown. Uredinia on adaxial leaf surface, in aecidioid groups, nearly colorless when old, doubtless bright yellow when fresh; spores (23-)25-30 (-36) x (17-)18-22(-23) μm, mostly obovoid, wall 1.5(-2) μm thick, colorless, echinulate, pores obscure but equatorial and apparently 3 or 4(5), without caps. Telia on adaxial surface, grouped as the uredinia, microcyclic in appearance, blackish brown, pulverulent; spores (32-)35-43(-46) x (23-) 25-30(-33) μm, ellipsoid or broadly so; wall uniformly 3-4 μm thick, verrucose with small rounded or conical verrucae spaced (2-)3-4(-5) μm, the spacing uneven, pale chestnut or golden brown, pore of upper cell apical or depressed 1/3, pore of lower cell mostly midway to hilum, without umbos, obscure; pedicels, colorless, usually broken near hilum.

Hosts and distribution: *Vernonia heydeana* Coult., *V. trifloseulosa* H.B.K.: Central America.

Type: on *Vernonia trifloseulosa,* Mazatenango, Guatemala, Holway No. 510 (PUR 37335).

The close aecidioid grouping of the sori and their deep seated location in the leaf are characteristic.

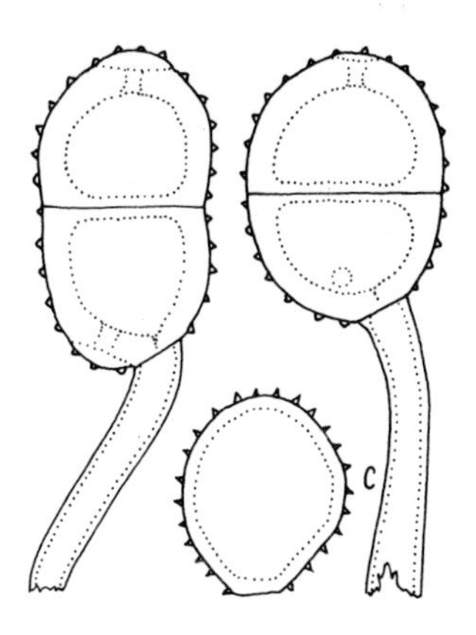

120a. *PUCCINIA NOTHA* H. S. Jack. & Holw. in Jackson, Bot. Gaz. 65:305. 1918 ssp. *NOTHA* var. *NOTHA*.

Spermogonia on adaxial leaf surface. Aecia few in groups on abaxial surface, peridium cylindrical, rupturing irregularly, whitish; spores 26-35 x 20-26 µm, irregularly ellipsoid or globoid, wall 1-1.5 µm thick, colorless, verrucose, the verrucae tending to be deciduous. Uredinia amphigenous, few and small, whitish when dry but doubtless bright yellow when fresh, spores (26-)28-32(-35) x (23-)26-29(-30) µm, broadly ellipsoid, broadly obovoid or nearly globoid, wall 1.5-2.5(-3) µm thick, colorless, echinulate, pores obscure, probably 3(4), equatorial. Telia mostly on adaxial surface, exposed, blackish brown, rather pulverulent; spores (35-)40-48(-52) x (24-)26-32(-34) µm, mostly broadly ellipsoid, wall (3-)3.5-5 µm thick at sides, 5-6.5(-8) µm over pores, clear chestnut brown or slightly paler over pores but not as defined umbos, echinulately verrucose with cones spaced (2-)3-4(-5) µm, pore of upper cell apical, of lower cell usually near hilum; pedicel yellowish or colorless, sometimes rugose basally, to 90 µm long.

Hosts and distribution: *Vernonia* spp.: Sinaloa, Mexico southward to Guatemala.

Type: on *Vernonia leiocarpa* DC., Solola, Guatemala, Holway No. 148 (PUR 33783).

120b. *PUCCINIA NOTHA* ssp. *NOTHA* var. *ATROCASTANEA* Urban, Acta Univ. Carolinae Biol. 1971:54. 1973.

This variety differs from var. *notha* because the teliospores are 46-55(-60) x 32-39(-41) µm, the wall is darker brown and the verrucae are spaced 2.5-6.5 µm.

Hosts and distribution: *Vernonia leiocarpa* DC., *V. standleyi* Blake: Guatemala and Honduras.

Type: on *Vernonia leiocarpa*, vicinity of Jutiapa, Guatemala, Standley No. 74968 (PUR F17897).

The variety is similar to *P. egregia* Arth. but has more deeply pigmented teliospores.

120c. *PUCCINIA NOTHA* ssp. *NOTHA* var. *PERAFFINIS* Urban, Acta Univ. Carolinae Biol. 1971:51. 1973.

This variety differs from var. *notha* because the teliospores are (43-)48-57(-61) x (30-)33-38(-40) µm and from var. *atrocastanea* because of the clear chestnut brown pigmentation of the wall.

Hosts and distribution: *Vernonia melancarpa* Blake, *V. pallens* Sch. Bip., *V. shannonii* Coult.: southern Mexico and Guatemala.

Type: on *Vernonia shannonii*, Quezaltenango, Guatemala, Holway No. 814 (PUR 33789).

Urban was doubtful of "The real existence of this variety ...."

Additionally, Urban described ssp. *socorrensis* on *Vernonia littoralis* Brand. from the Revilla Gigedo Islands because "The new subspecies differs chiefly in urediniospore germ pore number (mostly 2) and conspicuously more prominent and sparsely distributed spines." The subspecies is not known from the mainland.

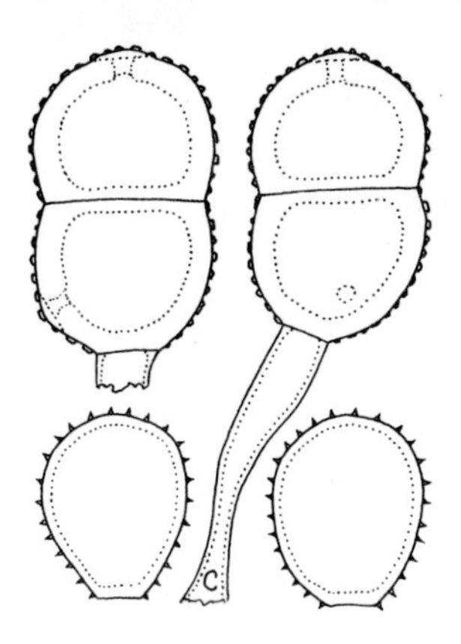

121. *PUCCINIA IDONEA* H. S. Jack. & Holw. in Jackson Bot. Gaz. 65:304. 1918.

Spermogonia and aecia unknown. Uredinia amphigenous, yellow when fresh fading to white; spores 23-28 x 18-21 µm, broadly ellipsoid or obovoid, wall 1-1.5 µm thick, colorless, echinulate, pores obscure but apparently equatorial. Telia mostly on abaxial surface, exposed, blackish brown, 3-4 µm thick at sides, to 7 µm over pores by a scarcely defined umbo, prominently verrucose with broad low verrucae, sometimes arranged in lines, pore of upper cell apical, of lower cell midway or below; pedicels colorless, to about 90 µm long, collapsing or not, rugose and slightly enlarged basally.

Hosts and distribution: *Vernonia heydeana* Coult., *V. trifloscutosa* H.B.K.: Costa Rica, Guatemala and El Salvador.

Type: on *Vernonia trifloscutosa*, Esquintla, Guatemala, Holway No. 499 (PUR 37322).

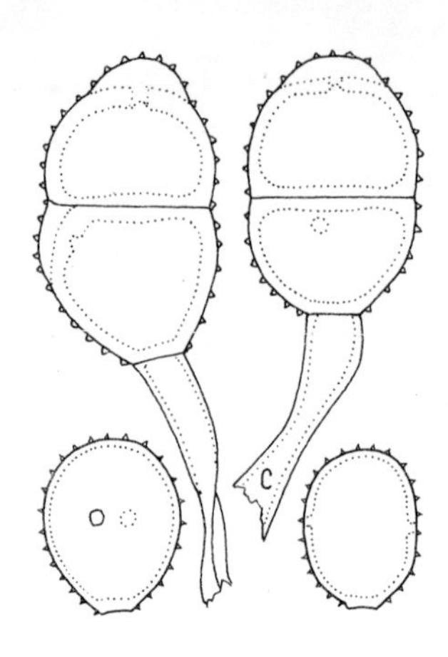

122a. *PUCCINIA CONOCLINII* Seym. ex Burrill, Bot. Gaz. 9:191. 1884 var. *CONOCLINII*.

Spermogonia mostly on abaxial leaf surface. Aecia amphigenous or mostly on abaxial surface, in small groups, without peridium but with catenulate spores; spores (20-)23-27(-31) x (15-)17-21(-23) µm, irregularly ellipsoid or nearly globoid, wall 1.5-2 µm thick, pale yellowish, echinulate. Uredinia mostly on abaxial surface, pale cinnamon brown; spores (20-)22-26(-33) x (18-)20-24(-28) µm, ellipsoid, obovoid or globoid, wall 1-1.5 µm thick, about golden brown, uniformly echinulate with spines spaced (1.5-)2-3 µm, pores 2, rarely 3, equatorial in only slightly or not flattened sides. Telia mostly on abaxial surface, exposed, relatively compact, blackish brown; spores (36-)40-48(-52) x (25-)28-33 (-36) µm, broadly ellipsoid or oblong ellipsoid, wall (2-) 2.5-3.5(-4) µm thick at sides, 6-9 µm over pores, from deep golden to chestnut brown but paler in the umbo over the pores, pore apical in each cell, verrucose with low conical warts spaced (1.5-)2-3.5(-4) µm; pedicels colorless, to 150 µm long but often much shorter.

Hosts and distribution: *Alomia microcarpa* (Benth.) B. L. Rob., and species of *Eupatorium* and *Ageratum*: the central United States south to Costa Rica; also in South America.

Type: on *Eupatorium coelestinum* L., Union County, Illinois, Earle No. 5034 (ILL 19970; isotype PUR 61218).

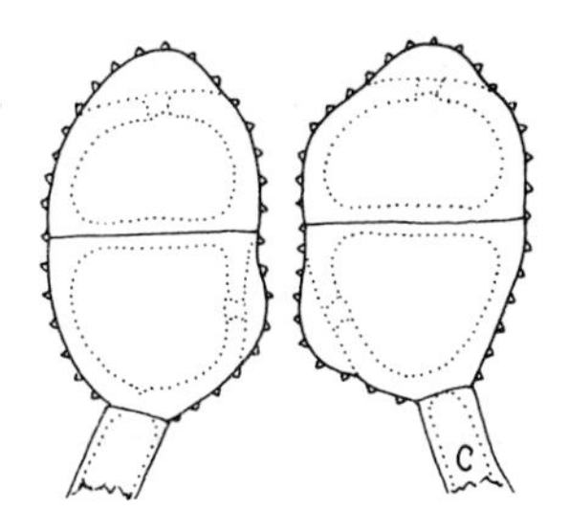

122b. *PUCCINIA CONOCLINII* var. *DEPRESSIPORA* Cumm., M. P. Brit. & J. W. Baxt. Mycologia 61:941. 1969.

Spermogonia and aecia unknown. Urediniospores as in var. *conoclinii*; teliospores as in var. *conoclinii* except 33-42 x 21-26 µm and with the pore of the lower cell midway between septum and hilum.

Type: on *Piqueria standleyi* B. L. Rob., Dept. Jutiapa, Guatemala, Standley No. 75496 (PUR 49996).

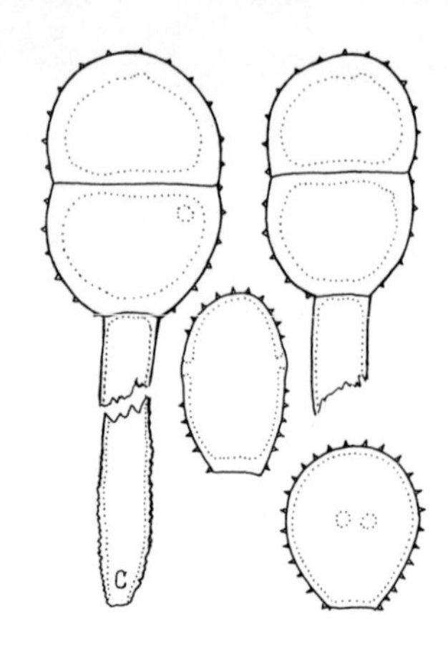

123. *PUCCINIA ECHINULATA* Joerst. Ark. Bot. 3:465. 1956.

Spermogonia and aecia unknown. Uredinia not seen; urediniospores in telia (26-)30-34(-38) x 24-29 µm, broadly ellipsoid or obovoid with pores face view, wall 1-1.5 µm thick, pale cinnamon brown or golden, echinulate, pores 2, equatorial in smooth, flattened sides. Telia amphigenous, exposed, blackish brown, pulverulent; spores (33-)36-44(-51) x (24-)27-29(-32) µm, ellipsoid, wall (2.5-)3-3.5(-4) µm thick at sides, 4-5 µm apically, uniformly chestnut brown, papillate with short cones spaced 2.5-5 µm, pore apical in each cell; pedicels colorless, rugose basally, to 90 µm long.

Hosts and distribution: *Dyssodia serratifolia* DC., *D. grandiflora* DC.: southern Mexico.

Type: on *Dyssodia serratifolia*, Las Hoyas cañon, Oaxaca, Oax., Pringle (S).

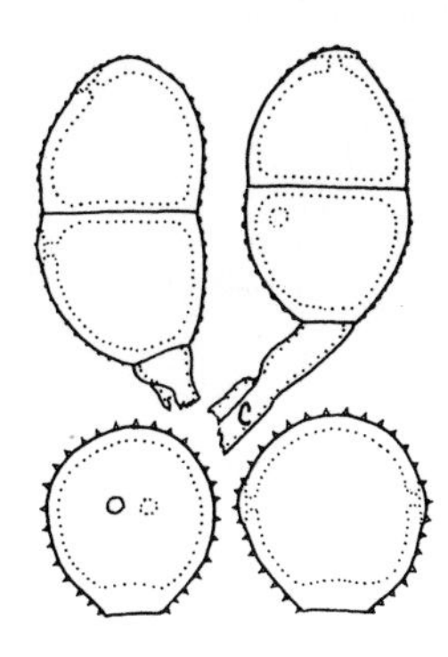

124. *PUCCINIA ALTISSIMORUM* Savile, Can. J. Bot. 48:1579. 1970.

Spermogonia and aecia unknown. Uredinia mostly on abaxial leaf surface, cinnamon brown; spores 21-25(-27) x (18-)20-23(-25) µm, nearly globoid, wall 1.5-2 µm thick, pale cinnamon brown or golden, uniformly echinulate, pores 2, rarely 1 or 3, equatorial or slightly above, with small caps. Telia on abaxial surface, exposed, blackish brown, pulverulent; spores (25-)28-38(-42) x (16-)18-22(-25) µm, mostly ellipsoid, wall (1-)1.5-2 µm thick, slightly thicker over pores by a low, illdefined umbo, clear chestnut or golden brown, punctately verrucose with verrucae spaced about 1.5-2 µm, pore of upper cell apical or somewhat depressed, of lower cell near septum; pedicels colorless, to 70 µm long, commonly broken short.

Hosts and distribution: *Cirsium altissimum* (L.) Spreng., *C. iowense* (Pamm.) Fern., *C. discolor* (Muhl.) Spreng.: the Mississippi Valley and eastward.

Type: on *Cirsium discolor*, Falkland, Del., Common No. 137 (holotype DAOM; isotypes Ellis & Ever. N. Amer. F. No. 2253).

Savile also established var. *horriduli* based on a specimen from Louisiana (Kellerman No. 5999 = PUR 37958) on *C. horridulum* Michx. that has urediniospores 25-35 µm long and the lower pore in the teliospore depressed to midway.

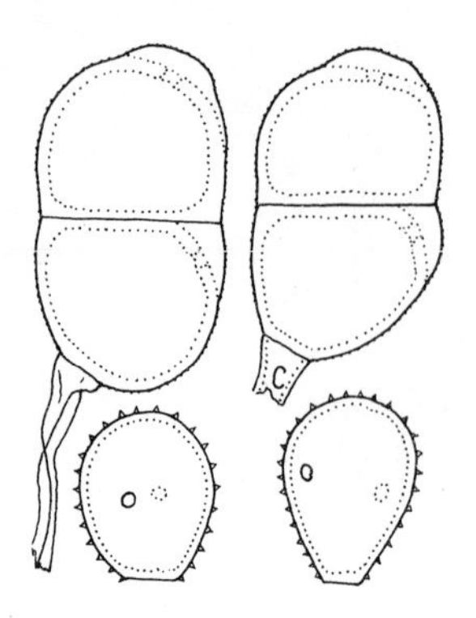

125. *PUCCINIA BACCHARIDIS-HIRTELLAE* Diet. & Holw. in Holway, Bot. Gaz. 31:331. 1901.

Spermogonia, aecia and uredinia unknown. Urediniospores in telia 21-26(-28) x (17-)19-22 µm, obovoid or broadly ellipsoid, wall 1 µm thick, pale cinnamon brown or golden, echinulate with short spines spaced (1.5-)2(-2.5) µm, pores 2, equatorial or approximately so, in smooth, slightly flattened areas. Telia on abaxial leaf surface, exposed, dark cinnamon brown, compact; spores (35-)38-46 (-54) x (23-)25-30(-32) µm, mostly ellipsoid or broadly ellipsoid, wall 1.5-2(-2.5) µm thick at sides, (3-)4-6(-7) µm thick apically, clear chestnut or golden brown, paler over the pores, minutely verrucose with the verrucae spaced about 1 µm, pore of upper cell apical or nearly so, of lower cell next to septum; pedicels colorless, to 75 µm long, often broken shorter.

Hosts and distribution: *Archibaccharis hirtella* (DC.) Heer.: southern Mexico.

Type: on *Baccharis hirtella* (= *A. hirtella*), Amecameca, Mex., Holway No. 375b (S; isotype PUR 33794).

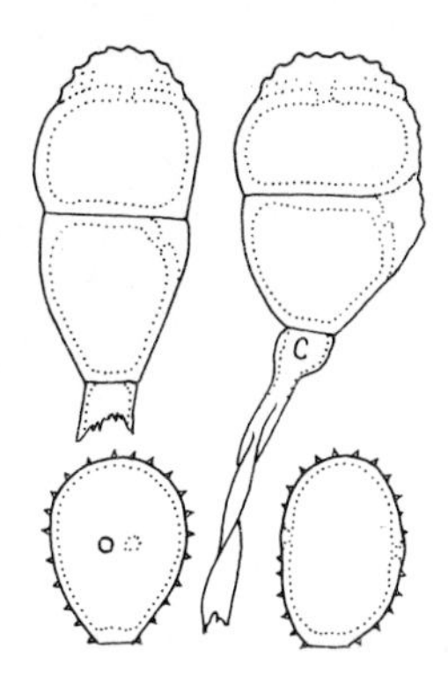

126. *PUCCINIA GUATEMALENSIS* J. Parm. Can. J. Bot. 45:2303. 1967.

Spermogonia amphigenous. Aecia amphigenous, around the spermogonia, uredinoid, pale cinnamon brown; spores (18-)20-24(-28) x (13-)16-20(-21) µm, obovoid or broadly ellipsoid, wall 1-1.5 µm thick, yellowish or golden, echinulate, pores 2, equatorial in slightly or not flattened sides. Uredinia scattered, otherwise similar to the aecia. Telia amphigenous or mostly on abaxial leaf surface, exposed, brown becoming gray from germinating, rather compact; spores (25-)30-40(-43) x (16-)18-23(-28) µm, broadly ellipsoid or obovoid, wall (1-)1.5(-2.5) µm thick at sides, clear chestnut brown, (3-)4-7(-8) µm and golden over pores, reticulate or striately reticulate on the apex becoming faintly so or smooth below, pore apical in each cell, each with a golden, defined umbo; pedicels colorless, and usually broken near hilum; germinating without dormancy.

Hosts and distribution: *Zexmenia* and *Wedelia* spp.: central Mexico to Honduras and El Salvador.

Type: on *Zexmenia frutescens* (Mill.) Blake, Quirigua, Guatemala, Holway No. 601 (PUR 37621).

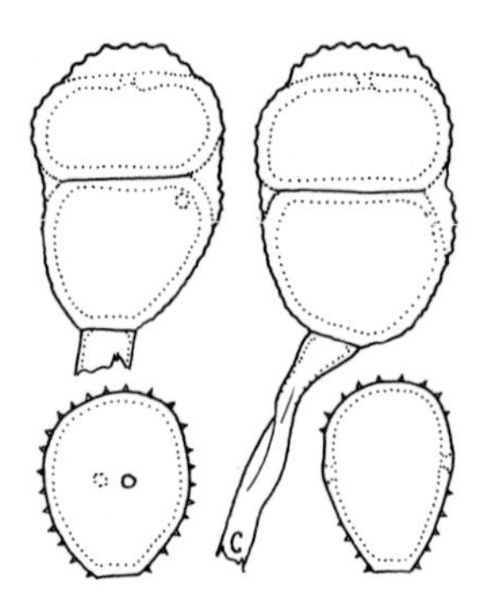

127. *PUCCINIA PROBA* H. S. Jack. & Holw. in Arthur, Mycologia 10:143. 1918.

Spermogonia amphigenous. Aecia amphigenous in tight groups, peridium cupulate or cylindrical, whitish; spores 16-23 x 14-21 µm, angularly globoid or ellipsoid, wall 1 µm thick, colorless, verrucose with small rods. Uredinia amphigenous, pale cinnamon brown; spores (18-)20-26(-29) x (15-)16-19(-21) µm, broadly ellipsoid or obovoid, wall 1 µm thick, yellowish, echinulate except around pores, pores 2, equatorial in slightly or not flattened sides. Telia amphigenous or mostly on abaxial leaf surface, exposed, blackish brown, pulverulent; spores of two kinds, clear chestnut brown or golden brown and germinating without dormancy, and deep chestnut brown and nongerminating, (24-)28-40(-45) x (16-)19-22(-25) µm, broadly ellipsoid or slightly obovoid, wall 1-2 µm thick at sides, (3.5-)5-7(-8) µm over pores, reticulate, especially apically, sometimes nearly smooth basally, pore of each cell apical, each with a paler, defined umbo; pedicels colorless, to 50 µm long but often broken near spores.

Hosts and distribution: *Zexmenia* spp.: central Chihuahua, Mexico southward to Costa Rica.

Lectotype: on *Zexmenia frutescens* (Mill.) Blake var. *villosa* (Polak) Blake, San José, Costa Rica, Holway No. 247 (PUR 37622).

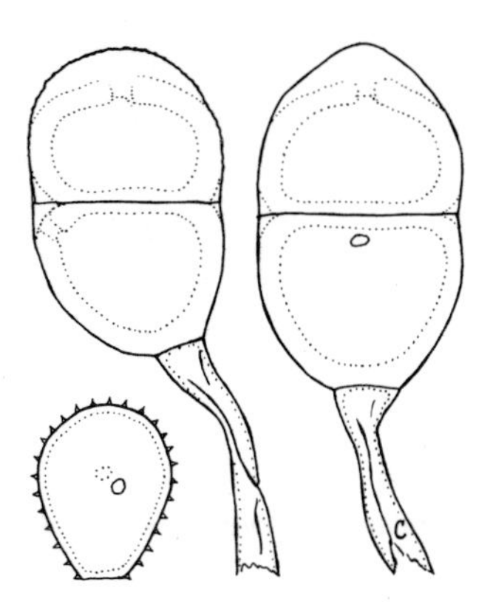

128a. *PUCCINIA CONCINNA* Arth. Bot. Gaz. 40:205. 1905 var. *CONCINNA*.

Aecia unknown. Uredinia not seen; spores in telia (22-)23-27(-30) x (18-)20-24(-26) µm, broadly ellipsoid or obovoid, wall 1-1.5(-2) µm thick, cinnamon brown or paler, uniformly echinulate, pores 2(3), equatorial, with slight or no caps. Telia mostly on abaxial leaf surface, sometimes on stems, exposed, blackish brown, compact; spores (37-)40-52(-60) x (21-)25-32(-38) µm, ellipsoid or oblong ellipsoid, wall (2.5-)3-4(-5) µm thick at sides, chestnut brown, (4-)5-9(-12) µm over pores by slightly paler but scarcely defined umbos, minutely rugose or rugosely reticulate or sometimes with discrete wartlets, pore apical in each cell; pedicel colorless, to 170 µm long but usually less than 125 µm.

Hosts and distribution: *Eupatorium greggii* Gray: southeastern Arizona and (probably) northwestern Chihuahua, Mexico.

Type: on *Conoclinium greggii* (Gray) Small (= *Eupatorium greggii*), Sierra Madre, Mexico, Nelson (PUR 37434).

The following variety differs in having dimorphic spores.

128b. *PUCCINIA CONCINNA* var. *DURANII* (Hennen, Leon.-Gall. & Cumm.) Cumm. Mycotaxon 5:403. 1977.
Puccinia duranii Hennen, Leon.-Gall. & Cumm. Southw. Nat. 16:367. 1972.

Aecia unknown. Uredinia and spores as in var. *concinna*. Telia on abaxial leaf surface, early exposed, blackish brown, compact; spores dimorphic, large type (36-)40-48 (-52) x (25-)27-29 µm, ellipsoid or broadly so, wall (2-) 2.5-3.5(-4) µm thick at sides, 5.5-9 µm at apex, chestnut brown or nearly so, small type 29-35 x 21-25 µm, ellipsoid or somewhat obovoid, wall 2-3 µm thick at sides, 2.5-4(-5.5) µm at apex, both types minutely rugose or rugosely reticulate, pore apical in each cell; pedicel colorless or tinted next the hilum, thin walled, mostly collapsing, to 125 µm long.

Type: on *Eupatorium greggii* Gray, 88 mi south of Hidalgo del Parral, Dgo., Mexico, Hennen No. 69-152 (PUR 63275). Not otherwise known.

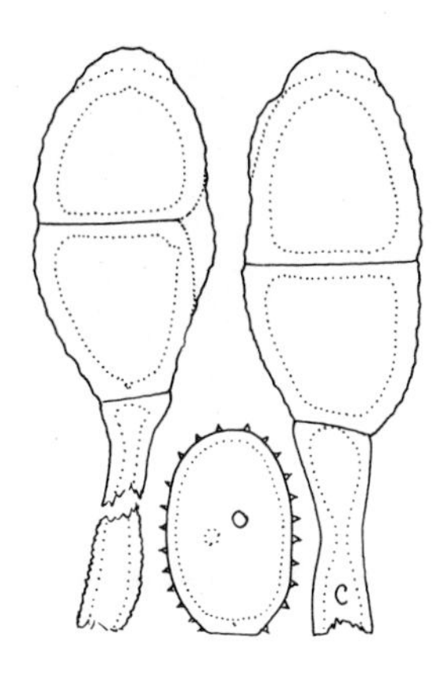

129. *PUCCINIA ZALUZANIAE* Arth. Bot. Gaz. 40:205. 1905.

Spermogonia and aecia unknown. Uredinia mostly on adaxial leaf surface, cinnamon brown; spores (20-)25-28(-32) x (16-)18-22(-24) µm, ellipsoid or narrowly obovoid, wall (1-) 1.5 µm thick, cinnamon brown or paler, echinulate except around pores, pores 2, equatorial in slightly flattened sides. Telia mostly on adaxial surface, exposed, small, blackish brown, pulverulent; spores (30-)34-48(-54) x (17-) 21-28(-31) µm, ellipsoid or oblong ellipsoid, wall tending to be bilaminate, (1.5-)2.5-3.5(-4) µm thick at sides, (3-) 4-6(-7) µm over pores, chestnut brown, finely rugose in reticulate or labyrinthiform patterns, appearing almost undulate in silhouette, pore of each cell apical, each covered by a pale brown, defined umbo; pedicels brownish next hilum, colorless below, rugose basally.

Hosts and distribution: *Zaluzania* spp.: from central to southern Mexico.

Type: on *Zaluzania montagnaefolia* Sch. Bip. (as *Z. asperrima*), Tehuacán, Pue., Holway No. 5347 (PUR 42613).

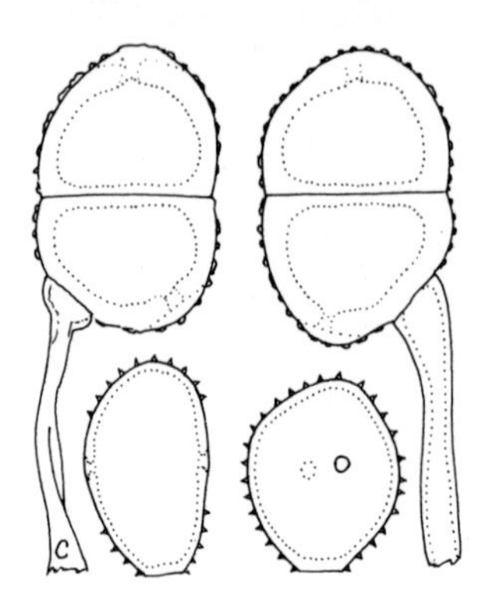

130. *PUCCINIA ZEXMENIAE* Diet. & Holw. in Holway, Bot. Gaz. 24:26. 1897.

Spermogonia amphigenous. Aecia amphigenous, in small groups, often along veins, peridium white, cylindrical; spores (16-)20-30(-36) x (13-)15-19(-21) µm, ellipsoid, oblong ellipsoid or globoid, often angular, wall (1-)1.5-2 µm thick, verrucose with short rods, colorless. Uredinia amphigenous or commoner on abaxial surface, cinnamon brown; spores (16-)20-28(-31) x (14-)16-23 µm, obovoid or nearly globoid with pores face view, ellipsoid with pores lateral, wall 1(-1.5) µm thick, pale cinnamon brown, echinulate except around each pore, pores 2, equatorial in usually flattened sides. Telia amphigenous, exposed, blackish brown, pulverulent; spores (26-)32-44(-49) x (20-)23-28(-32) µm, ellipsoid or broadly so, wall (2-)2.5-4(-5) µm thick at sides, 4-6(-9) µm over pores but the umbos not clearly defined, dark chestnut brown, reticulate with meshes mostly 2-3 µm diam with the delimiting ridges narrow or the ridge areas marked by beads and elongate warts, pore in apical cell mostly apical, in lower cell usually about midway; pedicels mostly colorless, to 150 µm long but usually 100 µm or less.

Hosts and distribution: *Zexmenia* spp.: central Mexico northwestward to southern Arizona.

Type: on *Zexmenia podocephala* Gray, Guadalajara, Jal., Mexico, Holway (S).

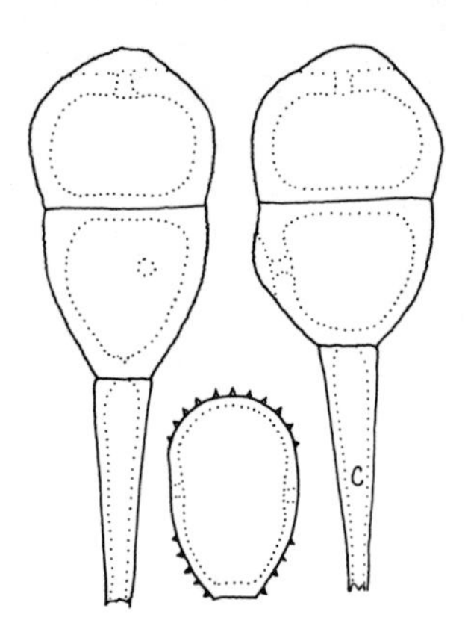

131. *PUCCINIA SUBDECORA* Syd. & Holw. in Sydow, Ann. Mycol. 1:17. 1903.

Spermogonia on adaxial leaf surface and on stems. Aecia amphigenous and on stems, around the spermogonia, uredinoid, dark brown; spores (23-)25-30(-34) x 20-25(-27) µm, broadly ellipsoid or obovoid with pores face view, ellipsoid or oblong ellipsoid and (14-)16-21(-23) µm wide with pores lateral, wall (1-)1.5-2 µm thick, cinnamon brown, echinulate except around pores, pores 2, equatorial in smooth, flattened sides. Uredinia amphigenous, scattered, otherwise similar to the aecia. Telia mostly on the abaxial surface, exposed, pulverulent, chocolate brown; spores (27-)30-44(-53) x (16-)18-28(-30) µm, ellipsoid or obovoid, wall (2-)2.5-3.5 (-5) µm thick at sides, 5-8 µm over pores, minutely rugosely verrucose or rugosely reticulate, clear chestnut brown except the pale umbo over each pore, pore apical in upper cell, mostly midway to hilum in lower cell; pedicel colorless, tapering downward, to 30 µm long.

Hosts and distribution: *Brickellia grandiflora* Nutt.: Colorado, Wyoming, Utah and California, U.S.A.

Type: Georgetown, Colorado, Holway No. 1102 (S; isotype PUR 37516).

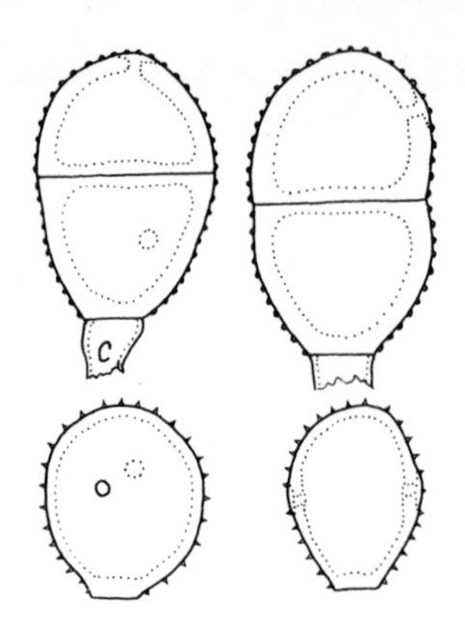

132. *PUCCINIA CYANI* Pass. in Rabenh. F. Europ. No. 1767. 1874.

Spermogonia amphigenous, systemic. Aecia uredinoid, systemic, amphigenous, cinnamon brown; spores like the urediniospores. Uredinia localized, mostly on abaxial leaf surface, cinnamon brown; spores (21-)23-26(-28) x (18-)20-24(-26) µm, broadly ellipsoid or globoid with pores face view, ellipsoid with pores lateral, wall 1.5-2 µm thick or thicker on pore bearing sides, cinnamon brown, echinulate, pores 2, equatorial, with small or no caps. Telia systemic and amphigenous when associated with spermogonia and aecia, later localized on abaxial surface, exposed, blackish brown, pulverulent; spores (30-)35-42(-46) x (23-)25-28(-32) µm, ellipsoid or broadly so, wall (2-)2.5-3.5(-4) µm at sides, slightly thicker over pores by poorly defined umbos, chestnut brown, punctately verrucose with verrucae spaced (1-)1.5-2(-2.5) µm, pore of upper cell apical or depressed, pore of lower cell midway to hilum or below; pedicels colorless, always broken near hilum; paler teliospores with thinner walls usually intermixed.

Hosts and distribution: on *Centaurea cyanus* L.: southern Canada south at least to California and Tennessee, also in Guatemala; widely distributed in other areas of the world.

Type: on *Centauria cyanus*, Parma, Italy (PARM; isotype Rabenh. F. Europ. 1767).

Savile (29) named var. *sublevis* for some Pacific Coast collections that have finely rugose to smooth teliospores.

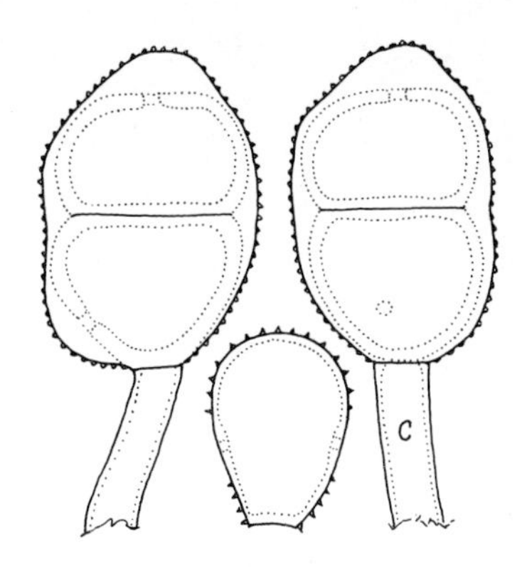

133. *PUCCINIA IOSTEPHANES* Diet. & Holw. in Holway, Bot. Gaz. 31:334. 1901.
Puccinia nanomitra P. Syd. & H. Syd. Monogr. Ured. 1: 182. 1902.

Spermogonia and aecia unknown. Uredinia amphigenous, cinnamon brown; spores 24-30(-32) x (19-)20-25(-27) µm, obovoid or broadly ellipsoid with pores face view, triangularly obovoid with pores lateral, wall (1-)1.5(-2) µm thick, cinnamon brown, echinulate except around pores, pores 2, equatorial, in flattened sides. Telia mostly on the adaxial leaf surface, exposed, blackish brown, pulverulent; spores (30-)35-50(-57) x (24-)28-34(-37) µm, broadly ellipsoid, wall bilaminate, the inner chestnut brown wall 2.5-3 µm thick, the outer golden layer 1-4 µm thick at sides and 4-9 (-11) µm over pores, verrucose with low cones spaced 2-3(-4) µm, pore apical in upper cell, in lower half of lower cell, each covered by a large, defined umbo; pedicels colorless, to 120 µm long, usually broken shorter, rugose basally.

Hosts and distribution: *Iostephanes heterophylla* Benth., *Viguiera dentata* (Cav.) Spreng., *V. eriophora* Greenm.: southern Mexico.

Type: on *Iostephanes heterophylla*, Cuernavaca, Mor., Holway No. 3488 (S; probable isotypes Barth. N. Amer. Ured. 455; Barth. F. Columb. 3849).

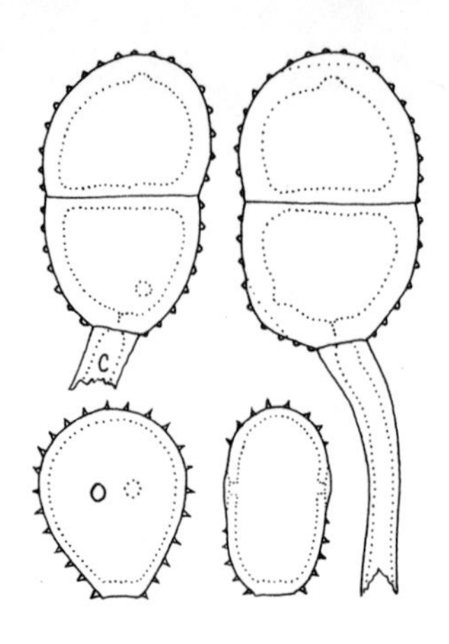

134a. *PUCCINIA ELECTRAE* Diet. & Holw. in Holway Bot. Gaz. 31:333. 1901 var. *ELECTRAE*.

Spermogonia and aecia unknown. Uredinia amphigenous or mostly on adaxial leaf surface, cinnamon brown; spores (24-)27-33(-35) x (18-)22-26(-28) μm, obovoid or broadly ellipsoid with pores face view, wall (1-)1.5-2 μm thick, cinnamon brown, echinulate except around the pores, pores 2, equatorial in flattened sides. Telia amphigenous or mostly on adaxial surface, early exposed, blackish brown, more or less pulverulent; spores (38-)42-48(-50) x (22-)25-32(-35) μm, mostly ellipsoid or broadly so, wall (2-)2.5-3.5(-4) μm thick at sides, chestnut brown, 4-6 μm over pores by pale, low, scarcely defined umbos that are part of a thin, illdefined outer layer, verrucose with low, rounded or cone shaped verrucae spaced about 2-4 μm, pore apical in upper cell, near pedicel in lower cell; pedicels colorless except near hilum, to 100 μm long.

Hosts and distribution: *Coreopsis mutica* DC.: southern Mexico and Guatemala.

Type: on *Coreopsis galeottii* Gray (= *C. mutica*), Oaxaca, Oax., Mexico, Holway No. 3664 (S; isotypes Barth. N. Amer. Ured. 1439).

Parmelee recognized the following varieties because of slight differences in the thickness of spore walls and in the size and spacing of the verrucae of the teliospores. The two varieties occur along the Rio Grande River, thus far from and in more arid conditions than var. *electrae*.

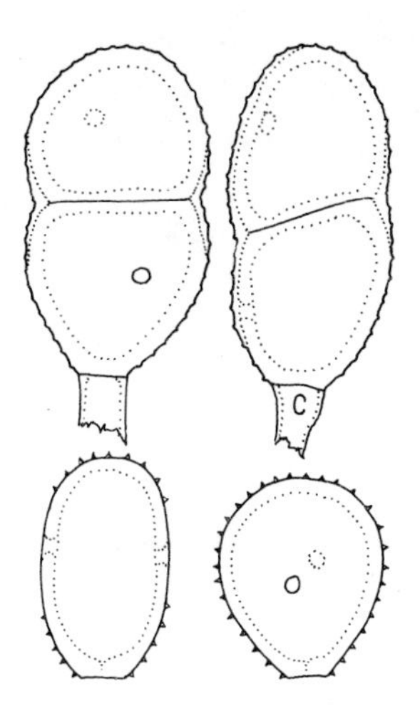

134b. *PUCCINIA ELECTRAE* var. *DEPRESSIPOROSA* J. Parm. Can. J. Bot. 45:2307. 1967.

Urediniospores (27-)30-38(-42) x (18-)20-27(-30) µm, wall (1.5-)2.5-3(-3.5) µm thick, golden to chestnut brown, pores 2, equatorial. Teliospores (37-)40-48(-54) x (23-)26-30(-35) µm, wall (2-)2.5-3.5(-4) µm, punctately verrucose, verrucae spaced about 1.5-2.5 µm, deep chestnut brown, pore in each cell usually depressed to near midway; pedicels colorless, broken near hilum.

Type: on *Zexmenia brevifolia* Gray, Big Bend National Park, Texas, Cummins No. 61-415 (PUR 58913). One other collection (Cummins 70-34) from the same region is known.

134c. *PUCCINIA ELECTRAE* var. *ROBUSTA* J. Parm. Can. J. Bot. 45:2306. 1967.

Urediniospores 26-29 x 19-25 µm, ellipsoid, wall 2-2.5 µm thick, near chestnut brown, pores 2. Teliospores 35-43 x 29-33 µm, wall (2.5-)3-4 µm thick, slightly thicker over pores, chestnut brown, verrucose with low verrucae spaced 1.5-3.5 µm, pore apical in upper cell, in lower half in lower cell; pedicel to 85 µm long.

Type: on *Zexmenia brevifolia* Gray, Presidio, Texas, J. H. Russell (PUR 49948). Not otherwise known.

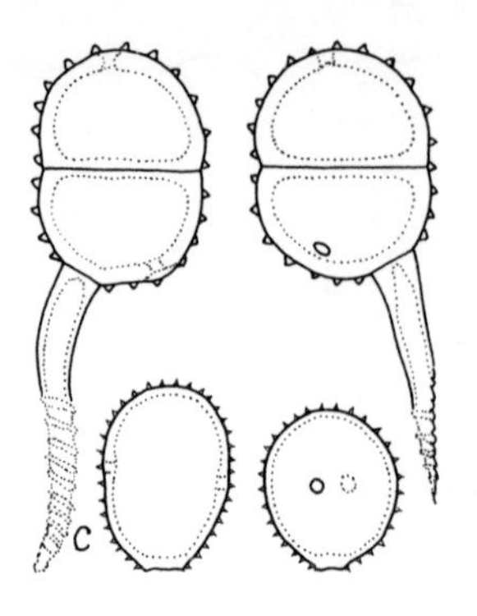

135. *PUCCINIA GLOBULIFERA* Arth. Bot. Gaz. 40:200. 1905.

Spermogonia and aecia unknown. Uredinia on abaxial leaf surface, cinnamon brown; spores (16-)18-23(-25) x (15-)17-21 µm, broadly ellipsoid with pores face view, triangular or strongly obovoid with pores lateral, wall 1-1.5 µm thick, cinnamon brown, echinulate except an area around each pore, pores equatorial. Telia amphigenous or mostly on abaxial surface, exposed, blackish brown, pulverulent; spores (25-)27-32(-34) x (20-)22-24(-26) µm, broadly ellipsoid, wall uniformly (1.5-)2-2.5 µm thick, chestnut brown, verrucose echinulate with broad, rather blunt cones 1.5-2 µm high spaced (2.5-)3-5(-7) µm, pore apical in upper cell, near pedicel in lower cell, without umbos, obscure, diorchidioid spores common; pedicels pale brownish next to hilum, colorless basally, the lower half rugose, often in an annular or spiral pattern, to 75 µm long, seldom less than 50 µm long.

Hosts and distribution: *Otopappus pringlei* (Greenm.) Blake, *O.* sp., *Notoptera* (?) sp.: Nayarit, Jalisco and Guerrero, Mexico.

Type: on *O. pringlei* (as *O. epaleaceus pringlei*), Iguala, Gro., Holway No. 5313 (PUR 42612; isotypes Bartholomew N. Amer. Ured. 143).

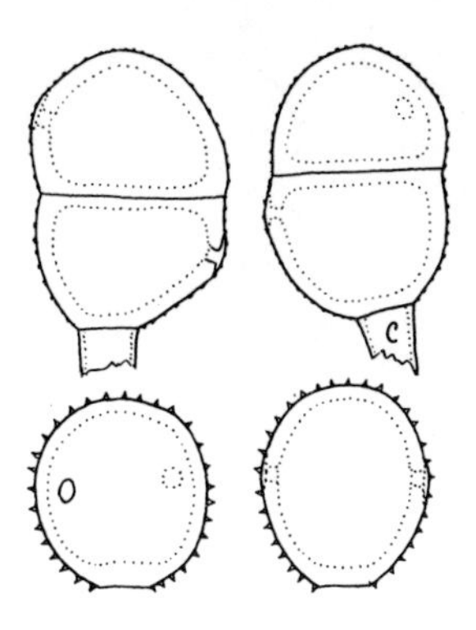

136. *PUCCINIA PINAROPAPPI* P. Syd. & H. Syd. Hedwigia 40 (Beibl.): 127. 1901.

Spermogonia and aecia unknown. Uredinia amphigenous, cinnamon brown; spores (21-)24-28(-30) x (19-)22-25 µm, broadly ellipsoid or globoid, wall (1.5-)2-2.5(-3) µm thick, uniformly and finely echinulate, about golden brown, the pore bearing sides not flattened, pores 2, equatorial, with inconspicuous or no caps. Telia amphigenous, exposed, blackish brown, more or less pulverulent; spores (28-)32-38 (-40) x (21-)24-28 µm, mostly ellipsoid or broadly so, wall uniformly 2-3(-3.5) µm thick, chestnut brown, minutely punctate with verrucae spaced about 2 µm, scarcely visible in silhouette, pore of upper cell nearly apical or usually depressed 1/3-1/2, pore of lower cell depressed 1/3-1/2, without differentiated caps; pedicels colorless, broken near hilum.

Hosts and distribution: *Pinaropappus roseus* (Less.) Less.: Texas.

Type: Austin, Texas, Long (S).

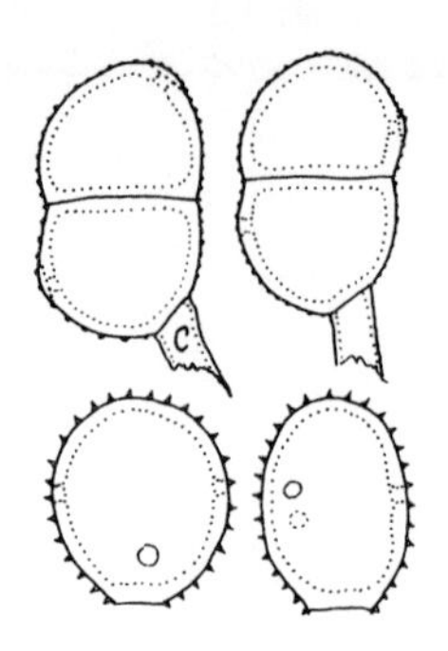

137a. *PUCCINIA VARIABILIS* Grev. Scot. Crypt. Flora pl. 75. 1824 var. *VARIABILIS.*

Spermogonia not seen. Aecia amphigenous or mostly on abaxial leaf surface, in small groups, peridium whitish; spores (16-)19-23(-25) x (13-)15-20 µm, broadly ellipsoid or globoid, wall 1(-1.5) µm thick, colorless, minutely verrucose. Uredinia amphigenous, cinnamon brown; spores (20-) 22-26(-28) x 19-23(-25) µm, broadly obovoid or globoid, wall 1.5(-2) µm thick, golden brown, uniformly echinulate, pores (2)3, equatorial, with slight or no caps. Telia amphigenous or mostly on abaxial surface, exposed, pulverulent, blackish brown; spores (21-)24-30(-33) x (15-)18-22(-24) µm, ellipsoid or broadly so, wall uniformly (1-)1.5(-2) µm thick, clear chestnut brown, punctately verrucose with verrucae spaced about 1.5-2 µm, pore of upper cell apical or depressed, pore of lower cell near septum or depressed 1/2 or more, each with a slight cap; pedicel colorless, always broken near hilum.

Hosts and distribution: *Taraxacum officinale* Wiggers: Nova Scotia and Quebec; also in Europe.

Type: on *Taraxacum officinale,* near Edinburgh, Scotland (E).

The following varieties differ mostly in having typically two pores in the urediniospores.

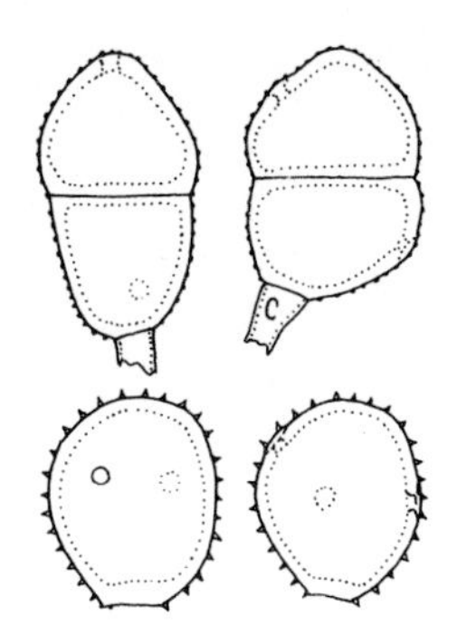

137b. *PUCCINIA VARIABILIS* var. *INSPERATA* (H. S. Jack.) Cumm. Mycotaxon 5:406. 1977.
Puccinia insperata H. S. Jack. Brooklyn Bot. Gard. Mem. 1:253. 1918.

Aecia and aeciospores as in var. *variabilis*. Uredinia and urediniospores as in var. *variabilis* except the pores 2, rarely 3. Telia as in var. *variabilis*; spores (23-)26-33 (-35) x (15-)17-21(-23) µm, mostly obovoid or ellipsoid, sometimes diorchidioid, wall uniformly 1.5-2 µm thick, punctately verrucose with verrucae spaced (1.5-)2(-3) µm, clear chestnut brown, pore of upper cell apical or depressed 1/4 to 1/2, pore of lower cell depressed 1/3 to 2/3, with slight or no caps, pedicel colorless, always broken near hilum.

Hosts and distribution: *Prenanthes alata* (Hook.) Griseb.: the Pacific Coast from Oregon to Alaska.

Type: on *Nabalus hastatus* (= *Prenanthes alata*), Hood River Valley, Oregon, Jackson No. 3265 (PUR 34812).

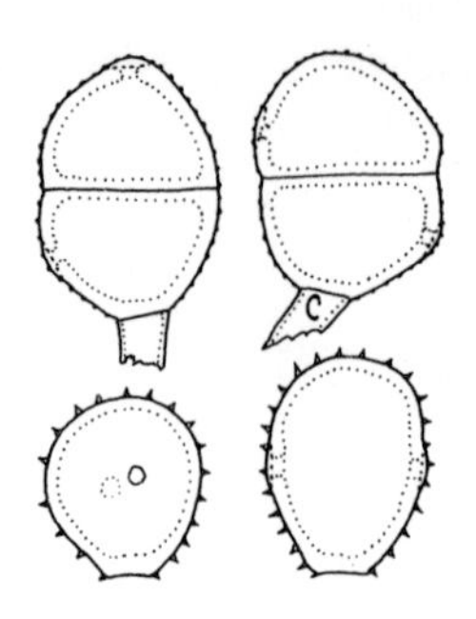

137c. *PUCCINIA VARIABILIS* var. *LAPSANAE* (Fckl.) Cumm. Mycotaxon 5:406. 1977.
Puccinia lapsanae Fckl. Jahr. Nass. Ver. Nat. 15:13. 1860.

Aecia and spores as in var. *variabilis*. Uredinia amphigenous; spores (17-)19-22(-24) x (15-)17-20(-22) µm, mostly broadly ellipsoid or obovoid, wall (1-)1.5(-2) µm thick, golden brown, uniformly echinulate, pores 2(3), equatorial. Telia amphigenous; spores (23-)26-29(-33) x (17-)19-23(-26) µm, mostly ellipsoid, wall uniformly 1.5-2 µm thick, clear chestnut brown, punctately verrucose with verrucae spaced (1-)1.5-2(-2.5) µm, pore of upper cell apical or depressed 1/3 to 2/3, pore of lower cell depressed 1/3 to 1/2, with slight caps; pedicel colorless, always broken near hilum.

Hosts and distribution: *Lapsana communis* L.: occasional in northern United States and southern Canada; also in Europe.

Type: on *Lapsana communis*, near Oestrich in Nassau, Germany (G).

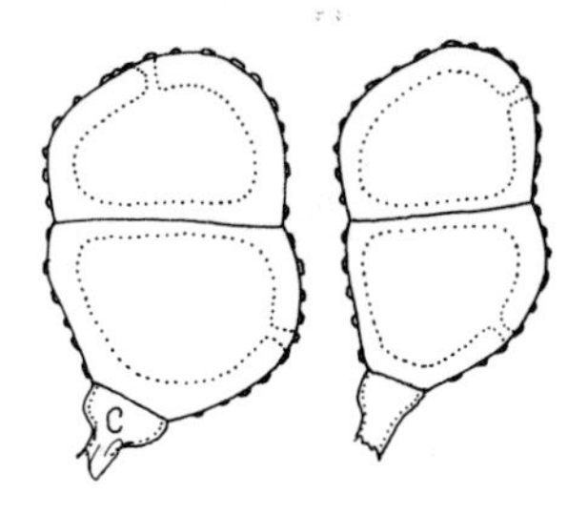

138. *PUCCINIA HYSTERIUM* (Str.) Roehl. Deutschl. Fl. Ed. 2. III, 3:131. 1813.
Uredo hysterium Str. Ann. Wetter. Ges. 2:102. 1810 (based on telia).

Spermogonia mostly amphigenous, from a systemic mycelium. Aecia amphigenous and on stems and inflorescence among the spermogonia, aecidioid, the peridium short cupulate; spores 23-30(-38) x 16-22 μm, globoid or broadly ellipsoid, wall 1-1.5 μm thick, colorless, minutely verrucose. Uredinia lacking. Telia amphigenous and on stems, among the aecia or separately, partially covered by the epidermis but soon exposed and pulverulent, chocolate brown; spores (29-)32-40 (-44) x (21-)24-30(-32) μm, variable in size and shape, mostly broadly ellipsoid, wall uniformly (2-)2.5-3(-3.5) μm thick, verrucose with flattish warts spaced 2.5-5 μm or these fused in short linear series or rarely in a pseudo-recticulate pattern, chestnut brown, pore of upper cell apical or approximately so, of lower cell midway or below; pedicels colorless, fragile and broken short.

Hosts and distribution: *Tragopogon pratensis* L.: southern Ontario and Quebec; also in Europe and Asia.

Type: on *Tragopogon pratensis*, Germany (not seen).

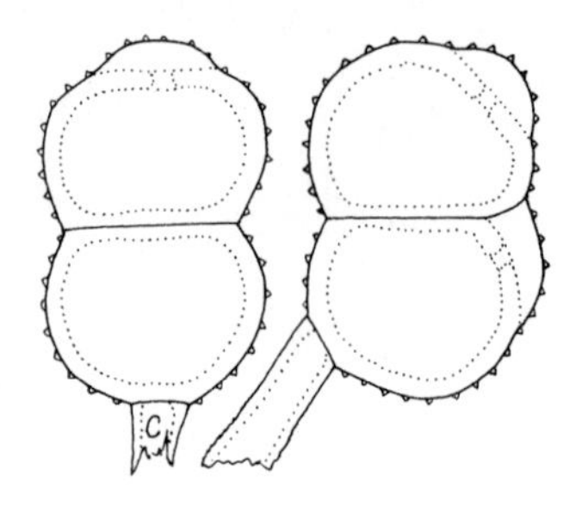

139. *PUCCINIA MCVAUGHII* Hennen, Leon-Gall. & Cumm. Southw. Nat. 16:359. 1972.

Spermogonia not seen. Aecia (old) on abaxial leaf surface, in small groups, peridium whitish; spores not seen. Uredinia not seen, probably wanting. Telia on abaxial surface, exposed, dark brown, more or less pulverulent; spores (36-)40-48(-52) x (23-)25-28(-30) µm, mostly ellipsoid, wall 1.5-2.5(-3) µm thick at sides, 5-7 µm thick over pores with clearly differentiated umbos, clear chestnut brown except the pale umbos, echinulate verrucose with low cones spaced (1.5-)2-3(-4) µm, pore of upper cell at or near apex, of lower cell at or near septum; pedicels colorless, rugose basally.

Type: on *Archibaccharis sescenticeps* (Blake) Blake, 2 km northeast of Campamento El Gallo, Gro., Mexico, Rzedowski & McVaugh No. 194 (PUR 61549). Not otherwise known.

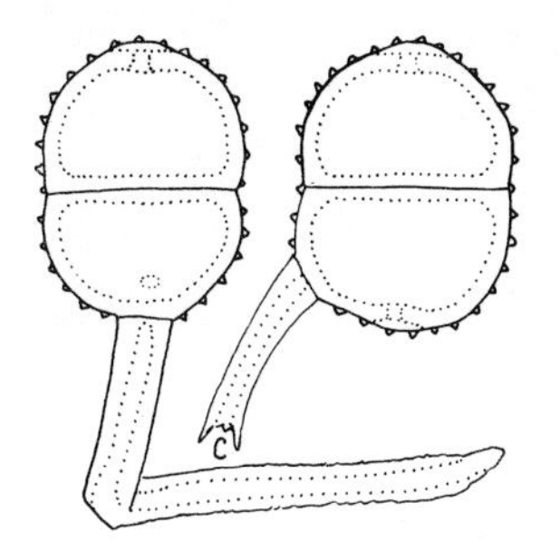

140. *PUCCINIA OTOPAPPICOLA* Joerst. Nytt. Mag. Bot. 6:137. 1958.
Puccinia cornuta H. S. Jack. & Holw. in Arthur, Amer. J. Bot. 5:533. 1918, not Hazslinszky 1877.

Spermogonia on adaxial leaf surface. Aecia along veins on lower surface in chlorotic areas, peridium brownish, cylindrical; spores 26-40 x 15-26 µm, angularly globoid or oblong, wall 1 µm thick and smooth basally, to 7 µm thick and verrucose above. Uredinia lacking. Telia mostly on abaxial leaf surface, exposed, more or less pulverulent, chocolate brown, mostly along the veins; spores (28-)30-38(-45) x (20-)22-28(-32) µm, broadly ellipsoid, wall 2.5-3(-3.5) µm thick, chestnut brown, verrucosely echinulate with cones about 1 µm high spaced 2-3.5 µm, pore apical in upper cell, near pedicel in lower cell, each with a paler, small umbo, the wall tending to be bilaminate elsewhere; pedicels brownish next to hilum, colorless basally, the lower half or third rugose in an annular or spiral pattern.

Hosts and distribution: *Notoptera brevipes* (B. L. Rob.) Blake, *N. scabridula* Blake: British Honduras and Guatemala.

Type: on *N. brevipes*, Guatemala City, Holway No. 846 (PUR 36485).

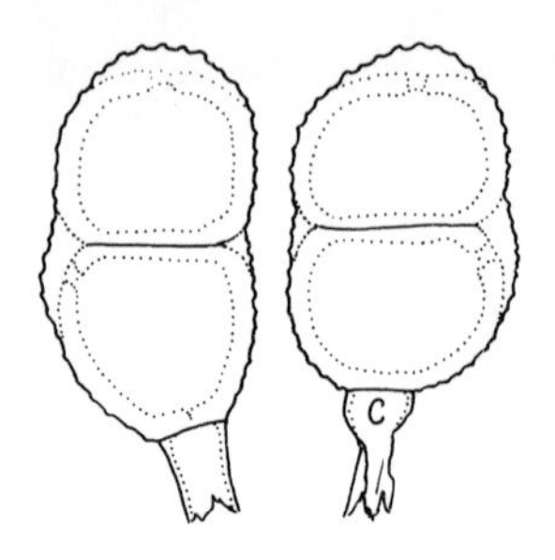

141. *PUCCINIA GHIESBRECHTII* J. Parm. Can. J. Bot. 45:2305. 1967.

Spermogonia amphigenous, in necrotic areas. Aecia and uredinia lacking. Telia amphigenous, around the spermogonia or radiating along veins, exposed, pulvinate, first formed sori cinnamon brown and with germinating spores then replaced by blackish brown sori with resting spores; spores (26-)30-45(-50) x (18-)22-27(-30) µm, ellipsoid or broadly so, wall 1-1.5 µm thick at sides, 6-9 µm over pores, near golden brown, coarsely and striately reticulate apically becoming minutely rugose to smooth basally in germinating type, 1.5-2.5(-3) µm thick at sides 4-7 µm over pores, deep chestnut brown, reticulate with meshes 1.5-3 µm wide, the delimiting ridges narrow and frequently incomplete in resting type, pore of each cell apical, each covered by a paler umbo which is clearly defined in germinating type, much less so in resting type; pedicels colorless, commonly broken near hilum, rugose basally when complete.

Hosts and distribution: *Zexmenia elegans* Sch. Bip., *Z. ghiesbrechtii* Gray: Sinaloa and Nayarit, Mexico.

Type: 15 miles east of Chupaderos, Sin., Cummins No. 63-710 (PUR 61637).

The species probably is a microcyclic derivative of *Puccinia proba*.

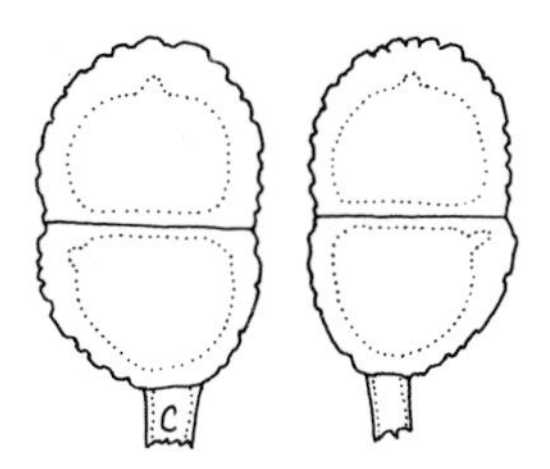

142a. *PUCCINIA DOVRENSIS* Blytt, Christ. Vidensk.-Selsk. Forh. 1896:64. 1896. var. *RUSSA* (Arth. & Cumm.) Hennen & Baxter, Mycologia 66:554. 1974.
Puccinia russa Arth. & Cumm. Ann. Mycol. 31:44. 1933.

Telia amphigenous, exposed, confluent from the beginning forming sori to as much as 5 mm diam, chocolate brown; spores (30-)32-37(-40) x (20-)22-24(-25) μm, mostly oblong ellipsoid, wall (2-)2.5-3(-3.5) μm thick at sides including sculpturing, 4-5(-6) μm over the pores but without defined umbos, verrucose with irregular warts as much as 5 μm diam, these often anastomosing in various patterns, pore apical in each cell but inconspicuous; pedicels brownish near hilum, usually broken near spore.

Type: on *Erigeron salsuginosus* Gray (? = *E. howellii* Gray), Garnet Lake, California, Blasdale No. 1337 (PUR 47223). Not otherwise known.

142b. *PUCCINIA DOVRENSIS* var. *LEPTOTICHA* (Hennen) Hennen in Hennen & Baxter, Mycologia 66:554. 1974.

Similar to the above but with spores (31-)32-43(-45) x 16-23 μm.

Type: on *Erigeron simplex* Greene, Independence Pass, east of Aspen, Colorado, Hennen No. 63-116 (PUR 60496). Also from Cottonwood Pass, Colo. and Breccia Peak, east of Moran, Wyo.

Puccinia dovrensis var. *dovrensis* is not known to occur in North America.

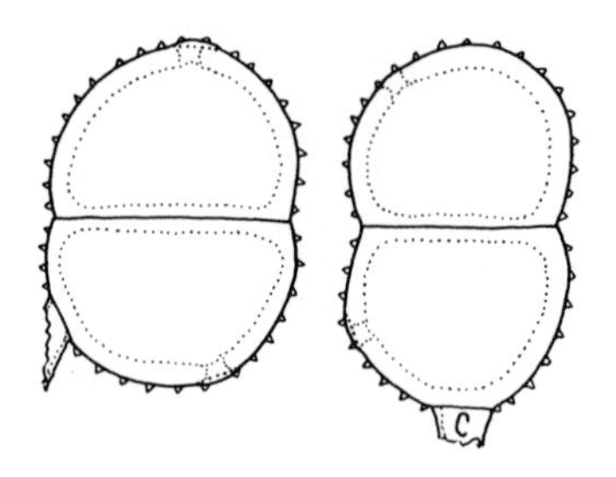

143. *PUCCINIA SUKSDORFII* Ellis & Ever. J. Mycol. 7:130. 1892.

Spermogonia amphigenous, systemic. Aecia and uredinia lacking. Telia amphigenous, systemic, exposed, blackish brown; spores (30-)35-45(-55) x (25-)27-35(-38) µm, ellipsoid, wall uniformly 2-2.5(-3) µm thick, echinulately verrucose with small conical verrucae 1-1.5 µm high spaced 2-3 (-4) µm, chestnut brown, pore of upper cell apical or depressed about 1/4 to the septum, pore of lower cell 1/2 or more toward the hilum, with slight or no caps; pedicels colorless, always broken near hilum.

Hosts and distribution: *Agoseris* spp.: Montana to Washington south to Arizona.

Lectotype: on *Troximon glaucum* (= *Agoseris glauca* (Pursh) D. Dietr.), Washington (without locality or date), Suksdorf (NY). Designation of a lectotype is necessary because a Kelsey specimen from Helena, Montana also was cited in the original.

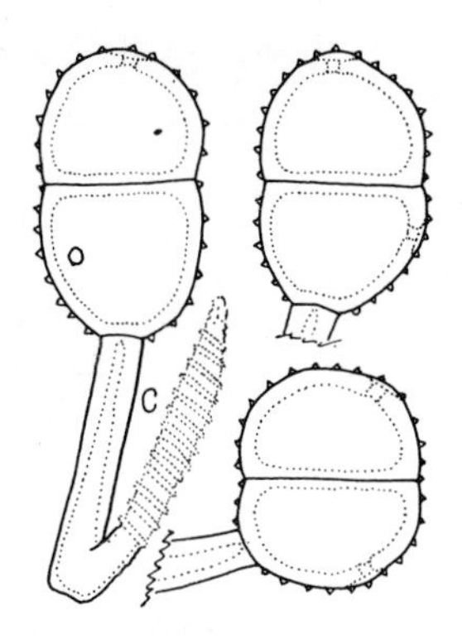

144. *PUCCINIA ANNULATIPES* Cumm. Mycologia 66:892-93. 1974.

Spermogonia mostly on adaxial leaf surface. Aecia and uredinia lacking. Telia on abaxial surface below the spermogonia, grouped in conspicuous chlorotic areas, exposed, blackish brown; spores (28-)30-38(-42) x (20-)22-25(-27) µm, ellipsoid or broadly ellipsoid, wall 2-3(-3.5) µm thick except over pores, chestnut brown, verrucosely echinulate with cones 1-1.5 µm high spaced (1.5-)2-3(-4) µm, pore apical in upper cell, about midway in lower cell, each with a slightly paler, umbo about 1.5 µm thick; pedicel brownish next to hilum, colorless below, thick walled, the lower half rugose in an annular or spiral pattern, to 85 µm long but usually about 50-60 µm long.

Type: on *Notoptera* sp., south of Puerto Vallarta, Jal., Mexico, Cummins No. 71-523 (PUR 64862). Not otherwise known.

The species differs from *P. noptopterae* Arth. (Jamaica) especially in having larger spores with thick walled, persistent pedicels. It differs from *P. ottopappicola* Joerst. in lacking aecia.

145. *PUCCINIA FEROX* Diet. & Holw. in Holway, Bot. Gaz. 31: 333. 1901.

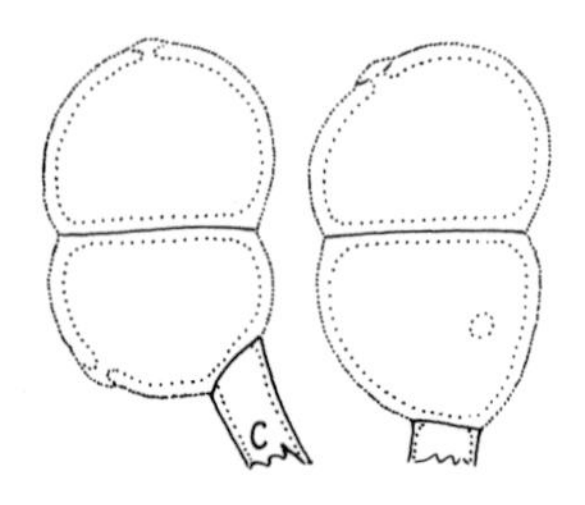

Telia caulicolous and amphigenous along the veins, exposed, dark cinnamon brown, confluent, pulverulent; spores (29-)33-46(-50) x (19-)22-28 (-32; -38) µm, ellipsoid, wall 1-1.5 µm thick, pale cinnamon or golden brown, minutely and closely verrucose, pore usually apical in upper cell, midway in lower cell, each with a small colorless papilla before germination, germinating without dormancy; pedicels fragile, broken near hilum.

Hosts and distribution: *Verbesina* spp.: southern Mexico to Guatemala and Costa Rica.

Type: on *Verbesina diversifolia* DC., Oaxaca, Oax., Mexico, Holway No. 3704 (S; probable isotypes Barth. F. Columb. 4569).

146. *PUCCINIA ABSICCA* H. S. Jack. & Holw. in Arthur, Mycologia 10:144. 1918.

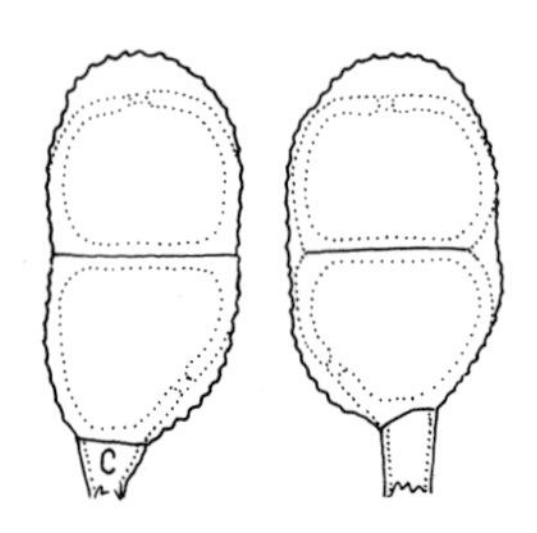

Spermogonia on adaxial surface of leaves. Aecia and uredinia lacking. Telia amphigenous, exposed, about cinnamon brown, pulverulent; spores (25-) 28-40(-43) x (17-)19-24(-26) µm, ellipsoid or oblong ellipsoid, wall 1.5 -2(-2.5) µm thick at sides, (3-)4-6 (-7) µm over pores, tending to be bilaminate, about golden brown or paler, finely rugose in an irregular pattern or the markings uniting in reticulate patterns, pore apical in upper cell, near the hilum in lower cell, each with a pale, clearly defined umbo; pedicels colorless, broken near the spore.

Type: on *Zexmenia frutescens* (Mill.) Blake var. *villosa* (Polak) Blake, San José, Costa Rica, Holway No. 239 (PUR 42002). Not otherwise known.

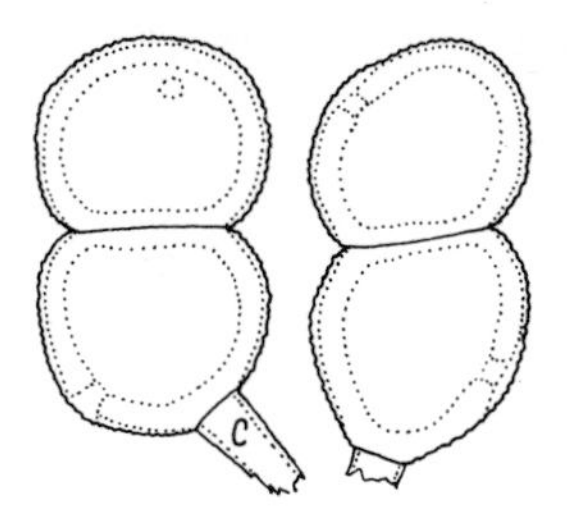

147. *PUCCINIA DISCRETA* H. S. Jack. & Holw. in Jackson, Bot. Gaz. 65:308-309. 1918.

Spermogonia on adaxial leaf surface, in small groups. Aecia and uredinia lacking. Telia mostly on adaxial surface, closely grouped around the spermogonia, exposed, about cinnamon brown, pulverulent; spores (31-)35-43(-46) x (19-) 21-25(-29) µm, ellipsoid, constricted and the cells separating easily; wall uniformly (2-)2.5-3 µm thick or slightly thicker at pores, about golden brown with a thin colorless outer layer, finely rugose or reticulately rugose with discrete wartlets or these usually fused in various patterns, pore of upper cell apical or mostly depressed about 1/3, pore of lower cell midway or below, with distinct umbos; pedicels colorless, always broken near hilum.

Hosts and distribution: *Vernonia stellaris* Llave ex Lex., *V.* sp.: San Luis Potosi, Mexico to Costa Rica; also in South America.

Type: on *V. deppeana* Less. (= *V. stellaris*), San José, Costa Rica, Holway No. 260 (PUR 41369).

Collectors have noted that the infections always are on young terminal growth.

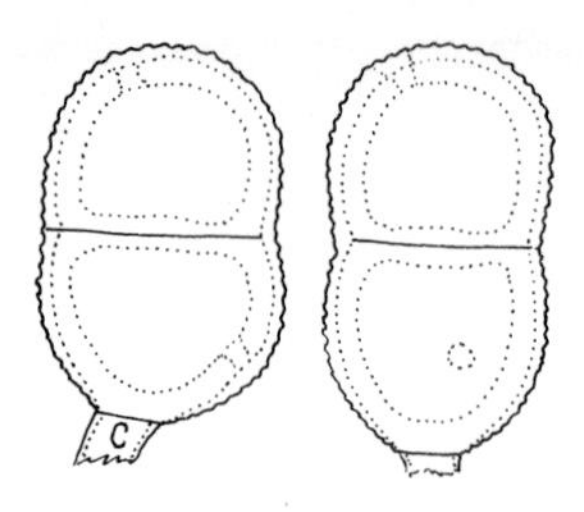

148. *PUCCINIA NEOROTUNDATA* Cumm. Mycologia 48:606. 1956.
Puccinia rugosa Speg. An. Soc. Cient. Argent. 17:92-93. 1884, not Billings 1871.
Puccinia rotundata Diet. Hedwigia 36:32. 1897, not Bonorden 1860.

Spermogonia, aecia and uredinia lacking. Telia amphigenous, in groups on slightly hypertrophied areas or along the veins, exposed, pulverulent, dark cinnamon brown; spores (30-)35-42(-45) x (20-)22-26(-28) µm, broadly ellipsoid, wall uniformly 3-4 µm thick or slightly thicker over pores, scarcely bilaminate but decidedly paler externally, rugose with small verrucae and ridges of various lengths, these tending to fuse in pseudoreticulate patterns, golden brown or paler, often almost lemon yellow, pore of upper cell apical or nearly so, of lower cell midway or below; pedicels colorless, fragile, always breaking near hilum.

Hosts and distribution: on *Vernonia* spp.: Costa Rica; also in South America.

Type: on Compositae (*Verbesina* ?), now considered to be *Vernonia*, Paraguari, Paraguay, Balansa No. 3433 (LPS; isotype PUR F7971). Urban (30) erroneously listed the type of *P. rotundata* as type of the species.

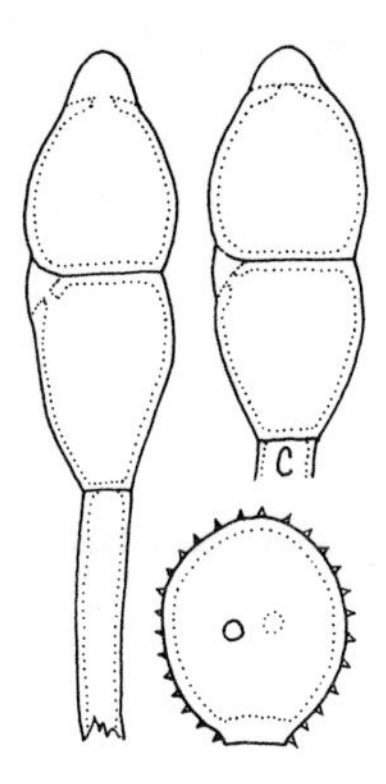

149a. *PUCCINIA ARACHIDIS* Speg. An. Soc. Cient. Argent. 17: 90. 1884 var. *ARACHIDIS*.

Spermogonia and aecia unknown. Uredinia mostly on abaxial leaf surface, about cinnamon brown; spores (21-)23-29 x (16-)18-22(-24) µm, mostly broadly ellipsoid or obovoid, wall 1.5(-2) µm thick, cinnamon brown, echinulate with fine spines spaced about 1.5-2 µm, pores 2, equatorial in often flattened areas. Telia mostly on abaxial surface, exposed, pulvinate, about cinnamon brown becoming gray from germination; spores (33-)38-56(-60) x (12-)14-18 µm mostly ellipsoid or oblong ellipsoid, wall 0.5-1 µm thick at sides and pale golden, (3-)4-5 µm over pore at apex and septum as nearly colorless, clearly defined umbos which disappear during germination, smooth; pedicels colorless, to 65 µm long but usually broken shorter; germinating without dormancy.

Hosts and distribution: *Arachis hypogaea* L.; southern United States and southward into South America.

Type: on *A. hypogaea*, Caa-guazu, Paraguay (LPS).

For a review of the literature of peanut rust see Bromfield (6).

149b. *PUCCINIA ARACHIDIS* var. *OFFUSCATA* (Arth.) Cumm. Mycotaxon 5:402. 1977.
Puccinia offuscata Arth. Bull. Torrey Bot. Club 47: 469. 1920.

Differs from var. *arachidis* in having urediniospores with 2-4, commonly 3, germ pores and paler teliospores. There are no dependable size differences.

Hosts and distribution: *Zornia bracteata* J. F. Gmel., *Z. diphylla* (L.) Pers.: Texas to Florida; also in the West Indies and South America.

Type: on *Z. diphylla*, Herradura, Prov. Piñar del Rio, Cuba, Baker No. 2143 (PUR 36786).

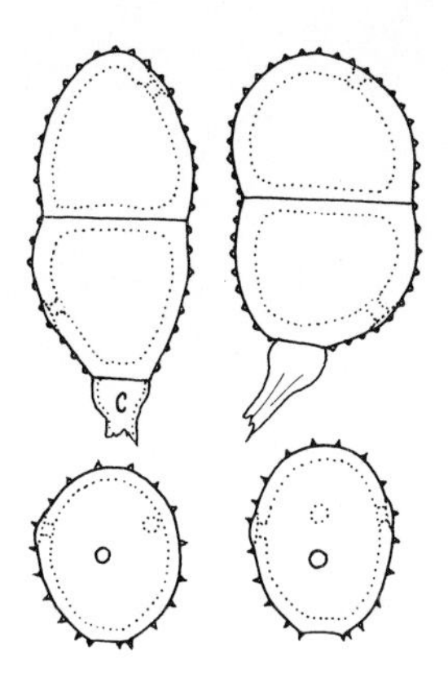

150. *PUCCINIA PAROSELAE* Cumm. Bull. Torrey Bot. Club 68:44. 1941.

Spermogonia and aecia unknown. Uredinia on abaxial leaf surface and on stems, cinnamon brown; spores (20-)22-28(-32) x (15-)17-20(-22) µm, variable but mostly obovoid or broadly ellipsoid, wall 1.5(-2) µm thick, pale cinnamon brown, uniformly echinulate, pores 3 or 4, equatorial or sometimes scattered, with small caps. Telia similar to uredinia but near chocolate brown, early exposed, pulverulent; spores (21-)22-26(-27) x (34-)37-42(-45;-50) µm, mostly ellipsoid, wall uniformly 2-2.5(-3) µm thick, chestnut brown or paler, echinulate verrucose with small cones spaced 1.5-2 µm and tending to be in lines, pore midway or below in lower cell, apical or slightly subapical in upper cell, with slight or no paler umbos; pedicels colorless, fragile and always broken near the spore.

Type: on *Dalea mollis* Benth. (*Parosela mollis*), Santa Rosa Canyon, Riverside County, California, Parks and Jordan No. 6430 (PUR 49188); not otherwise known.

2. UROMYCES Unger

Exantheme Pflanzen. p. 277. 1833.

Spermogonia subepidermal, globoid, type 4 (16). Aecia subepidermal in origin, erumpent, aecidioid with catenulate spores, or uredinoid with spores borne singly on pedicels. Uredinia subepidermal in origin, erumpent, with or without paraphyses; spores borne singly on pedicels. Telia subepidermal in origin, mostly erumpent; spores borne singly on pedicels, 1 celled, with 1 germ pore, wall mostly pigmented; basidium external.

Type species: *Uromyces appendiculatus* (Pers.) Unger.

KEY TO SPECIES OF *UROMYCES* ON COMPOSITAE

1. Teliospores with surface sculpture 2
1. Teliospores smooth 4

 2. Species microcyclic, only telia known .... *oblongisporus* (1)
 2. Species with uredinia and telia 3

3. Teliospore pedicel 50 µm or less long, usually broken short .... *martinii* (2)
3. Teliospore pedicel to 130 µm long, usually persistent .... *cucullatus* (3)

 4. Species with uredinia and telia 5
 4. Species microcyclic, with telia only 13

5. Teliospore wall uniformly thin 6
5. Teliospore wall thickened apically 7

 6. Urediniospore pores 3, equatorial ..... *pressus* (4)
 6. Urediniospore pores 2, equatorial .... *purus* (5)

7. Teliospore apical wall broadly thickened 8
7. Teliospore apical wall thickened as a differentiated umbo 10

8. Urediniospore wall nearly colorless, pores 3 .... *senecionicola* (6)
8. Urediniospore wall cinnamon brown, pores 2 9

9. Urediniospores mostly 22-25 µm long, pores subequatorial .... *polymniae* (7)
9. Urediniospores mostly 32-40 µm long, pores equatorial .... *compactus* (8)

10. Urediniospores mostly less than 27 µm long 11
10. Urediniospores mostly more than 27 µm long 12

11. Urediniospores mostly 19-21 µm long; teliospores mostly 28-33 µm long .... *columbianus* (9)
11. Urediniospores mostly 24-27 µm long; teliospores mostly 32-42 µm long .... *montanoae* (10)

12. Urediniospores mostly 26-33 µm long; teliospores mostly 32-42 µm long ...... *bidenticola* (11)
12. Urediniospores mostly 30-36 µm long; teliospores mostly 38-50 µm long .... *salmeae* (13)

13. Teliospore apical wall thickened as a differentiated pale umbo .... *bidentis* (12)
13. Teliospore apical wall broadly thickened 14

14. Teliospore apical wall 3.5-5 µm thick *amoenus* (14)
14. Teliospore apical wall thicker 15

15. Teliospores mostly 23-30 x 12-17 µm .... *rudbeckiae* (15)
15. Teliospores mostly 27-33 x 17-23 µm .. *sommerfeltii* (16)

KEY TO SPECIES OF *UROMYCES* ON LEGUMES

Teliospore wall smooth Section I
Teliospore wall sculptured Section II

SECTION I

Uredinia produced; macrocyclic or potentially so Section IA
Uredinia lacking; demicyclic or microcyclic ..... Section IB

Section IA

1. Uredinia with paraphyses 2
1. Uredinia lacking paraphyses 3

2. Paraphyses thin walled, straight..... *mexicanus* (17)
2. Paraphyses thick walled, incurved .... *lespedezae-procumbentis* (18)

3. Urediniospore pores scattered 4
3. Urediniospore pores equatorial or above 5

4. Teliospore wall much thickened at apex *lupini* (19)
4. Teliospore with only a papilla at apex .... *trifolii-repentis* var. *fallens* (39)

5. Urediniospore pores superequatorial .... *vignae* (20)
5. Urediniospore pores equatorial or nearly so 6

6. Aecia systemic, uredinoid 7
6. Aecia localized or unknown 8

7. Aecio-urediniospore pores 2 .... *glycyrrhizae* (21)
7. Aecio-urediniospore pores 4 or 5 .... *hyalinus* (22)

8. Teliospores golden or paler, side wall 1 µm thick 9
8. Teliospores chestnut, side wall 1.5-2 µm or more 11

9. Teliospores globoid, 18-22 µm diam *yurimaguasensis* (23)
9. Teliospores ellipsoid or oblong ellipsoid 10

10. Urediniospore pores 3 .... *dolicholi* (24)
10. Urediniospore pores 2 .... *neurocarpi* (25)

11. Teliospores globoid, pedicel to 150 µm long .... *tenuistipes* (26)
11. Teliospores more or less obovoid, pedicel 100 µm or less 12

12. Teliospores with papilla at apex .... *trifolii-repentis* (39)
12. Teliospores with broadly thickened apical wall 13

13. Urediniospore pores mostly 4, mostly equatorial .... *viciae-fabae* (27)
13. Urediniospore pores mostly 3, equatorial 14

14. Teliospore pedicel brown .... *ervi* (28)
14. Teliospore pedicel colorless ...... *indigoferae* (29)

Section IB

1. Species demicyclic; producing aecia and telia 2
1. Species microcyclic; producing telia only 4

2. Aecia systemic .... *psoraleae* (30)
2. Aecia localized 3

3. Teliospores mostly 22-30 µm long .... *ervi* (28)
3. Teliospores mostly 29-40 µm long .... *montanus* (31)

4. Telia systemic; spores mostly 28-36 µm long .... *abbreviatus* (32)
4. Telia localized; spores mostly 22-27 µm long .... *trifolii* (54)

SECTION II

Uredinia produced; macrocyclic or potentially so Section IIA
Uredinia lacking; demicyclic or microcyclic Section IIB

Section IIA

1. Uredinia with paraphyses 2
1. Uredinia lacking paraphyses 3

2. Paraphyses thin walled, mostly straight .... *hedysari-paniculati* (33)
2. Paraphyses thick walled, strongly incurved .... *antiguanus* (34)

3. Urediniospore pores 6 or more, scattered 4
3. Urediniospore pores 5 or fewer 6

4. Teliospores wall with thick, pale outer layer .... *crotalariae* (35)
4. Teliospore wall simple, chestnut brown 5

5. Urediniospore pores (4-)6 or 7(8), spore wall mostly 2-3 µm thick .... *ciceris-arietinus* (36)
5. Urediniospore poees (6)7-9(-11), spore wall mostly 1.5-2 µm .... *occidentalis* (37)

6. Teliospore side wall smooth, apex with striae or rugosity 7
6. Teliospores with distinct verrucae 8

7. Teliospore side wall 1 µm thick, apex faintly striate .... *yurimaguasensis* (23)
7. Teliospore side wall 3-4.5 µm thick, apex faintly rugose .... *tenuistipes* (26)

8. Urediniospore pores 2(3) 9
8. Urediniospore pores 3 or 4(5) 11

9. Teliospores uniformly verrucose .... *clitoriae* (38)
9. Teliospores with few irregularly arranged verrucae 10

10. Teliospores mostly 22-26 µm long, side wall 1.5-2 µm thick .... *trifolii-repentis* (39)
10. Teliospores mostly more than 28 µm long, side wall 2.5-3.5 µm thick .... *appendiculatus* (40)

11. Teliospore pedicel persistent, to 150 µm long, spores faintly rugose .... *tenuistipes* (26)
11. Teliospore pedicel usually broken short, not more than 50 µm long 12

12. Teliospores verrucose or striately ridged 13
12. Teliospores obviously reticulate 17

13. Teliospores striately ridged .... *striatus* (41)
13. Teliospores discretely verrucose or with some merging verrucae 14

14. Urediniospores mostly 17-20 µm long *calopogonii* (42)
14. Urediniospores more than 20 µm long 15

15. Urediniospores mostly 20-24 µm long; teliospores mostly 18-23 x 15-19 µm .... *cologaniae* (43)
15. Urediniospores mostly 22-28 µm long 16

16. Teliospores mostly 25-29 µm long, with often merging, elongate verrucae .... *illotus* (44)
16. Teliospores mostly 19-21 µm long, with discrete verrucae .... *punctatus* (45)

17. Urediniospores mostly 22-24 µm long, pores 3-5, more or less equatorial .... *bauhiniae* (46)
17. Urediniospores mostly 27-30 µm long, pores 3, rarely 2 or 4, equatorial .... *imperfectus* (47)

Section IIB

1. Species demicyclic; producing aecia and telia 2
1. Species microcyclic; producing telia only 6

  2. Aecia systemic 3
  2. Aecia localized 5

3. Teliospores with a few ridges or verrucae .... *lapponicus* (48)
3. Teliospores with more or less uniform verrucae or interrupted ridges or combinations 4

  4. Teliospores with an inconspicuous papilla over pore .... *elegans* (50)
  4. Teliospores with a broad pale umbo over the pore .... *coloradensis* (51)

5. Teliospore with a lens like papilla over the pore .... *minor* (52)
5. Teliospore with a broad pale umbo over the pore .... *hedysari-obscuri* (53)

  6. Telia systemic; spores linearly ridged .... *phacae-frigidae* (49)
  6. Telia localized 7

7. Teliospores with few scattered verrucae or smooth .... *trifolii* (54)
7. Teliospores reticulate .... *bauhiniicola* (55)

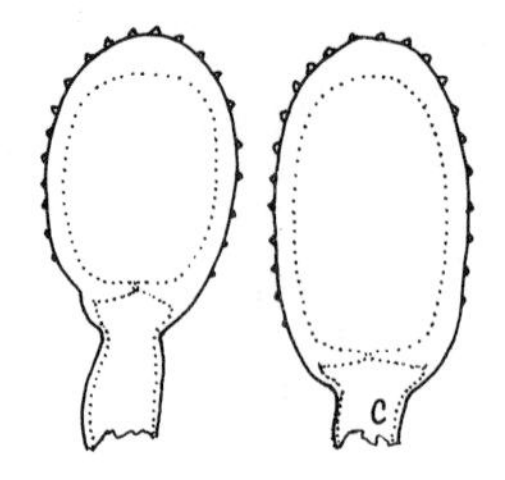

1. *UROMYCES OBLONGISPORUS* Ellis & Ever. Bull. Torrey Bot. Club 25:507. 1898.

Spermogonia, aecia and uredinia unknown. Telia amphigenous, exposed, scattered, blackish brown, more or less pulverulent, with abundant sterile, cylindrical, thin walled or often thick walled, colorless hyphae (paraphyses) among the spores; spores (26-)30-37(-40) x (17-)18-21(-24) μm, mostly ellipsoid or oblong ellipsoid, wall (1.5-)2(-2.5) μm thick at sides, 4-6(-7) μm apically, uniformly chestnut brown or slightly paler apically but without a differentiated umbo, verrucose echinulate with small cones mostly spaced 2-3 μm, conspicuous apically but much less so basally; pedicels colorless, thin walled, appearing to be a continuation of the spore wall, to about 40 μm long but usually broken short.

Type: on *Artemisia tridentata* Nutt., Sweetwater County, Wyoming, Nelson No. 3546 (NY; isotype PUR 42425). Not otherwise known.

The telia are not arranged as in most microcyclic species but other spore forms are not present.

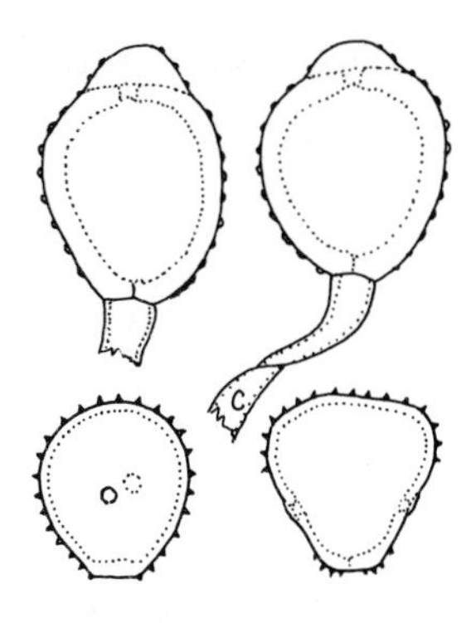

2. *UROMYCES MARTINII* Farl. Proc. Amer. Acad. Arts Sci. 18: 79. 1883.

Spermogonia on adaxial leaf surface. Aecia on abaxial surface, few in a group, peridium yellowish, erose; spores (17-)19-23(-25) x (15-)16-19(-21) µm, globoid, broadly ellipsoid or ellipsoid, wall 1-1.5 µm thick, pale yellowish, moderately verrucose with rod like verrucae that may be united in short series. Uredinia on abaxial surface, pale cinnamon brown; spores (18-)20-24 x (17-)20-22 µm, broadly ellipsoid or obovoid with pores face view, triangularly obovoid with pores lateral, wall 1-1.5 µm thick, pale cinnamon or golden brown, echinulate except around pores, pores 2, equatorial or slightly below, with small caps. Telia on abaxial surface, exposed, chocolate brown, pulverulent; spores (27-)29-33(-35) x (20-)22-26 µm, broadly ovoid or broadly ellipsoid, wall 2-3 µm thick at sides, chestnut brown, (6-) 7-9(-10) µm at apex by a nearly colorless, defined umbo, verrucose with low, rounded verrucae spaced mostly about 2-3 µm; pedicels colorless, to 50 µm long but usually broken short.

Hosts and distribution: *Melanthera hastata* Michx., *M. parvifolia* Small: southern Florida.

Type: on *Melanthera hastata*, Florida, 1880, Martin (FH; probable isotypes Ellis N. Amer. Fungi 1067).

3. *UROMYCES CUCULLATUS* H. Syd. & P. Syd. Ann. Mycol. 2: 349. 1904.
Uromyces pianhyensis P. Henn. Hedwigia 47:266. 1908.

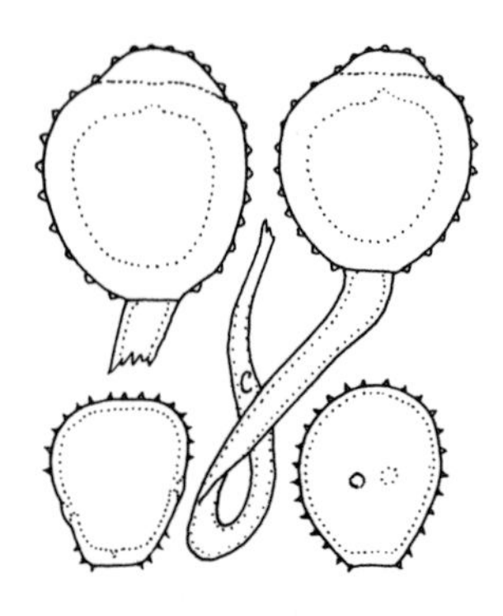

Spermogonia on adaxial leaf surface. Aecia on abaxial surface, few in a group, peridium cylindrical becoming lacerated, whitish; spores 22-26 x 19-24 µm, broadly ellipsoid or globoid, commonly angularly so, wall 1 µm thick, pale yellowish, prominently verrucose. Uredinia amphigenous or mostly on adaxial surface, about cinnamon brown; spores (16-)18-21 x (17-) 18-22(-23) µm, broadly ellipsoid or obovoid with pores in face view, mostly depressed globoid or more or less triangular with pores lateral, wall 1-1.5 µm thick but usually thicker at hilum, about cinnamon brown, echinulate except around the pores, pores 2, equatorial or usually somewhat subequatorial, with slight or no caps. Telia amphigenous or often mostly on adaxial surface, exposed, blackish brown, somewhat pulverulent; spores (24-)28 -33(-35) x (22-)24-28(-30) µm, mostly globoid or nearly so, wall (3.5-)4-4.5(-5) µm thick at sides and deep chestnut brown, (6-)7-10(-12) µm apically with a golden, defined umbo, the side wall sometimes tending to be bilaminate, conspicuously verrucose with low conical verrucae spaced (2-)2.5-3(-5) µm; pedicels colorless, to 120 µm long, often flexuous or shorter and straight.

Hosts and distribution: *Baltimora recta* L., *Wedelia acapulcensis* H.B.K. and species of *Perymenium*, and *Zexmenia*: central Mexico to Panama.

Type: on *Zexmenia aurantiaca* Klatt: Costa Rica, Tonduz No. 9836 (S; isotype PUR 17327).

Uromyces cucullatus differs from *U. blainvilleae* Berk. in having much longer and persistent pedicels on the teliospores and an umbo that is about 1/3 narrower. Aecia are not known in *U. blainvilleae* nor has the species been recorded in Continental North America. Possibly, *U. cucullatus* could be considered a variety of *U. blainvilleae*.

4. *UROMYCES PRESSUS* Arth. & Holw. in Arthur, Mycologia 10: 125. 1918.
Maravalia pressa (Arth. & Holw.) Mains, Bull. Torr. Bot. Club 66:177. 1939.

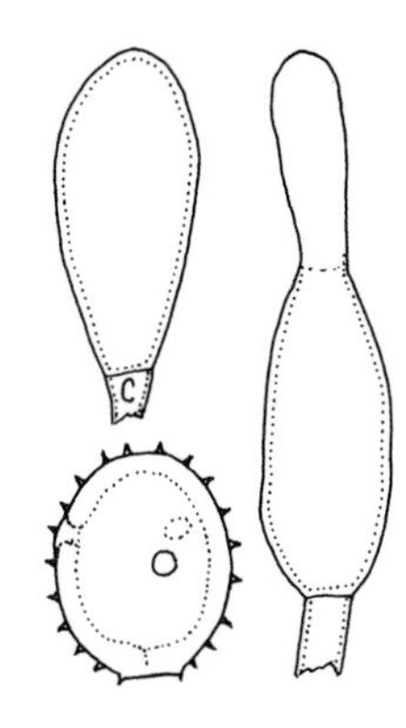

Spermogonia on adaxial surface in small groups. Aecia uredinoid, associated with the spermogonia, otherwise and the spores as the uredinia. Uredinia mostly on abaxial surface, pale yellowish brown; spores 26-29(-33) x (22-)23-26 µm, broadly obovoid or broadly ellipsoid, wall (2.5-)3-3.5(-4) µm thick, uniformly echinulate, yellowish or pale golden, pores 3, equatorial, with slight or no caps. Telia mostly on abaxial surface, exposed, whitish, compact; spores (26-) 30-37(-40) x (15-)16-19(-21) µm, ovoid, ellipsoid or oblong ellipsoid, wall uniformly 0.5 µm thick, colorless, smooth, germination occurs by continued growth of the apex to form a basidium; pedicels colorless, to about 30 µm long but usually broken short.

Hosts and distribution: *Vernonia stellaris* Llave ex Lex.: Costa Rica and Guatemala.

Type: on *Vernonia deppeana* (= *V. stellaris*), San José, Costa Rica, Holway No. 361 (PUR 36672).

The type species of the genus *Maravalia*, *M. pallida* Arth. & Thaxt., is known only in the telial stage. Its host plant is *Pithecellobium latifolium* (L.) Benth., a member of the subfamily Mimosoideae of the Leguminosae. It is probable that the spermogonia of *M. pallida* will prove to be type 7 (16) as in *Chaconia* and most species of *Ravenelia*. The spermogonia are type 4 in *Uromyces pressus* and *U. purus*, hence these species are probably correctly excluded from *Maravalia*.

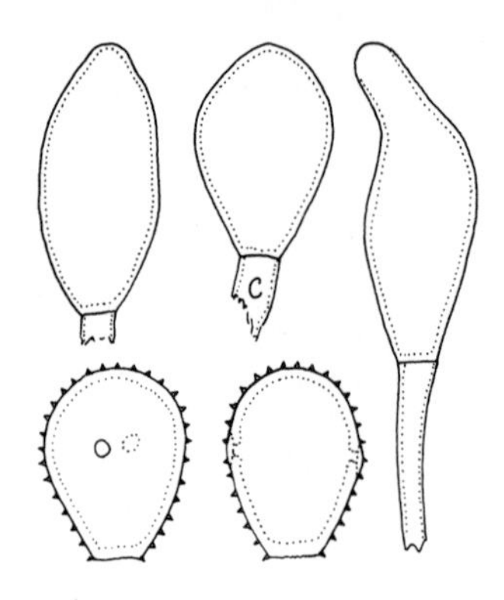

5. *UROMYCES PURUS* (H. Syd.) Cumm. Mycotaxon 5:407. 1977.
Argomycetella pura H. Syd. Ann. Mycol. 23:313. 1925.
Maravalia pura (H. Syd.) Mains, Bull. Torrey Bot. Club 66:178. 1939.

Spermogonia mostly on adaxial leaf surface. Aecia uredinoid, grouped about the spermogonia, otherwise like the uredinia. Uredinia mostly on abaxial surface, scattered, pale cinnamon brown; spores (25-)27-32(-37) x 18-23 µm, obovoid or broadly ellipsoid, wall 1.5-2 µm thick, cinnamon brown or paler, uniformly echinulate, pores 2, equatorial, with slight caps. Telia on abaxial surface, exposed, whitish, compact; spores (25-)28-34(-37) x (15-)16-18(-20) µm, ellipsoid, ovoid or more or less oblong, wall uniformly 0.5 µm or less thick, colorless, smooth, germinating immediately by continued elongation of the apex, without a differentiated germ pore; pedicels colorless, to 30 µm long.

Hosts and distribution: *Vernonia patens* H.B.K., *V.* sp.: Costa Rica and San Luis Potosí (on *V.* sp.).

Type: on *Vernonia patens*, San José, Costa Rica, 6 January 1925, Sydow No. 3 (holotype lost; isotype PUR F2269).

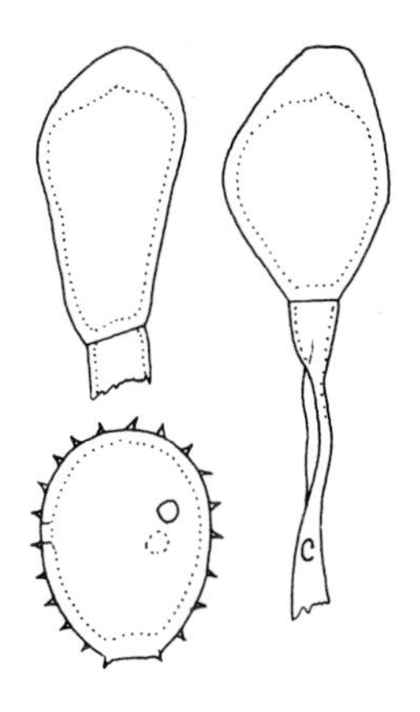

6. *UROMYCES SENECIONICOLA* Arth. Bot. Gaz. 40:198. 1905.

Spermogonia on adaxial leaf surface. Aecia in groups opposite the spermogonia, peridium whitish, fragile; spores too old for description. Uredinia mostly on abaxial leaf surface, yellowish when old, doubtless bright yellow; spores (25-)28-35(-39) x (18-)20-23(-25) µm, ellipsoid or obovoid, wall 1.5-2 µm, yellowish to colorless, uniformly echinulate, pores 3(4?), approximately equatorial, obscure. Telia on the abaxial surface, grayish black, covered by host epidermis, without paraphyses; spores (25-)28-40(-44) x (16-)18-23(-25) µm, variable but mostly ellipsoid or obovoid, wall 1.5-2 µm thick at sides, (3-)4-7(-8) µm apically, uniformly clear chestnut or deep golden brown, smooth; pedicels golden, to 50 µm long but commonly shorter.

Hosts and distribution: *Senecio* spp.: from Durango, Mexico to Guatemala.

Type: on *Senecio roldana* DC., Amecameca, Mex., Mexico Holway No. 5183 (PUR 17234).

The colorless urediniospores were mentioned by Arthur (*loc. cit.*), but in the N. Amer. Flora (1) they were described as cinnamon brown, probably by mistake.

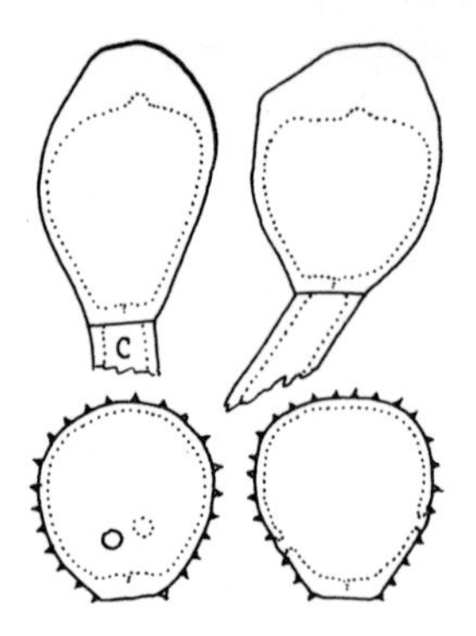

7. *UROMYCES POLYMNIAE* Diet. & Holw. in Holway, Bot. Gaz. 31:327. 1901.

Spermogonia in small groups in adaxial leaf surface. Aecia in circles opposite the spermogonia, peridium delicate, erose; spores (18-)20-28(-30) x (15-)17-21(-23) µm, globoid or broadly ellipsoid, wall 1.5-3 µm thick, colorless, verrucose. Uredinia mostly on abaxial leaf surface, cinnamon brown; spores (20-)22-25(-27) x (21-)23-27 µm, broadly obovoid, or globoid, often slightly wider than high, wall 1-1.5 µm thick but the base slightly thicker, uniformly echinulate, cinnamon brown, pores 2, subequatorial or near hilum, without discernible caps. Telia mostly on abaxial surface, blackish brown, exposed, compact; spores (26-)30-36(-38) x (20-)22-26(-28) µm, broadly ellipsoid or broadly obovoid, wall 1.5-2(-2.5) µm thick at sides, (5-)7-10(-12) µm apically, uniformly chestnut brown or somewhat paler apically; smooth; pedicels slightly yellowish, to about 60 µm long.

Hosts and distribution: *Polymnia maculata* Cav.: central Mexico and Guatemala; also in Brazil.

Lectotype: Rio Hondo, near Mexico City, Holway (S; isotypes Barth. N. Amer. Ured. No. 395). Lectotype selected here.

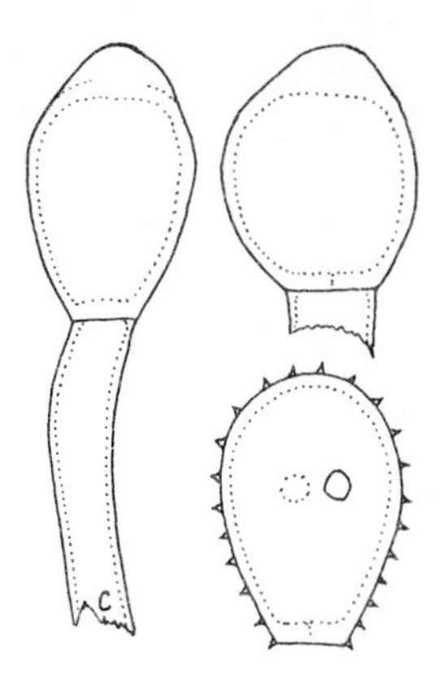

8. *UROMYCES COMPACTUS* Peck, Bot. Gaz. 7:56. 1882.

Spermogonia and aecia on stems in elongate groups; peridium colorless or pale yellowish, becoming lacerated, spores (26-)28-35(-40) x (20-)22-27(-30) µm, mostly ellipsoid or broadly ellipsoid, wall 1-1.5 µm thick, colorless, finely and inconspicuously verrucose. Uredinia on stems, about cinnamon brown; spores (30-)32-40(-46) x (20-)23-28 (-31) µm, ellipsoid or obovoid, wall 1.5-2 µm thick, cinnamon brown, echinulate, pores 2, equatorial in smooth, often flattened sides, with low caps. Telia on stems, exposed, compact, blackish brown; spores (28-)33-42(-48) x (19-)22-28(-30) µm, obovoid, ellipsoid or oblong ellipsoid, wall 1.5-2(-2.5) µm thick at sides, (4-)5-7(-8) µm apically, about chestnut brown, the apical thickening only slightly paler, smooth; pedicels persistent, colorless or pale golden, to 95 µm long.

Hosts and distribution: *Aster spinosus* Benth.: the southwestern United States and northern Mexico.

Type: on dead stems of some "Composite Plant". Pringle (NYS).

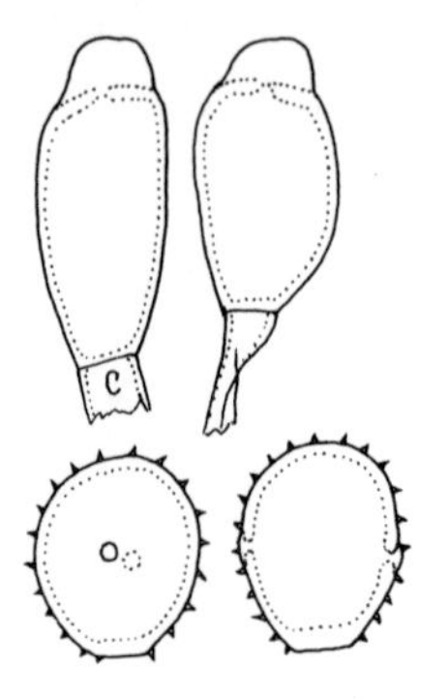

9. *UROMYCES COLUMBIANUS* Mayor, Mem. Soc. Neuch. Sci. Nat. 5:467. 1913.

Spermogonia mostly on adaxial surface. Aecia on abaxial surface in groups along nerves, blister like, the peridium poorly formed, not exserted; spores (15-)18-23(-24) x (14-)16-18(-20) µm, variable but mostly globoid or broadly ellipsoid, wall 1.5(-2) µm thick, including rod like verrucae, colorless. Uredinia on abaxial surface, cinnamon brown; spores (18-)19-21(-23) x (16-)17-19(-20) µm, obovoid or broadly ellipsoid with pores face view, wall 1-1.5 µm thick, cinnamon brown, uniformly echinulate, pores 2, equatorial, with small caps. Telia on abaxial surface, exposed, cinnamon brown becoming gray from germination, compact; spores (25-)28-33(-38) x (12-)15-18(-20) µm, mostly obovoid or oblong ellipsoid, wall 0.5-1 µm thick basally, to 1.5 µm toward apex, golden brown, (5-)6-8(-9) µm at apex by a nearly colorless, cork like umbo, smooth; pedicels colorless, to 55 µm long but usually broken much shorter.

Hosts and distribution: *Melanthera* spp.: Florida and eastern Mexico to Costa Rica; also in South America and the Caribbean.

Type: on *Melanthera aspera* (Jacq.) Steud. var. *canescens* (Kuntze) Thell., near Supia, Dept. Cauca, Colombia, Mayor No. 148 (NEU).

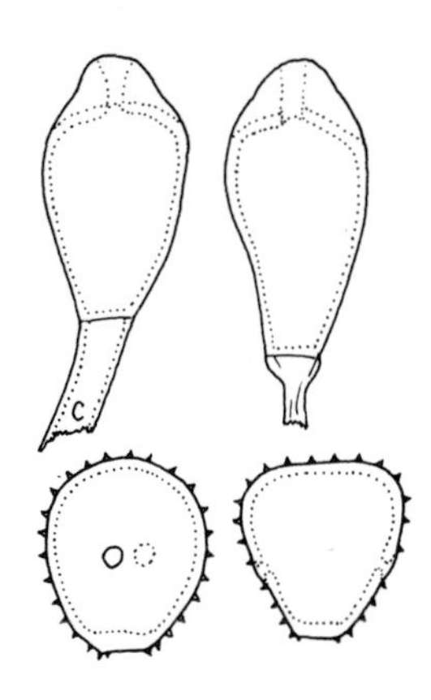

10. *UROMYCES MONTANOAE* Arth. & Holw. in Arthur, Mycologia 10:127. 1918.

Spermogonia on adaxial leaf surface in tight groups on small, slightly hypertrophied areas. Aecia uredinoid, grouped opposite the spermogonia, otherwise as the uredinia. Uredinia on abaxial surface, cinnamon brown; spores (22-)24-27(-29) x (20-)22-26 µm, obovoid or broadly ellipsoid with pores face view, strongly obovoid triangular with pores lateral; wall 1-1.5 µm thick, cinnamon brown, echinulate except directly over pores, pores 2, equatorial or slightly below, with slight or no caps. Telia on abaxial surface, exposed, compact, cinnamon brown becoming gray from germination; spores (28-)32-42(-46) x (17-)18-22(-24) µm, mostly obovoid, wall 1 µm thick and golden or pale cinnamon brown at sides, (4-)6-8(-9) µm thick apically by a nearly colorless, defined umbo, smooth; pedicels nearly colorless, to about 50 µm long.

Hosts and distribution: *Montanoa dumosa* Klatt, *M. hibiscifolia* Benth., *M. pittieri* B. L. Rob. & Greenm.: Costa Rica and Guatemala.

Type: on *M. pittieri*, San Lucas Tolimán, Dept. Solola, Guatemala, Holway No. 176 (PUR 36684).

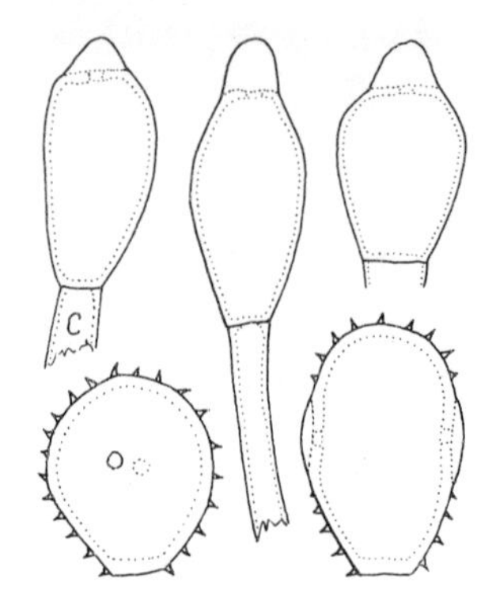

11. *UROMYCES BIDENTICOLA* Arth. Manual Rusts U.S. & Canada p. 342. 1934.
Uromyces bidenticola Arth. Mycologia 9:71. 1917, *nomen nudum*.
Klebahnia bidentis Arth. N. Amer. Flora 7:481. 1922, not *Uromyces bidentis* Lagerh. 1895.

Spermogonia amphigenous, few in a group. Aecia amphigenous, around the spermogonia, uredinoid; spores like the urediniospores. Uredinia amphigenous or mostly on abaxial leaf surface, cinnamon brown or darker; spores (22-)26-33 (-37) x (19-)21-24(-26) µm, broadly ellipsoid or obovoid, wall 1.5-2.5 µm thick, cinnamon brown, echinulate except around pores, pores 2, equatorial or slightly above, with obvious caps. Telia mostly on abaxial surface, exposed, compact, cinnamon brown but becoming gray with basidia; spores (30-)32-40(-45) x (15-)17-20(-23) µm, ellipsoid, oblong ellipsoid or elongately obovoid, wall 1 µm thick at sides, golden brown, (3-)4-8(-10) µm thick at apex with a colorless umbo; pedicels colorless, to 55 µm long.

Hosts and distribution: on *Bidens* spp.: the southern United States southward to Central America; also in South America, Africa and Asia.

Neotype: on *Bidens squarrosa* H.B.K., Guatemala City, Guatemala, 31 Dec. 1914, Holway No. 4 (PUR 36759).

A neotype is required because no specimens are cited in the Manual and no available United States specimen had telia. The neotype has all spore stages and has measurements of urediniospores and teliospores made in 1915, hence doubtless they were used in writing the diagnosis.

12. *UROMYCES BIDENTIS* Lagerh. in Patouillard & Lagerheim, Bull. Soc. Mycol. France 11:213. 1895.

Spermogonia, aecia and uredinia lacking. Telia on the abaxial surface of leaves in close, circinate groups, exposed, compact, cinnamon brown but soon gray from germination; spores (30-)32-40(-45) x (15-)17-20(-23) μm, mostly oblong ellipsoid or elongately obovoid, wall 1 μm thick at sides, golden brown, 4-9 μm thick at apex with a nearly colorless defined umbo, smooth; pedicel to about 65 μm long, colorless.

Hosts and distribution: *Bidens* spp.: Florida, San Luis Potosí, Mexico, Guatemala and Costa Rica; also in Puerto Rico and South America.

Type: on *Bidens andicola* H.B.K., Chillo pres Quito, Ecuador, Lagerheim (S).

This species is a microcyclic derivative of *U. bidenticola* and differs from it only in life cycle.

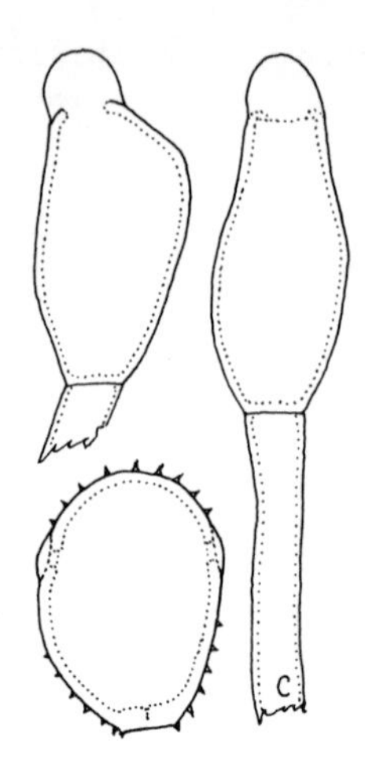

13. *Uromyces salmeae* Arth. & Holw. in Arthur, Amer. J. Bot. 5:445. 1918.

Spermogonia amphigenous. Aecia amphigenous, few in a group, the peridium breaking easily; spores (29-)32-45(-50) x (20-)23-26(-28) µm, more or less ellipsoid or oblong, wall (1.5)2-2.5 µm thick at sides including verrucae, usually thickened apically to 3-6 µm, yellowish or pale brownish, coarsely verrucose with small rods, short ridges or irregular patterns. Uredinia mostly on abaxial leaf surface, cinnamon brown; spores (28-)30-36(-40) x (22-)24-27(-29) µm, broadly ellipsoid or obovoid, wall 1.5(-2) µm thick, pale cinnamon brown, echinulate except around and below the pores, pores 2, above the equator, with small caps. Telia on abaxial surface, exposed, dark cinnamon brown, compact; spores (35-)38-50(-55) x 17-22 µm, mostly ellipsoid or oblong, wall 0.5-1 µm thick at sides, golden or yellowish, thickened at apex 5-8(-10) µm by a nearly colorless, cork like, umbo, smooth; pedicels colorless, to 65 µm long, usually broken shorter.

Hosts and distribution: *Salmea scandens* (L.) DC.: Guatemala; also reported in Puerto Rico.

Type: San Lucas Tolimán, Solola, Guatemala, Holway No. 188 (PUR 17239).

Germination of the teliospores probably occurs without extended dormancy.

14. *UROMYCES AMOENUS* H. Syd. & P. Syd. Ann. Mycol. 4:28. 1906.

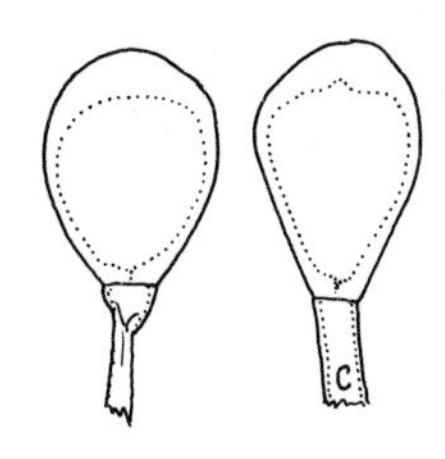

Telia on abaxial surface, exposed, closely grouped and often confluent, blackish brown; spores (20-)22-28(-31) x (14-)16-21(-23) µm, mostly globoid or broadly obovoid, wall 1.5-2(-2.5) µm thick at sides, (2.5-)3.5-5(-7) µm apically, uniformly golden or clear chestnut brown, smooth; pedicels yellowish, to 60 µm long but usually broken shorter.

Hosts and distribution: *Anaphalis margaretacea* (L.) Benth. & Hook. *sensu lat.*: Wyoming to northern California and British Columbia.

Lectotype: Mts. of Skamania County, Wash., Suksdorf (S); isotypes Ell. & Ever. F. Columb. No. 1795 as *Uromyces gnaphalii*). Lectotype designated here.

15. *UROMYCES RUDBECKIAE* Arth. & Holw. in Arthur, Bull. Iowa Agr. Coll. Bot. 1884:154. 1885.

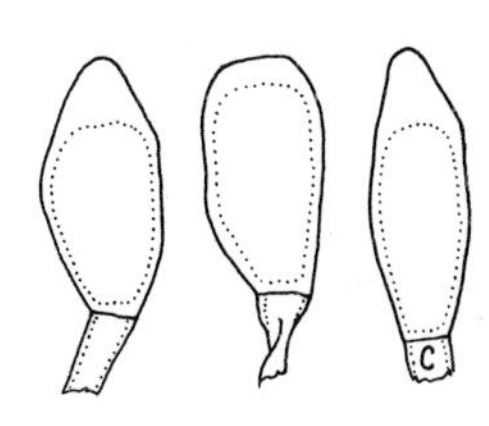

Telia mostly on abaxial leaf surface, exposed, few to many in tight groups, cinnamon brown becoming gray from germination, compact; spores (19-)23-30 (-33;40) x (10-)12-17 µm, mostly oval, ellipsoid or obovoid, wall 1 µm thick at sides, (3-)5-7(-8) µm at apex, pale golden or yellowish, smooth; pedicels colorless, to 55 µm long.

Hosts and distribution: *Rudbeckia laciniata* L.: Tennessee to the District of Columbia, Ontario, Manitoba, Montana and south to Texas and New Mexico; also in Japan.

Type: Decorah, Iowa, 1883, Holway (PUR 38748).

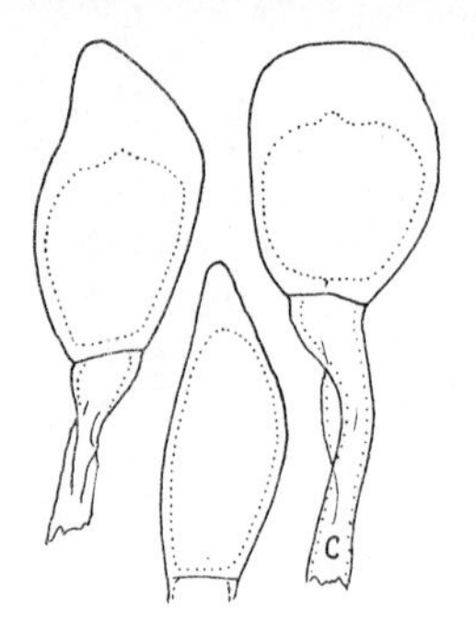

16. *UROMYCES SOMMERFELTII* Hyl., Joerst. & Nannf. Opera Bot. 1:96. 1953.
    Caeoma solidaginis Somm. Suppl. Fl. Lappl. p. 234. 1826.
    Uromyces solidaginis (Somm.) Niessl Verh. Naturf. Ver. Brünn 10:163. 1872, not Fckl. 1860.

Spermogonia, aecia and uredinia lacking. Telia on abaxial leaf surface and sometimes on stems, exposed, closely grouped, blackish brown, compact; spores (25-)27-33(-37) x (14-)17-23(-25) µm, mostly ovoid or obovoid or the longer spores elongately ellipsoid, wall (1-)1.5-2(-2.5) µm thick at sides, (6-)8-10(-12) µm at apex, clear chestnut brown but the longer spores mostly pale yellow, smooth, without a differentiated umbo; pedicels colorless, to 55 µm long.

Hosts and distribution: *Solidago* spp.: Wyoming and Colorado to Oregon and British Columbia; also in Europe.

Type: on *Solidago virga-aurea* L., Saltdal in Nordland, Norway, Sommerfelt (not seen).

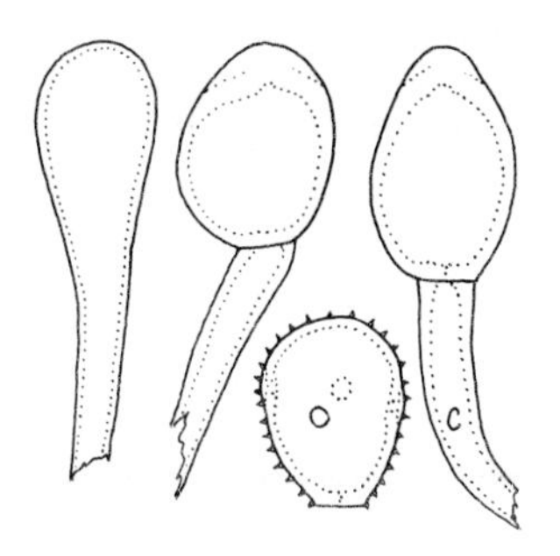

17. *UROMYCES MEXICANUS* Diet. & Holw. in Holway, Bot. Gaz. 24:24. 1897.

Spermogonia and aecia unknown. Uredinia on abaxial leaf surface, pale cinnamon brown, with peripheral, colorless, uniformly thin walled, cylindrical to clavate or capitate paraphyses 8-20 µm wide in upper part; spores (18-)21-24(-26) x (16-)17-20(-21) µm, mostly broadly ellipsoid, wall 1(-1.5) µm thick, uniformly echinulate, pores 3-5, equatorial, with slight caps. Telia mostly in elongate and often confluent groups on stems, exposed, blackish, compact; spores (21-)23-28(-29) x (15-)17-20(-22) µm, mostly ellipsoid or ovoid, wall (1.5-)2-3(-3.5) µm thick at sides, (3.5-)4.5-6(-7) µm apically, nearly uniformly chestnut brown, smooth; pedicels nearly colorless, to 100 µm long.

Hosts and distribution: *Desmodium* spp., especially *D. procumbens* (Mill.) A. S. Hitchc.: southern Arizona and New Mexico to South Central Mexico.

Lectotype: on *Desmodium* sp., City of Mexico, 9 Oct. 1896, Holway (S; isotype PUR 14796).

The paraphyses have been overlooked previously.

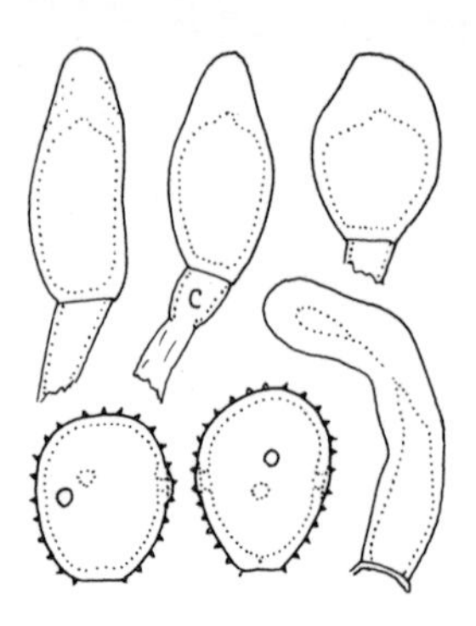

18. *UROMYCES LESPEDEZAE-PROCUMBENTIS* (Schw.) Curt. Cat. Pl. N. Carolina p. 123. 1867.
Puccinia lespedezae-procumbentis Schw. Schr. Nat. Ges. Leipzig 1:73. 1822.

Spermogonia few on adaxial leaf surface. Aecia on abaxial surface in small groups, peridium short, yellowish, recurved; spores (14-)16-19 x (13-)15-18 µm, mostly globoid, wall 1 µm or less thick, colorless, minutely verrucose. Uredinia mostly on abaxial surface, small, pale cinnamon brown, with peripheral, incurved, yellowish, thick walled paraphyses to about 55 µm long; spores (18-)19-22(-23) x (14-)16-18 µm, broadly ellipsoid or obovoid, wall 1-1.5 µm thick, yellowish or golden, uniformly echinulate, pores 3 or 4, equatorial, with slight caps. Telia amphigenous or mostly on abaxial surface, and on stems, exposed, blackish brown, compact; spores (19-)24-32(-37) x (12-)14-17 µm, ellipsoid or oblong ellipsoid, wall 1.5 µm thick at sides, pale chestnut or golden brown at sides, (5-)7-10(-12) µm thick and becoming paler apically, smooth; pedicels colorless, to about 85 µm long.

Hosts and distribution: *Lespedeza* spp.: Ontario and the eastern half of the United States.

Neotype: on *Lespedeza hirta* (L.) Ell., Murphy, North Carolina, 1920, Bartholomew (PUR 14361; isotypes Barth. N. Amer. Ured. No. 2392).

Arthur and Bisby (3) report that the holotype is "Represented only by an empty packet, ...".

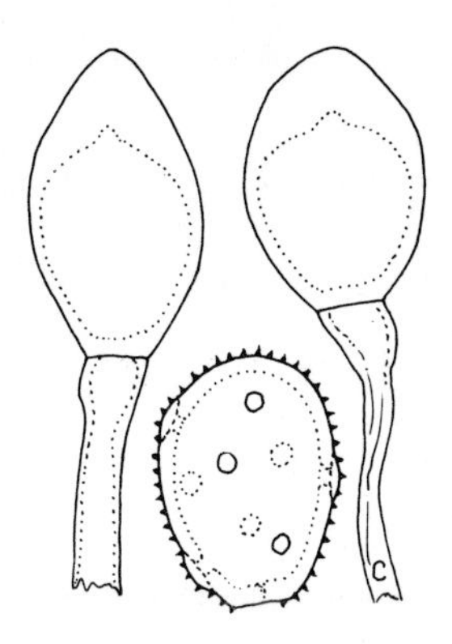

19. *UROMYCES LUPINI* Berk. & Curt. Proc. Amer. Acad. Sci. 4: 126. 1858.

Spermogonia mostly in small groups on adaxial leaf surface. Aecia on abaxial surface, peridium cupulate, short, whitish, recurved; spores 21-29 x 18-24 µm, broadly ellipsoid or globoid, verrucose. Uredinia amphigenous or mostly on abaxial surface, yellowish brown; spores (25-)28-33(-37) x (19-)22-26(-28) µm, variable but mostly ellipsoid or broadly so, wall (1.5-)2-2.5(-3) µm thick, yellowish or golden, uniformly echinulate, pores scattered, (6-)7-10 (-11), with obvious caps. Telia amphigenous or mostly on abaxial surface, exposed, blackish brown, compact; spores (26-)30-36(-40;-46) x (19-)21-28(-31) µm, mostly broadly ellipsoid, broadly obovoid or ovoid, wall (1.5-)2-3 µm thick at sides, (5-)7-11(-13) µm apically, uniformly chestnut brown or becoming paler apically, smooth; pedicels colorless or essentially so, to 80 µm long.

Hosts and distribution: *Lupinus* spp.: Nebraska to British Columbia southward to Baja California and south central Mexico.

Type: on *Lupinus* sp., California, Wright (K).

20. *UROMYCES VIGNAE* Barcl. J. Asia. Soc. Bengal 60:211. 1891.
Uromyces phaseoli (Pers.) Wint. *vignae* (Barcl.) Arth. Man. Rusts U.S. and Canada p. 297. 1934.

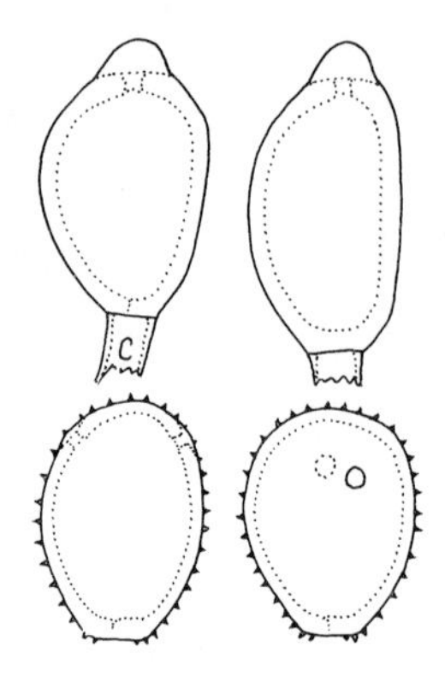

Spermogonia on the adaxial surface of leaflets in small groups. Aecia mostly on the abaxial surface and on petioles in groups, peridium whitish, lacerate; spores 20-26 x 16-20 μm, ellipsoid or oblong ellipsoid, wall 1-1.5 μm thick, minutely verrucose. Uredinia amphigenous and on petioles, cinnamon brown; spores (23-)25-30(-32) x 20-23 μm, broadly ellipsoid or obovoid, wall 1.5 -2 μm thick, pale golden, uniformly echinulate, pores 2 near the apex, with slight caps. Telia distributed as the uredinia, exposed, blackish brown, pulverulent; spores (28-)30-38(-42) x (17-)19-24 μm, obovoid or ellipsoid, wall 2-2.5(-3) μm thick at sides, (4-)4.5-7(-8) μm at apex by a pale, defined umbo, smooth; pedicel colorless, to about 40 μm long; the spores germinate without dormancy.

Hosts and distribution: *Vigna* spp., especially *V. sinensis* (L.) Endl.: widely distributed where cowpeas are grown.

Type: on *Vigna vexillata* (L.) A. Rich., near Simla, India, Barclay (K). Dr. Derek Reid has informed me that a specimen in K is stamped "Herb. Arthur Barclay. Recd. 1892" and suggests that the specimen is properly to be considered as the holotype.

The description of the aecia is adapted from Fromme (12).

The species differs from *U. appendiculatus* because of the strongly superequatorial pores of the urediniospores and the smooth, germinating teliospores.

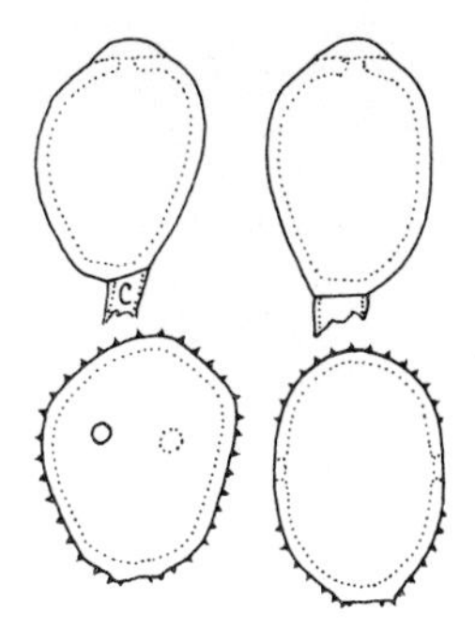

21. *UROMYCES GLYCYRRHIZAE* Magn. Ber. Dtsch. Bot. Ges: 8: 380, 383. 1890.

Spermogonia mostly on abaxial leaf surface, systemic. Aecia among the spermogonia, uredinoid, cinnamon brown; spores (24-)26-30(-32) x (21-)24-27(-30) µm with pores face view, broadly obovoid or nearly globoid, wall 1.5 µm thick, cinnamon brown, echinulate, pores 2, equatorial in smooth areas of flattened sides, without caps. Uredinia uncertain, when systemic not distinguishable from aecia, when localized (rare or rarely collected) associated with localized telia; spores as the aeciospores. Telia mostly on abaxial leaf surface, from a systemic mycelium or localized (when more likely to be on both surfaces), exposed, chocolate brown, pulverulent; spores (23-)25-30(-32) x (15-)17-21(-23) µm, ellipsoid or obovoid, wall 1.5-2(-2.5) µm thick, clear chestnut or dark golden brown except the apex with a pale umbo, smooth; pedicel colorless, always broken near hilum.

Hosts and distribution: *Glycyrrhiza lepidota* Pursh: Texas to Manitoba, Saskatchewan and California; also in Europe and northern Africa to China.

Type: Colorado Springs, Colorado, 14 Aug. 1889, Holway (HBG; probable isotypes Sydow Ured. No. 503).

Magnus' text leaves no doubt that the teliospores were described from the Holway specimen.

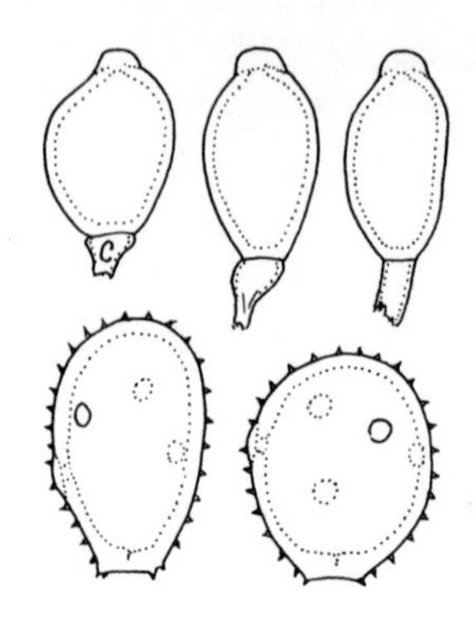

22. *UROMYCES HYALINUS* Peck, Bot. Gaz. 3:34. 1878.

Spermogonia amphigenous, systemic. Aecia mostly on the abaxial leaf surface, pale cinnamon brown, distributed from a systemic mycelium, uredinoid; spores (23-)25-30(-33) x (18-)20-24(-25) μm, obovoid, ellipsoid or broadly so, wall (1.5-)2(-2.5) μm thick, golden brown, uniformly echinulate, pores mostly 4 or 5, rarely 3 or 6, scattered or strongly tending to be in the equatorial zone, with caps. Uredinia lacking or, if not, then indistinguishable from the aecia. Telia mostly on abaxial surface, systemic, exposed, dark cinnamon or nearly chestnut brown, pulverulent, teliospores (20-)22-20(-35) x (13-)16-19(-22) μm ellipsoid or ovoid, wall (1-)1.5(-2) μm thick at sides, dark golden brown, 3-4 (-5) μm thick at apex by a pale, narrow umbo, smooth; pedicels colorless, always broken near the spore.

Hosts and distribution: *Sophora nuttalliana* B. L. Turn., *S. stenophylla* Gray: South Dakota and Wyoming to Arizona and Chihuahua, Mexico.

Type: on *Sophora sericea* (= *S. nuttalliana*), Cañon City, Colo., Brandegee (NYS; isotypes PUR 36620).

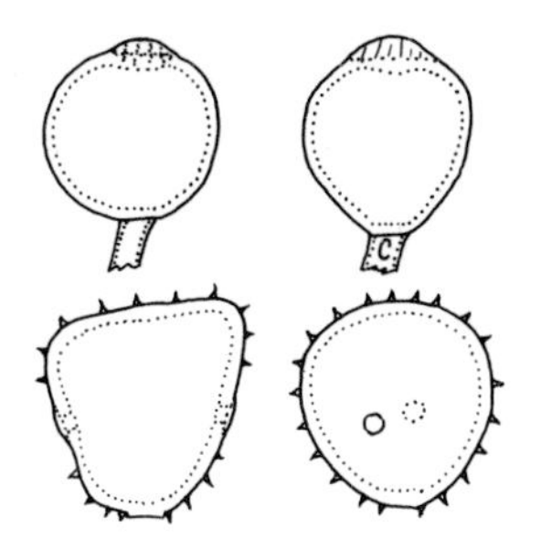

23. *UROMYCES YURIMAGUASENSIS* P. Henn. Hedwigia 43:157. 1904.

Spermogonia on adaxial leaf surface, few in a group. Aecia on abaxial surface opposite the spermogonia, uredinoid, cinnamon brown, spores (23-)24-27(-29) µm high, (22-) 23-27 µm wide, triangularly obovoid with pores lateral, broadly ellipsoid or broadly obovoid (and less often seen) with pores face view, wall (1-)1.5 µm thick, cinnamon brown, echinulate except around pores, pores 2 (rarely 3) in flattened sides, slightly subequatorial, with small caps. Uredinia like the uredinoid aecia except scattered; spores like the aeciospores. Telia on abaxial surface, exposed, about cinnamon brown, moderately compact; spores (17-)18-22(-23) µm diam, globoid, wall 1 µm thick at sides, golden brown, smooth, 3 µm thick over the pores with a nearly colorless, indistinctly striate umbo; pedicels colorless, to 65 µm long but commonly shorter.

Hosts and distribution: *Clitoria arborescens* Ait.: Panama; also in South America.

Type: on *Clitoria* sp., Rio Huallaga, Yurimaguas, Peru, Ule No. 3224 (B).

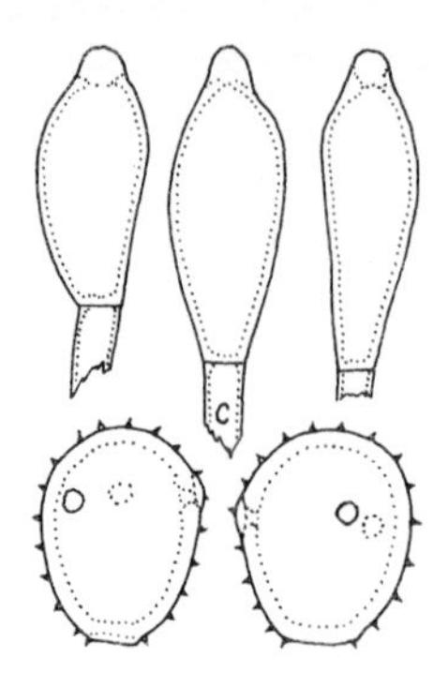

24. *UROMYCES DOLICHOLI* Arth. Bull. Torrey Bot. Club 33:27. 1906.

Spermogonia and aecia unknown. Uredinia most abundant on abaxial leaf surface, cinnamon brown or darker; spores (19-)21-26(-28) x (17-)18-22(-24) µm, broadly ellipsoid or obovoid, wall (1.5-)2(-2.5) µm thick, about cinnamon brown, finely and uniformly echinulate, pores (2)3(4), equatorial or slightly above, with obvious caps. Telia mostly on abaxial surface, exposed, yellowish brown, relatively pulverulent; spores (22-)27-35(-38) x (9-)11-14(-18) µm, narrowly ellipsoid, oblong ellipsoid or elongately obovoid, wall 0.5 -1 µm thick at sides, pale golden, 3-6 µm at apex by a nearly colorless umbo, smooth; pedicels colorless, to about 30 µm long but usually broken near hilum.

Hosts and distribution: *Cajanus indicus* Spreng., *Rhynchosia texana* (Nutt.) Torr. & Gray: Florida, Texas, eastern Mexico, Honduras and Nicaragua; also in the West Indies and South America.

Type: on *Dolicholus texanus* (= *R. texana*), San Angelo, Texas, Shear (PUR 15963; isotype Barth. F. Columb. 4091).

The teliospores are similar to those of *U. neurocarpi* but the urediniospores differ in size and shape.

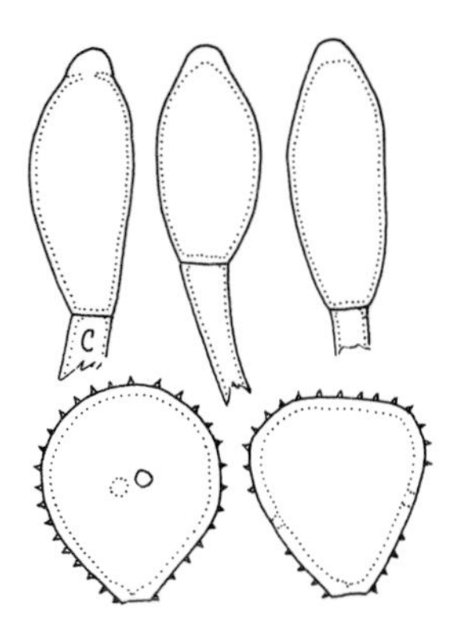

25. *UROMYCES NEUROCARPI* Diet. Hedwigia 34:292. 1895.

Spermogonia on adaxial leaf surface, few. Aecia on abaxial surface opposite the spermogonia, often associated with veins, uredinoid; spores similar to the urediniospores. Uredinia mostly on abaxial surface, dark cinnamon brown; spores (18-)22-26 µm high, (17-)21-25 µm wide, triangularly obovoid with pores lateral, more or less globoid with pores face view, wall (1-)1.5 µm thick, cinnamon brown, echinulate except around pores, pores 2, in flattened sides, without caps, slightly subequatorial. Telia mostly on abaxial surface, exposed, compact, yellowish brown becoming gray from germination; spores (24-)28-36(-40) x (9-)11-16(-18) µm, mostly narrowly ellipsoid or narrowly oblong ellipsoid, wall 0.5-1 µm thick at sides, 2-4 µm at apex, uniformly pale yellowish, smooth; pedicels colorless, to 65 µm long.

Hosts and distribution: *Clitoria* spp., especially *C. rubiginosa* Juss.: Nayarit and Oaxaca, Mexico to Panama; also in the islands of the Caribbean and in South America.

Type: on *Neurocarpum cajanifolium* (= *Clitoria cajanifolia* (Presl) Benth.); Bahia, Brazil, Lhotsky (S).

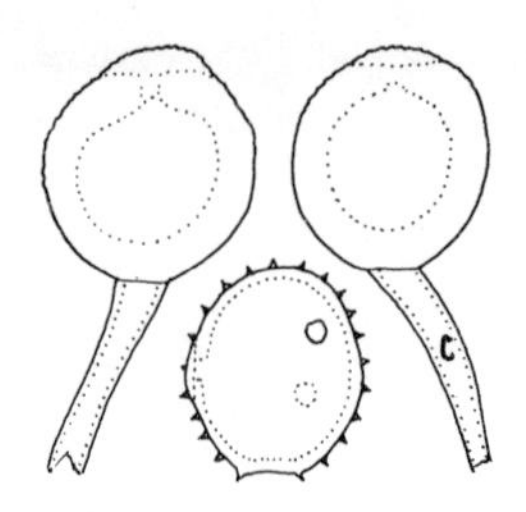

26. *UROMYCES TENUISTIPES* Diet. & Holw. in Holway, Bot. Gaz. 24:25. 1897.

Spermogonia and aecia unknown. Uredinia mostly on abaxial leaf surface, about cinnamon brown; spores (18-)20-24 (-26) x 18-23 µm, broadly ellipsoid or globoid, wall 1.5-2 µm thick, pale golden brown, echinulate, pores 3 or 4, more or less equatorial, with slight caps. Telia mostly on abaxial surface, exposed, pulverulent, blackish brown; spores (22-)24-27(-29) x (20-)21-24(-26) µm, globoid or nearly so, wall (3-)3.5-4.5 µm thick at sides, chestnut brown, 4.5-6 µm at apex with a paler, low umbo over the pore, finely rugose but this often obvious only on the umbo; pedicels colorless, usually remaining more or less terete, commonly flexuous and tapering, to 155 µm long.

Hosts and distribution: *Desmodium strobilaceum* Schl., *D.* sp.: central and southern Mexico.

Type: on *Desmodium* sp., Eslava, near City of Mexico, Holway (S; isotype PUR 14755).

27. *UROMYCES VICIAE-FABAE* Schroet. Hedwigia 14:161. 1857.
Uredo viciae-fabae Pers. Syn. Meth. Fung. p. 221. 1801 (telia present but not described).
Uromyces fabae deB. Ann. Sci. Nat. Bot. IV. 20:80. 1863, *nom. nudum.*

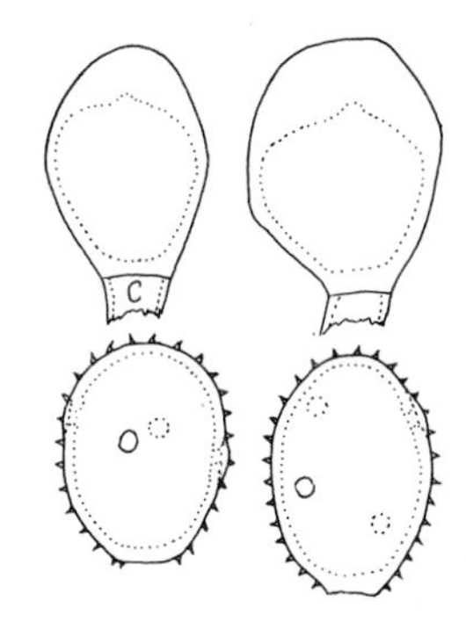

Spermogonia mostly on abaxial leaf surface. Aecia mostly on abaxial surface in small groups, mostly along veins, peridium cupulate, whitish; spores 18-26 x 15-21 µm, broadly ellipsoid, wall 1(-1.5) µm thick, verrucose. Uredinia amphigenous, yellowish brown; spores (22-)24-29(-32) x (17-)19-22(-24) µm, broadly ellipsoid, wall 1.5(-2) µm thick, pale golden, uniformly echinulate, pores (3)4(5), equatorial or variously distributed, with small caps. Telia sometimes on adaxial surface or sometimes amphigenous and on stems, exposed, blackish brown, compact; spores (24-)27-35(-39) x (17-)19-23(-25) µm, mostly oval or obovoid, wall (1.5-)2-2.5(-3) µm thick at sides, (5-)6-9(-10) µm at apex, smooth, uniformly chestnut brown; pedicels brownish at least apically, to about 60 µm long.

Hosts and distribution: species of *Lathyrus*, *Pisum* and *Vicia*: circumglobal.

Neotype: on *Vicia faba* L., collector and locality not given but part of the Persoon Herbarium (L 910.264-550).

Joerstad (18) designated this specimen as lectotype. Schroeter did not cite specimens, although he described teliospores, nor is it known that he examined Persoon specimens. For these reasons, I designate the above number as neotype.

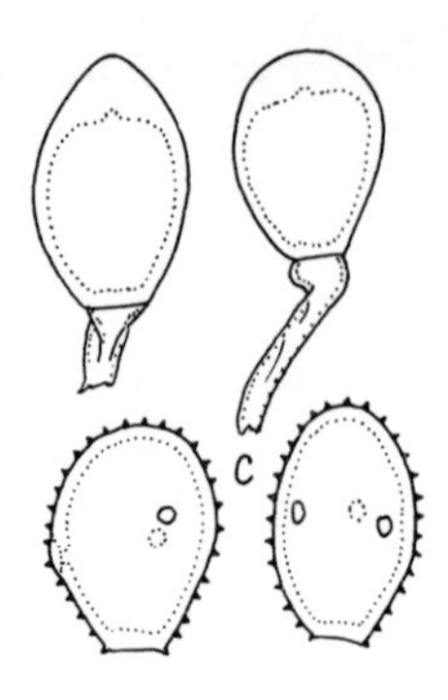

28. *UROMYCES ERVI* West. Bull. Acad. Belg. 21:234. 1854.

Spermogonia in small groups on the adaxial surface of leaves. Aecia mostly on the abaxial leaf surface, grouped around the spermogonia or scattered singly (resulting from infection by aeciospores), with short cupulate peridium; spores (15-)18-22(-25) x (12-)15-19(-21) µm, more or less globoid, wall 1(-1.5) µm thick, pale yellowish, verruculose. Uredinia rare, on either leaf surface or on petioles and stems, pale cinnamon brown; spores (20-)23-26(-30) x (17-) 19-22(-24) µm, broadly ellipsoid or obovoid, wall 1.5-2 µm thick, yellowish or golden, uniformly echinulate, pores equatorial (2)3(4), with small caps. Telia mostly on petioles and stems, becoming more or less confluent, somewhat slowly exposed, compact, blackish brown; spores (20-)22-30 (-33) x (14-)17-21(-23) µm, variable, mostly obovoid or oblong ellipsoid, wall at sides (1-)1.5-2(-2.5) µm thick, wall broadly thickened in apex (3-)4-6(-8) µm, chestnut brown or the longer narrower spores usually paler, smooth; pedicel persistent, brownish, to 80 µm long.

Hosts and distribution: *Vicia hirsuta* (L.) S. F. Gray: Nova Scotia, Canada; also from Europe to China and Japan.

Type: on *Ervum hirsutum* L. (= *Vicia hirsuta*), Selzinne near Namur, Belgium (BR; isotypes West. Herb. Crypt. Belg. No. 849).

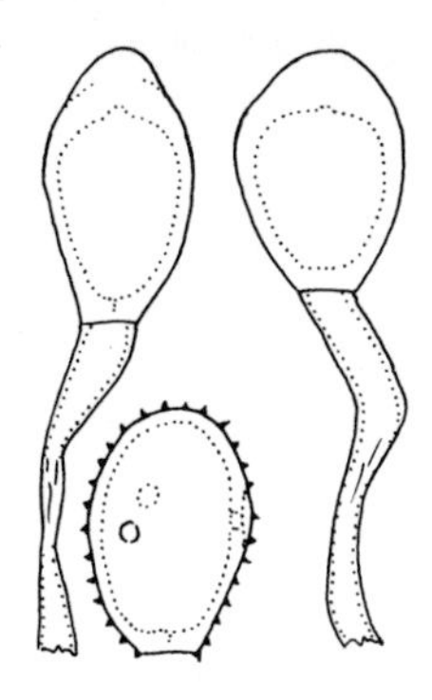

29. *UROMYCES INDIGOFERAE* Diet. & Holw. in Holway, Bot. Gaz. 31:328. 1901.

Spermogonia and aecia unknown. Uredinia amphigenous, cinnamon brown; spores (20-)23-28 x (16-)18-20(-22) µm, mostly ellipsoid or broadly so, wall 1.5(-2) µm thick, cinnamon or golden brown, uniformly echinulate, pores mostly 3, less commonly 4, rarely 2, equatorial, with obvious caps. Telia amphigenous and on stems, exposed, blackish brown, compact; spores (22-)25-30(-33) x (15-)17-21(-23) µm, mostly broadly ellipsoid or obovoid, wall (1.5-)2-3(-3.5) µm thick at sides, (4-)6-8(-10) µm apically, chestnut brown becoming slightly paler apically but not as a defined umbo, smooth, some spores have thinner, paler walls; pedicels colorless, to 100 µm long but usually shorter.

Hosts and distribution: *Indigofera* spp.: Florida and Texas to Panama; also in South America.

Type: on *Indigofera mexicana* Benth., Oaxaca, Oax., Mexico, Holway No. 3722 (S; isotype PUR 14758).

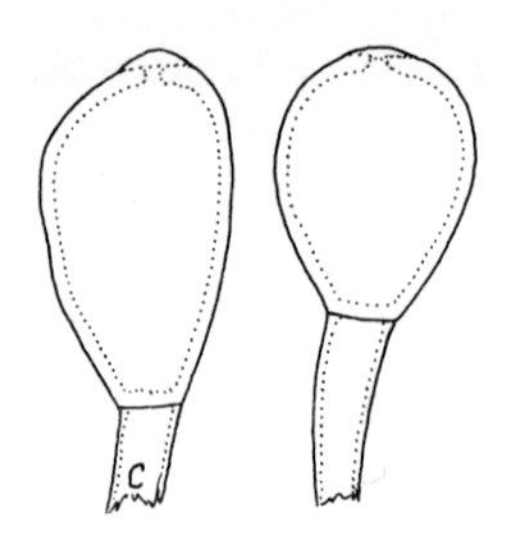

30a. *UROMYCES PSORALEAE* Peck, Bot. Gaz. 6:239. 1881 var. *PSORALEAE*.

Spermogonia on abaxial leaf surface, distributed from a systemic mycelium. Aecia on abaxial surface among the spermogonia, peridium relatively short, yellowish, the margin erose or recurved; spores (18-)21-26(-28) x (16-)18-22 (-23) µm, mostly globoid or broadly ellipsoid, wall 1-1.5 µm thick, finely verrucose. Uredinia lacking. Telia amphigenous and on stems, exposed, pulverulent, chocolate brown; spores variable in size, (26-)29-40(-42) x (17-)20-25(-27) µm, mostly obovoid or ellipsoid, wall 1.5-2 µm thick, clear chestnut brown, slightly thicker at pore with an illdefined, low, papilla, smooth; pedicels colorless, to 65 µm long but usually broken shorter.

Hosts and distribution: *Psoralea* spp., especially *P. lanceolata* Pursh: Saskatchewan to Idaho, Nebraska and Arizona.

Type: on *Psoralea lanceolata*, Salt Lake City, Utah, 1880 Jones (NYS; isotype PUR 35008).

The following variety differs in having narrower teliospores.

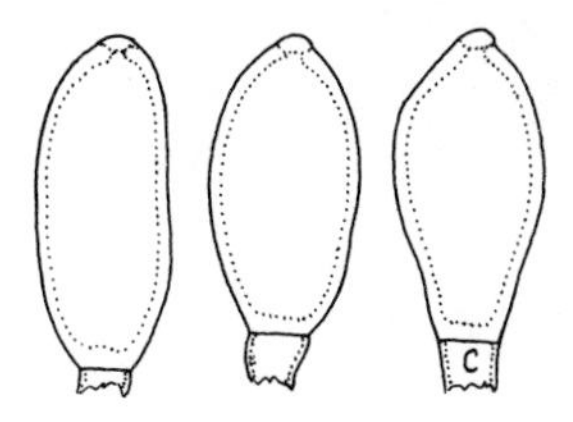

30b. *UROMYCES PSORALEAE* Peck var. *ARGOPHYLLAE* (Seym.) Arth. Man. Rust of U.S. and Canada p. 245. 1934.
Uromyces argophyllae Seym. Proc. Boston Soc. Nat. Hist. 24:185. 1889.

Spermogonia, aecia and aeciospores as in var. *psoraleae*; uredinia lacking; teliospores variable, (24-)29-38(-42) x (11-)13-19(-21) µm, mostly ellipsoid or oblong ellipsoid, wall 1.5-2 µm thick, about cinnamon or golden brown, with a pale, low papilla over the pore, smooth; pedicels colorless, usually broken at the hilum.

Hosts and distribution: *Psoralea* spp.: Saskatchewan to Illinois, Texas and Arizona.

Lectotype: on *Psoralea argophylla* Pursh, Bismark, Dakota (North Dakota), Seymour (FH; isotypes Ellis & Ever. N. Amer. F. No. 1862). Two localities are listed in the original, hence this designation of a lectotype. It is the logical selection because the diagnosis is preceded by "*Uromyces argophylli*, Seymour, in Ell. and Ev. N. Am. Fung. 1862."

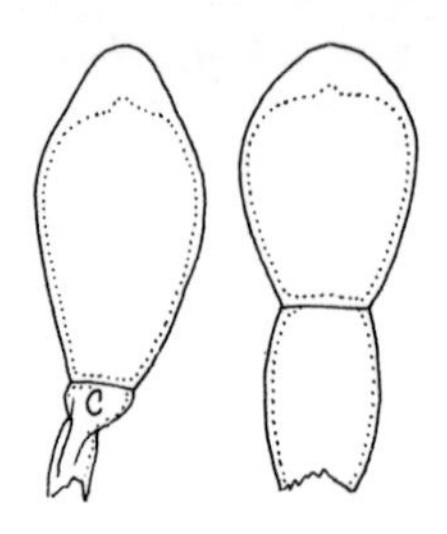

31. *UROMYCES MONTANUS* Arth. Bot. Gaz. 39:386. 1905.

Spermogonia amphigenous, in small groups. Aecia mostly on abaxial leaf surface, in small groups, peridium short, the margin erose, pale yellowish; spores (24-)30-38(-42) x (17-)20-26(-29) µm, mostly broadly ellipsoid or ellipsoid, wall 2-3 µm thick, yellowish, finely verrucose. Uredinia lacking. Telia on abaxial surface, exposed, in congested groups, about cinnamon brown, becoming gray from germination, compact; spores (26-)29-40(-44) x (15-)17-22(-24) µm, mostly ellipsoid or oblong ellipsoid, wall 0.5-1 µm thick at sides, (3-)4-7(-8) µm at apex, pale golden or the apex nearly colorless, smooth; pedicels colorless, often nearly as wide as the spore, to 65 µm long.

Hosts and distribution: *Lupinus* spp.: southern Mexico, Guatemala and Costa Rica.

Type: on *L. mexicanus* H.B.K.: Nevada de Toluca, Mex., Mexico, Holway (PUR 34998).

All collections are from relatively high altitudes, the type locality being 10,400 ft; one other is from 13,000 ft.

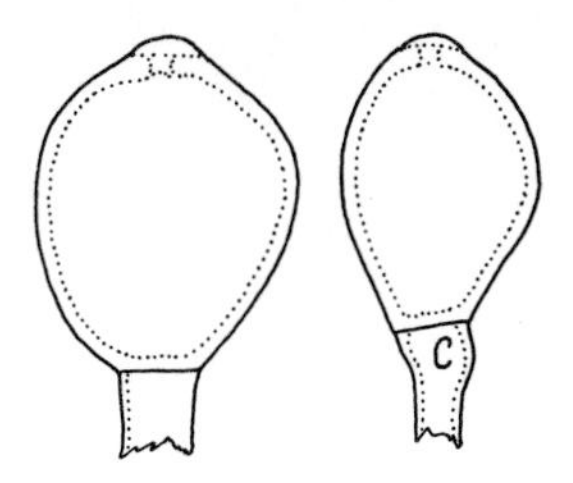

32. *UROMYCES ABBREVIATUS* Arth. Bull. Torrey Bot. Club 42: 587. 1915.

Spermogonia on the abaxial leaf surface among the telia. Telia mostly abaxial, closely and widely distributed from a systemic mycelium, exposed, chocolate brown; spores (26-)28-36(-38) x (20-)22-27(-29) µm and occasional odd size spores, mostly broadly ellipsoid or obovoid, wall (1-)1.5-2 µm thick at sides, (2-)2.5-3.5(-5) µm at apex, with a slight, pale, flattish umbo, clear chestnut brown, smooth; pedicel colorless, usually broken shorter.

Hosts and distribution: *Psoralea physodes* Dougl., *P. purshii* Vail: Idaho to Washington and California.

Type: on *P. purshii*, Winnemucca, Nevada, Griffiths and Morris (PUR 38657; isotypes Griffiths W. Amer. F. No. 390).

This species doubtless is a microcylic derivative of *U. psoraleae* Peck var. *psoraleae*.

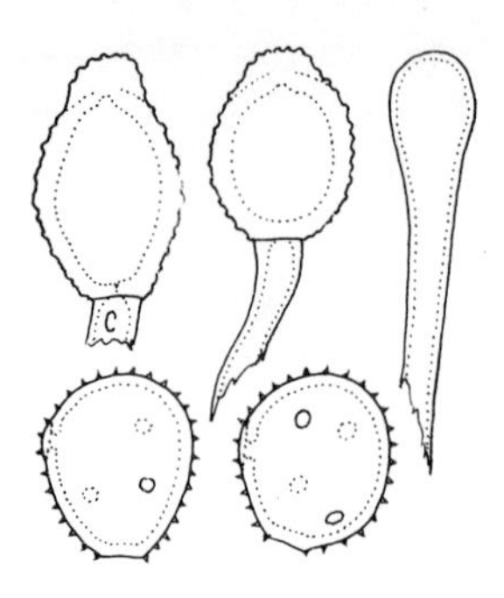

33. *UROMYCES HEDYSARI-PANICULATI* (Schw.) Farl. in Ellis N. Amer. Fungi No. 246. 1879.
Puccinia hedysari-paniculati Schw. Schr. Nat. Ges. Leipzig 1:74. 1822.

Spermogonia on adaxial leaf surface. Aecia on abaxial surface, peridium cupulate with erose margin, whitish; spores 22-23 x 21-24 µm, mostly globoid, wall 1-1.5 µm thick, colorless, verrucose. Uredinia mostly on abaxial surface, cinnamon brown or paler, with thin walled colorless, mostly clavate paraphyses from 15 to 25 µm wide apically; spores (19-)21-25(-27) x (16-)17-20(-22) µm, broadly ellipsoid, wall (1-)1.5(-2) µm thick, about golden brown, closely echinulate, pores (3)4-7, scattered, with low caps. Telia mostly on abaxial surface but adaxial on some hosts, exposed, blackish brown, relatively pulverulent; spores (20-)22-30(-33) x (16-)17-21(-23) µm, ovoid, ellipsoid or broadly ellipsoid, wall (2-)2.5-3 µm thick at sides, chestnut brown, 4-6(-8) µm thick at apex by a paler illdefined umbo, from labyrinthiformly rugose with united warts and ridges to reticulate with meshes about 1-2.5 µm wide; pedicels colorless except apically, to 60 µm long, often about 40 µm.

Hosts and distribution: *Desmodium* Massachusetts to Ontario and Minnesota south to Texas and Central America; also in South America.

Neotype: on *Desmodium* (det. to be *D. paniculatum* (L.) DC. by J. K. Small, 1908), Newfield, New Jersey (FH; isotypes Ellis N. Amer. F. No. 246. 1879) Neotype designated here.

According to Arthur and Bisby (3) there is only an empty packet in the Schweinitz herbarium (PH).

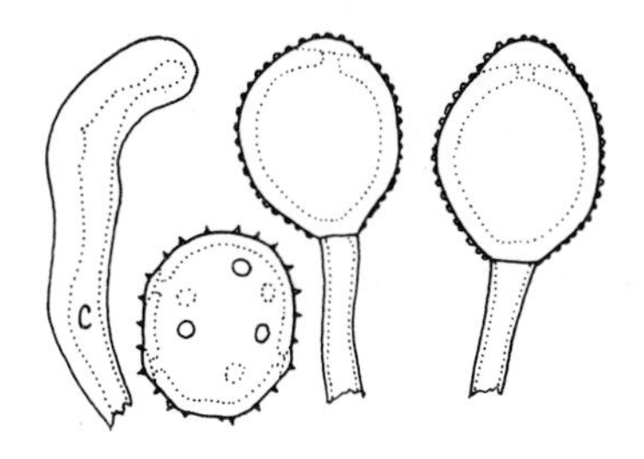

34. *UROMYCES ANTIGUANUS* Cumm. Bull. Torrey Bot. Club 67: 612. 1940.

Spermogonia and aecia unknown. Uredinia on abaxial leaf surface, yellowish brown, with abundant, colorless or yellowish, incurved, peripheral paraphyses, the inner wall 1-1.5 µm thick, the outer wall 2.5-5 µm thick; spores (18-) 20-22(-24) x (17-)18-20(-21) µm, broadly ellipsoid or obovoid, wall 1-1.5(-2) µm thick, yellowish or golden brown, uniformly echinulate, pores 7-9, scattered, with small caps. Telia mostly on abaxial surface, exposed, blackish brown, more or less pulverulent; spores (22-)24-29(-31) x 19-23 (-24) µm, mostly broadly ellipsoid, wall 2.5-3 µm thick at sides, 3-5 µm apically but without a defined umbo, dark chestnut brown, reticulate or pseudoreticulate with meshes 1-2 µm diam; pedicels colorless, to 65 µm long.

Hosts and distribution: *Desmodium orbiculare* Schl.: central Mexico to Guatemala.

Type: Cuesta de las Canas, above Antigua, Guatemala, Standley No. 58900 (PUR 49057).

35. *UROMYCES CROTALARIAE* (Arth.) J. W. Baxt. Mycologia 54: 437. 1962.
Uropyxis crotalariae Arth. Amer. J. Bot. 5:429. 1918.
Haplopyxis crotalariae (Arth.) H. Syd. & P. Syd. Ann. Mycol. 17:105. 1919.

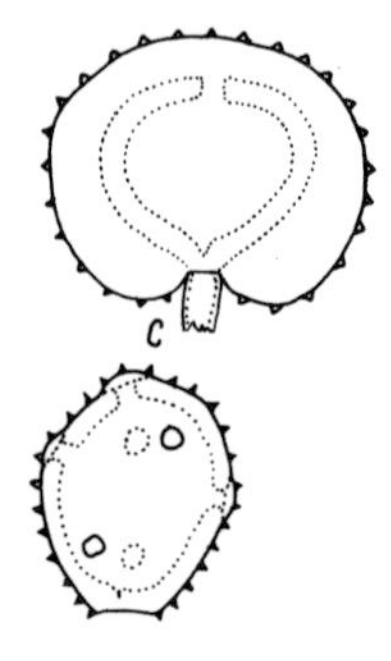

Spermogonia and aecia unknown. Uredinia mostly on abaxial leaf surface, pale cinnamon brown; spores (22-)24-28(-32) x 20-25 µm broadly ellipsoid or obovoid, wall 1.5-2(-2.5) µm thick, pale golden brown, echinulate, pores (5)6-8, scattered, with caps. "Telia hypophyllous, black or chocolate brown, pulverulent; teliospores oblate spheroid, bluntly conoid or ovoid, assuming a globoid shape when turned to present the apical pore or point of pedicel attachment in surface view, oblate spheroid teliospores 27-32 µm broad x 20-27 µm high, conoid or ovoid teliospores 20-27 µm broad x 24-30 µm high; wall bilaminate, inner layer 2-2.5 µm, thinner and papillate at the point of pedicel attachment, chestnut brown, pore apical, inconspicuous; outer layer swelling in water to a thickness of 3-10 µm, sharply indented at the point of pedicel attachment, hyaline, yellow or pale cinnamon brown, verrucose echinulate with short conical tubercles; pedicel thin walled, hyaline, usually breaking away at the point of attachment to the spore, occasionally persistent, 12-22 µm long." (from Baxter *loc. cit.*).

Hosts and distribution: *Crotalaria* spp.: Central Mexico to Costa Rica; also in Brazil and South Africa.

Type: on *Crotalaria vitellina* Ker., Laguna, Dept. Amatitlan, Guatemala, Kellerman No. 5397 (PUR 7357; the holotype is useless).

Included here are the Costa Rican specimens on *C. vitellina* that were referred to *Uromyces decoratus* Syd. in the N. Amer. Flora (1).

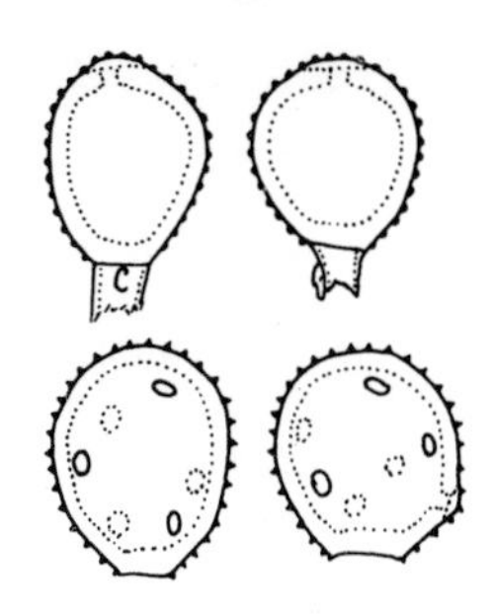

36. *UROMYCES CICERIS-ARIETINI* Jacz. in Boyer & Jaczewski, Bull. Soc. Bot. France 40:CCLXXII. 1893.

Spermogonia and aecia unknown. Uredinia on both sides of leaves, the rachis, and stems, abundant, appearing as if systemic, cinnamon brown; spores (20-)23-27(-29) x (16-)18-22 µm, broadly ellipsoid, wall variable (1.5-)2-3(-3.5) µm thick, golden or cinnamon brown, uniformly echinulate, pores scattered, (4-)6 or 7(8), with obvious caps. Telia about as the uredinia or usually less abundant, exposed, chocolate brown, more or less pulverulent; spores (18-)20-25(-27) x (16-)18-20(-21) µm, ellipsoid, broadly obovoid or globoid, wall uniformly 2-2.5(-3) µm thick except an umbo about 1-1.5 µm over pore, verrucose with small warts spaced 1.5-3 µm, or these merged into short ridges, or sometimes interspersed with ridges nearly the length of the spore, chestnut brown; pedicel colorless, always broken near hilum.

Hosts and distribution: *Cicer arietinum* L.: Central Mexico; also in Europe, Africa and east to India.

Type: on *Cicer arietinum*, Montpellier, France, Boyer (not seen).

The Mexican plants are so heavily infected as to suggest an autoecious species with the primary infection occurring in the seedling stage.

37. *UROMYCES OCCIDENTALIS* Diet. Hedwigia 42 (Beibl.):98. 1903.
Uromyces substriatus P. Syd. & H. Syd. Ann. Mycol. 4:30. 1906.

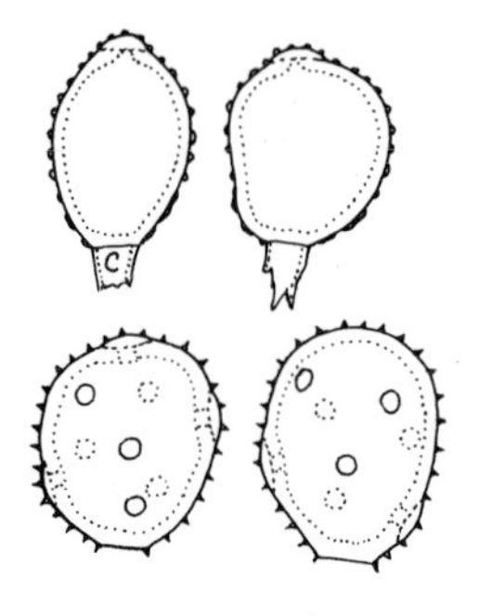

Spermogonia mostly on abaxial leaf surface, scattered on a systemic mycelium. Aecia on abaxial surface, similarly scattered, peridium short with recurved erose margin; spores 19-24 x 15-20 µm, globoid or broadly ellipsoid, wall 1.5(-2) µm thick, colorless, finely verrucose. Uredinia mostly on abaxial surface, yellowish brown; spores (20-)22-27(-29) x (16-)18-21 (-23) µm, obovoid, broadly ellipsoid or globoid, wall 1.5-2 µm thick, yellowish or golden brown, echinulate, pores (6)7-9(-11), scattered, with caps. Telia mostly on abaxial surface, exposed, often in circles, chocolate brown, pulverulent; spores (17-)19-25(-28;33) x (14-)16-20(-22) µm, variable in size and shape in some collections, mostly broadly ellipsoid or obovoid, wall 1.5-2 µm thick except for a small pale umbo over pore, verrucose with small verrucae more or less uniformly spaced (1-)1.5-2(-2.5) µm, or in reticulate patterns, or in lines, or fused to form ridges, especially toward hilum; pedicels colorless, always broken near hilum.

Hosts and distribution: of aecia, *Euphorbia* section *Tithymalus* ; of uredinia and telia, *Lupinus* spp.: Wyoming and Montana to California, Arizona and Mexico.

Lectotype: on *Lupinus latifolius* J. Agar., Sissons, California, July 27, 1894, Blasdale (S; isotype Barth. N. Amer. Ured. No. 96). Lectotype designated here following Arthur's (1) citation of the type locality as "Sissons, California, on *Lupinus latifolius*."

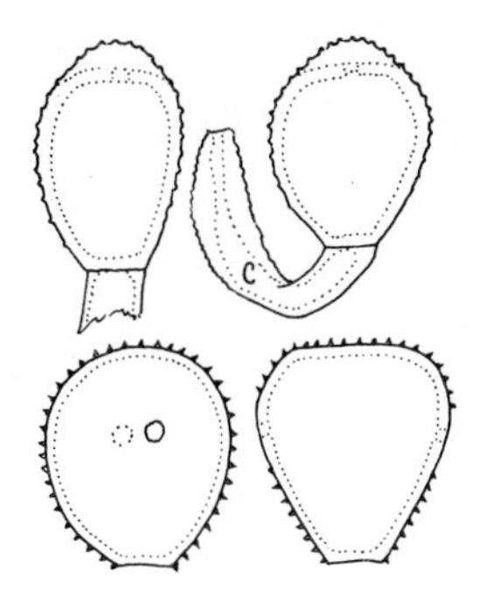

38. *UROMYCES CLITORIAE* Arth. Bot. Gaz. 39:389. 1905.

Spermogonia and aecia unknown. Uredinia on abaxial leaf surface, cinnamon brown; spores (25-)27-30 x 25-27 µm, mostly triangularly obovoid, wall (1.5-)2(-2.5) µm thick, cinnamon brown, echinulate except around each pore, pores 2, usually slightly above the equator, with slight or no caps. Telia on abaxial surface, exposed, pulverulent, blackish brown; spores (20-)22-27(-30) x (15-)17-20(-23) µm, from ellipsoid to globoid, wall (2-)2.5-3.5(-4) µm thick at sides, chestnut brown, 3.5-5 µm at apex with a pale, defined umbo, verrucose with discrete, small verrucae spaced mostly 2 µm apart; pedicels colorless, rugose basally, to 40 µm long but often broken short.

Hosts and distribution: *Clitoria mexicana* Link: central Mexico.

Type: Jalapa, Veracruz, Holway No. 3058 (PUR 15884). One other collection is known.

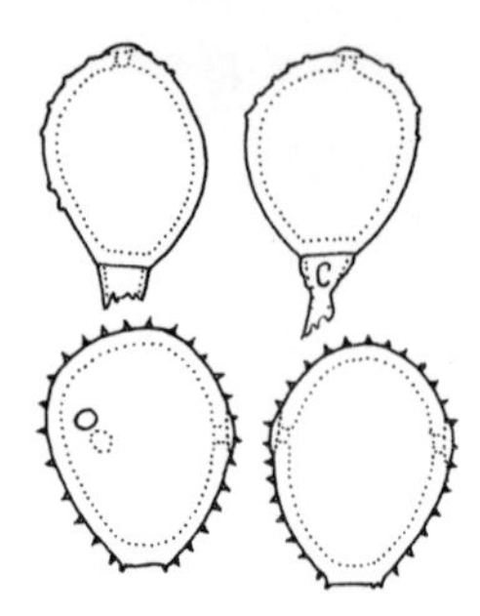

39a. *UROMYCES TRIFOLII-REPENTIS* Liro, Bidr. Kaenned. Finl. Nat. Folk 65:94. 1908 var. *TRIFOLII-REPENTIS*.
Uromyces trifolii auct. non (Hedw. ex DC.) Fckl. 1870.
Uromyces trifolii-hybridi H. Paul, Krypt. Forsch. 2:50. 1917.
Uromyces hybridi W. H. Davis, Mycologia 16:216. 1924.

Spermogonia mostly on adaxial leaf surface. Aecia on both leaf surfaces and petioles, in groups, peridium short, margin erose; spores (16-)18-22(-24) x (14-)15-18(-20) µm, nearly globoid or broadly ellipsoid, wall about 1 µm thick, colorless, finely verrucose. Uredinia amphigenous and on stems, pale cinnamon brown; spores (21-)24-27(-29) x (17-)19-22(-24) µm, broadly ellipsoid, broadly obovoid or globoid, wall (1.5-)2(-2.5) µm thick, golden brown, echinulate, pores 2 or 3(4), equatorial, with small caps. Telia mostly on abaxial leaf surface and on petioles and stems, exposed, pulverulent, chestnut brown; spores (20-)22-26(-29) x (15-)18-22(-23) µm, oval, broadly ellipsoid or globoid, wall 1.5-2 µm thick except a small, pale papilla over pore, pale chestnut brown, smooth or with a few scattered or linearly arranged verrucae; pedicel colorless, always broken near the spore.

Hosts and distribution: *Trifolium* spp., circumglobal, especially on *T. hybridum* L. and *T. repens* L.

Lectotype: on *T. repens*, Evo, Finland (H). Not seen.

The following variety has been treated as a species or merged with *U. trifolii-repentis* (often as *U. trifolii*). Differences in the number and arrangement of the urediniospore pores may justify varietal rank but scarcely specific rank.

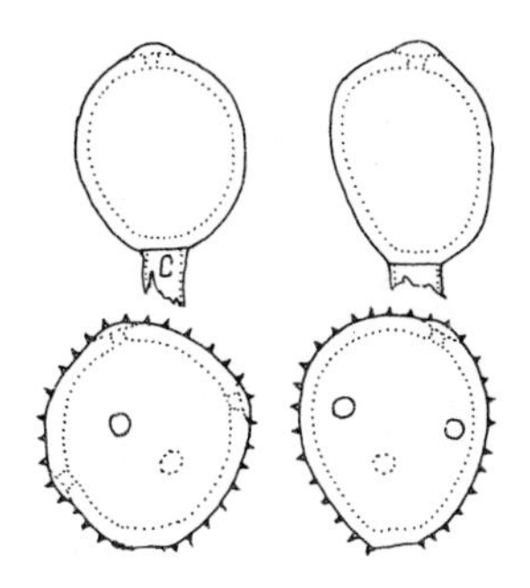

39b. *UROMYCES TRIFOLII-REPENTIS* Liro var. *FALLENS* (Arth.) Cumm. Mycotaxon 5:407. 1977.
Nigredo fallens Arth. N. Amer. Flora 7:254. 1912.
Uromyces fallens (Arth.) Barth. Handb. N. Amer. Ured. p. 61. 1928.

Generally similar to var. *trifolii-repentis*. Urediniospores (23-)24-28(-30) x (19-)21-24(-25) µm, wall (1.5-)2 (-3) µm thick, golden brown, uniformly echinulate, pores often 3 or 4, equatorial, usually 3 or 4 more or less equatorial and one in or near apex, less commonly 6 or 7, with small caps. Teliospores (20-)22-26(-29) x (17-)18-22(-23) µm, wall 1.5-2 µm thick, clear chestnut brown, mostly smooth.

Hosts and distribution: *Trifolium* spp., especially *T. pratense* L.: widely distributed where red clover is grown.

Neotype: on *Trifolium pratense*, Emporia, Kansas, 3 Nov. 1903, J. E. Bartholomew (PUR 15322; isotypes Barth. N. Amer. Ured. No. 685). Neotype designated here.

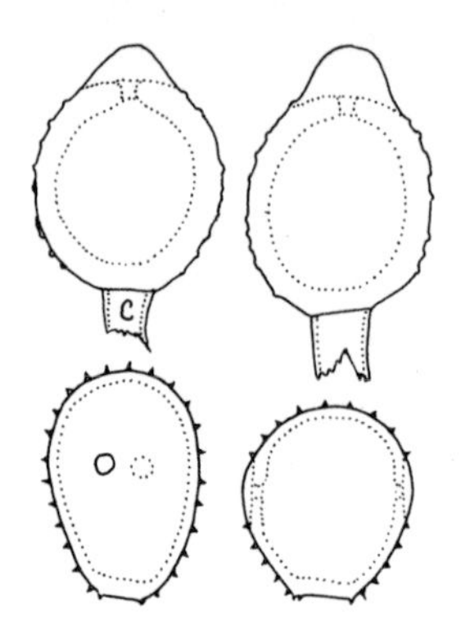

40a. *UROMYCES APPENDICULATUS* (Pers.) Unger, Einfl. Bodens p. 216. 1836 var. *APPENDICULATUS*.
Uredo appendiculata α. *Uredo phaseoli* Pers. Syn. Method. Fung. p. 222. 1801 (based on telia).
Uromyces phaseoli (Pers.) Wint. Hedwigia 19:37. 1880.

Spermogonia on adaxial leaf surface. Aecia on abaxial surface, peridium short, whitish, margin erose; spores (18-)20-28(-33) x (16-)18-20(-24) µm, ellipsoid, oblong ellipsoid or angularly globoid, wall 1-1.5 µm thick, colorless, finely verrucose. Uredinia amphigenous, cinnamon brown; spores (20-)24-30(-33) x (18-)20-27(-29) µm, mostly obovoid or broadly ellipsoid, wall 1.5-2 µm thick, cinnamon or golden brown, echinulate, pores 2, equatorial or slightly above with smooth caps. Telia amphigenous, circinate around uredinia or scattered, exposed, pulverulent, blackish brown; spores (24-)28-33(-35) x (20-)22-27(-29) µm, ovoid, broadly ellipsoid or globoid, wall (2-)2.5-3.5(-4) µm thick at sides, (5-)6-9(-11) µm thick at pore by a pale, defined, umbo, chestnut brown, with few to numerous verrucae randomly distributed or lineal or even fused into ridges, rarely smooth; pedicels colorless, to about 45 µm long.

Hosts and distribution: species of *Phaseolus* and *Strophostyles*: widely distributed where beans are cultivated and common on many native hosts, especially in warm climates.

Type: on *Phaseolus vulgaris* L.: locality not given, Europe (L). Not seen.

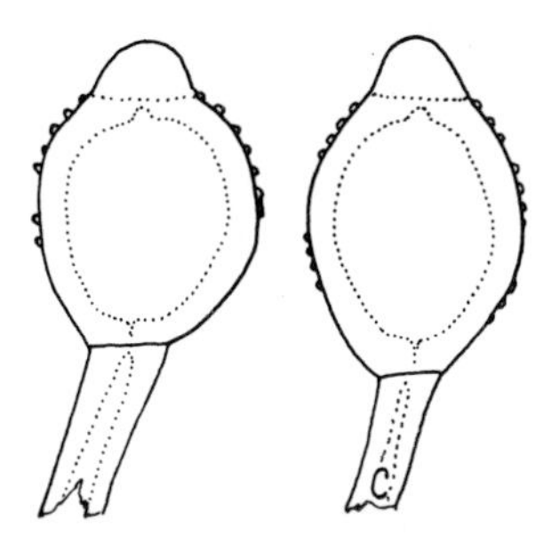

40b. *UROMYCES APPENDICULATUS* (Pers.) Unger var. *PUNCTIFORMIS* (P. Syd.) Cumm. Mycotaxon 5:407. 1977.
Uromyces punctiformis P. Syd. Ured. No. 1513. 1901.

Spermogonia, aecia and uredinia unknown. Urediniospores in the telia about 24-26 x 19-21 µm, mostly obovoid, wall 1-1.5 µm thick, golden brown, echinulate except around pores. Telia mostly on the adaxial leaf surface, exposed, small, blackish brown; spores (33-)35-40(-42) x (26-)28-31 (-33) µm, broadly ellipsoid or broadly obovoid, wall 3-4 µm thick, chestnut brown, verrucose with small warts variously distributed subapically, occasionally lineal, the lower half of spore commonly smooth, the pore covered by a pale, smooth, defined umbo (5-)7-10 µm high; pedicel persistent, to 80 µm long.

Type: on *Ramirezella strobilophora* (B. L. Rob.) Rose *(Vigna strobilophora* B. L. Rob.) Guadalajara, Jal., Mexico, Holway (S. isotypes Sydow, Ured. No. 1513). Not otherwise known.

The variety differs from var. *appendiculatus* mostly because of larger teliospores, whose pedicels are thick walled and persistent.

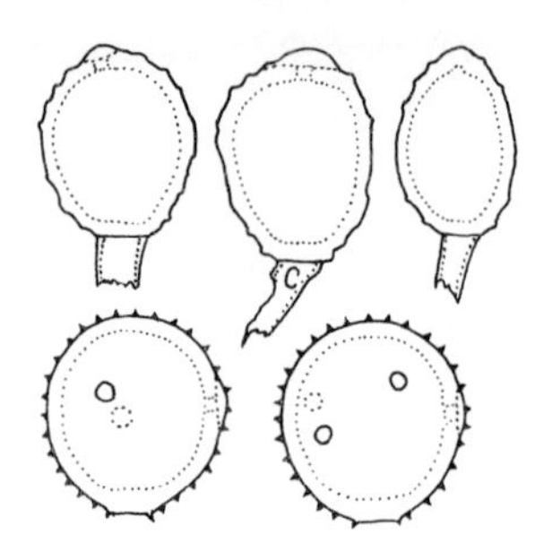

41. *UROMYCES STRIATUS* Schroet. Abh. Schles. Ges. Vaterl. Cult. Nat. Abth. 1869-72:11. 1870.

Spermogonia scattered on abaxial leaf surface from a systemic mycelium. Aecia abaxial among the spermogonia, peridium white with recurved margin; spores 14-25(-28) x 11-20 µm, globoid or broadly ellipsoid, wall about 1-1.5 µm thick, colorless, finely verrucose. Uredinia mostly on abaxial leaf surface, cinnamon brown; spores (17-)20-25(-27) x (16-)19-22(-23) µm, globoid or broadly ellipsoid, wall (1-) 1.5-2(-2.5) µm thick, pale cinnamon or golden brown, echinulate, pores 3 or 4 (rarely 2 or 5), equatorial, with small caps. Telia darker brown than uredinia, exposed, pulverulent; spores (17-)19-25(-29) x (13-)15-20(-21) µm, ellipsoid, obovoid or globoid, wall (1-)1.5-2 µm thick at sides, pale chestnut or dark golden brown, (2.5-)3-4(-5) µm at apex with a pale umbo, with longitudinal ridges about 1 µm wide spaced 1.5-2.5(-3) µm on centers with the ridges discrete, or anastomosed, or elongated warts sometimes replacing some ridges; pedicels colorless, usually broken short.

Hosts and distribution: of aecia, *Euphorbia cyparissias* L., *E. esula* L.; of uredia and telia, species of *Medicago* and less commonly *Hosakia*: southern Canada and southward; circumglobal on *Medicago* spp.

Lectotype: on *Medicago lupulina* L., Pirscham near Breslau, Silesia; designated by Hylander, Joerstad and Nannfelt (17).

The only verified records of the aecial stage in North America were recorded by Parmelee (24) who made inoculations.

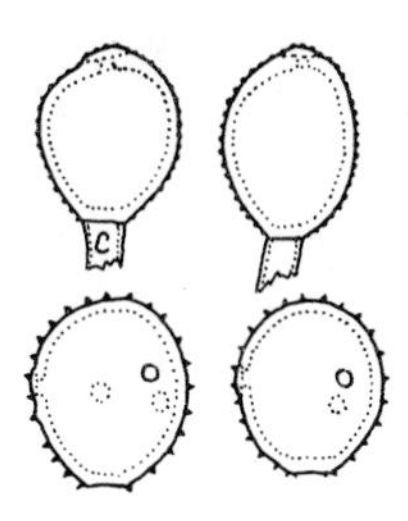

42. *UROMYCES CALOPOGONII* Cumm. Bull. Torrey Bot. Club 70:80. 1943.

Spermogonia and aecia unknown. Uredinia on abaxial leaf surface, yellowish brown; spores (16-)17-20(-22) x (15-)16-18 μm, broadly ellipsoid or globoid, wall 1-1.5 μm thick, pale cinnamon brown or golden, uniformly echinulate, pores 3 or 4, equatorial, with very small caps. Telia as the uredinia but chestnut brown, exposed, pulverulent; spores (16-)18-21(-23) x (14-)15-17(-18) μm, mostly ellipsoid or obovoid, wall 1.5(-2) μm thick at sides, the apex 2-2.5(-3) μm thick with a differentiated pale papilla or low umbo, pale chestnut brown, verrucose with mostly discrete, small, slightly irregularly shaped verrucae; pedicel colorless, always broken near the hilum.

Hosts and distribution: *Calopogonium galactioides* (H.B.K.) Benth.: Guatemala.

Type: near Chimaltenango, Standley No. 79808 (PUR 50320).

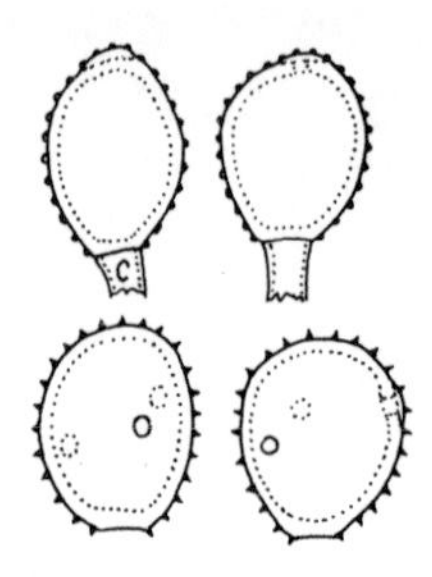

43. *UROMYCES COLOGANIAE* Arth. Bot. Gaz. 39:387. 1905.

Spermogonia and aecia unknown. Uredinia on abaxial leaf surface, yellowish brown; spores (18-)20-24(-25) x (16-)17-20(-22) µm, mostly broadly ellipsoid or obovoid, wall (1.5-)2 µm thick, golden, uniformly echinulate, pores 3(4), approximately equatorial, with small caps. Telia on abaxial surface, cinnamon brown or darker, exposed, pulverulent; spores (16-)18-23(-25) x (14-)15-19(-21) µm oval, broadly ellipsoid or globoid, wall 1.5(-2) µm thick except a low, pale umbo over the pore, cinnamon brown or pale chestnut brown, verrucose with small verrucae spaced 2-3 µm and often united basally by fine ridges, pedicels colorless, always broken near the hilum.

Hosts and distribution: *Cologania* spp.: central Mexico to Costa Rica; also in the West Indies.

Type: on *C. pulchella* H.B.K., Patzcuaro, Mich., Mexico, Holway No. 3192 (PUR 15530).

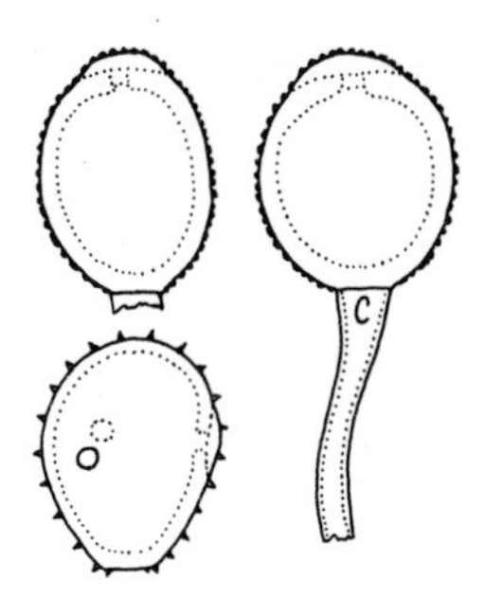

44. *UROMYCES ILLOTUS* Arth. & Holw. in Arthur, Amer. J. Bot. 5:441. 1918.

Uredinia on abaxial leaf surface, cinnamon brown; spores (23-)25-28(-30) x (17-)19-23(-24) µm, mostly obovoid, wall 1.5-2(-2.5) µm thick, cinnamon brown, uniformly echinulate, pores 3 or 4, equatorial, with obvious caps. Telia on abaxial surface, early exposed, blackish brown, moderately compact; spores (23-)25-29(-31) x (19-)20-24 µm, broadly ellipsoid or obovoid, wall 2-2.5 µm thick at sides, 3-5 µm apically with a slightly paler but not clearly defined broad, low umbo, chestnut brown, rugosely verrucose with slightly elongated verrucae which may be discrete or united in short series with a tendency to be oriented longitudinally; pedicel colorless, to about 45 µm long.

Type: on *Mucuna andreana* Mich., Chinaulta, Dept. of Guatemala, Guatemala, Holway No. 487 (PUR 42404). Not otherwise known.

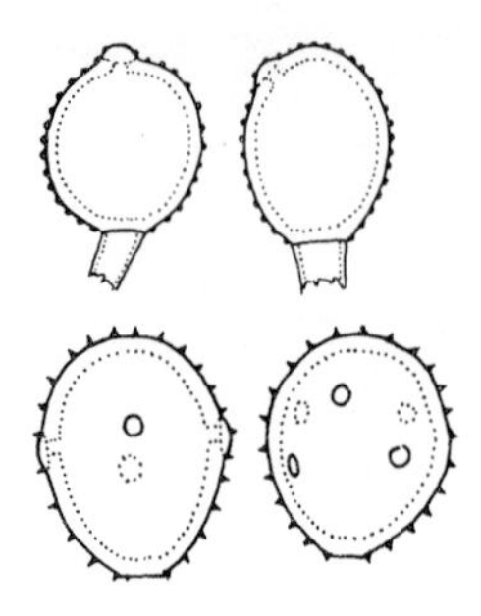

45. *UROMYCES PUNCTATUS* Schroet. Abh. Schles. Ges. Vaterl. Cult. Nat. Abth. 1869-72:10. 1870.

Spermogonia and aecia systemic in *Euphorbia cyparissias* L. and *E. virgata* W. & K. but not recorded in the Americas. Uredinia amphigenous, about cinnamon brown; spores (19-)22-28(-30) x (17-)19-23 µm, broadly ellipsoid, globoid or obvoid, wall 1.5-2 µm thick, light cinnamon brown, uniformly echinulate, pores 3-5, equatorial or approximately so in most spores, with small caps. Telia as the uredinia but about chestnut brown, exposed, pulverulent; spores (17-)19-21(-23) x (15-)17-19 µm, variable but mostly broadly obovoid or globoid, wall (1-)1.5(-2) µm thick, chestnut brown except a small, hyaline papilla over the pore, verrucose with small verrucae spaced 1.5-2(-3) µm; pedicel colorless, short, fragile.

Hosts and distribution: species of *Astragalus* and *Oxytropis*: Wisconsin to Washington and southward to Mexico and Guatemala; also in Europe.

Lectotype: on *Astragalus glycyphyllus* L., Breslau in Silesia; (lectotype designated by Hylander, Joerstad and Nannfelt (17) following Arthur's (1) citation of the type locality as "Breslau, Germany, on *Astragalus glycophyllus*.").

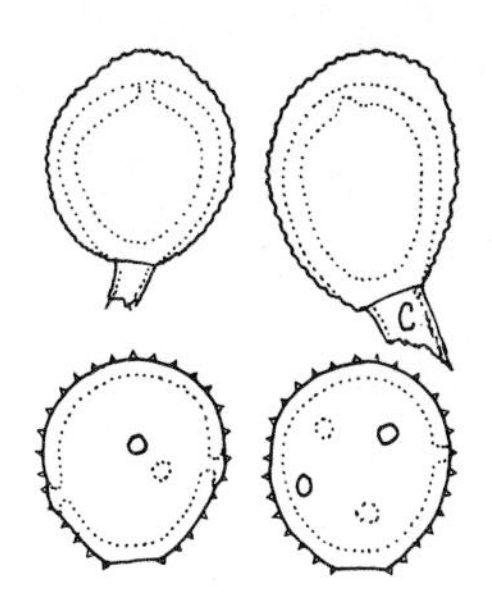

46. *UROMYCES BAUHINIAE* P. Henn. Hedwigia 34:90. 1895.
Uromyces guatemalensis Vest. Ark. Bot. 4:20. 1905.

Spermogonia and aecia unknown. Uredinia amphigenous or sometimes mostly or only on the adaxial surface, about cinnamon brown; spores (20-)22-24(-25) x (18-)20-22 µm, mostly broadly ellipsoid or globoid, wall 1.5-2 µm thick, cinnamon or golden brown, echinulate, pores 3-5(6), in the equatorial region or scattered, with slight or no caps. Telia as the uredinia but chocolate brown, exposed, pulverulent; spores (21-)23-26(-28) x 21-24 µm, obovoid, broadly ellipsoid or globoid, wall conspicuously bilaminate, the inner chestnut brown layer uniformly (1.5-)2(-2.5) µm thick, the outer layer increasing gradually from 0 thickness at the hilum to as much as 5 µm at the apex, thus forming an almost complete but sometimes inconspicuous outer envelop, uniformly reticulate with meshes (1-)1.5-2(-2.5) µm diam; pedicel colorless, usually broken near hilum.

Hosts and distribution: *Bauhinia* spp.: Sinaloa, Mexico to Costa Rica; also in South America.

Type: on *Bauhinia* sp., Paranaiba, Minas Gerais, Brazil, Ule No. 1906 (B).

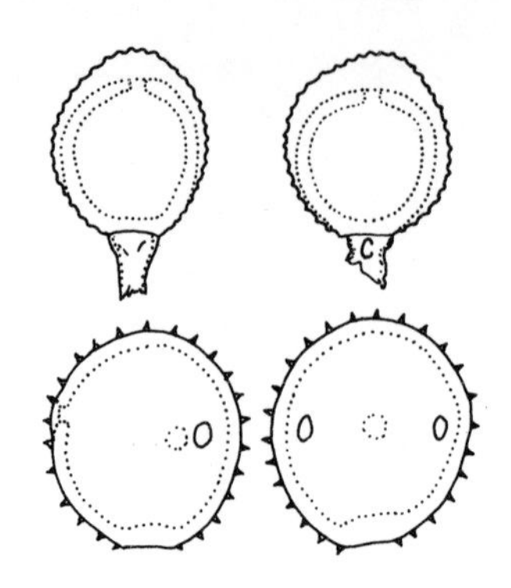

47. *UROMYCES IMPERFECTUS* Arth. Bull. Torrey Bot. Club. 47: 472. 1920.
Uromyces bauhiniae Vest. Ark. Bot. 4:21. 1905. Not *Uromyces bauhiniae* P. Henn. 1895.

Spermogonia and aecia unknown. Uredinia mostly on adaxial leaf surface, cinnamon brown; spores (25-)27-30(-32) x (22-)24-28(-29) μm, mostly globose, wall (1.5-)2 μm thick, cinnamon brown, echinulate, pores 3, occasionally 4, rarely 2, equatorial, with small caps. Telia as the uredinia but chocolate brown, exposed, pulverulent; spores (20-)22-24 (-27) x (18-)20-23(-24) μm, broadly ellipsoid or globoid, wall bilaminate, the inner chestnut brown layer 1.5-2 μm thick, the outer usually distinct layer increasing from 0 thickness at the hilum to as much as 4 μm at the apex, reticulate with meshes 1-1.5 μm diam; pedicel colorless, usually broken near the spore.

Hosts and distribution: *Bauhinia* spp.: Nicaragua; also reported in Jamaica and South America.

Type: on *Bauhinia* sp., Nicaragua (without locality or date but during 1853-56), Wright (K).

The above binomials, as well as *Uredo bauhiniae* Berk. & Curt., all are based on the type collection.

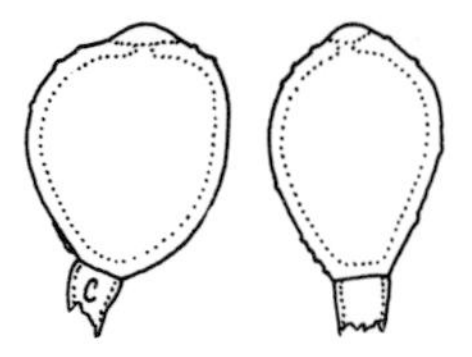

48. *UROMYCES LAPPONICUS* Lagerh. Bot. Notiser 1890:274. 1890.

Spermogonia scattered from a systemic mycelium. Aecia among the spermogonia on abaxial surface or on stems, cupulate, whitish; spores (17-)21-25(-28) x (14-)16-20(-22) µm, ellipsoid or globoid, wall (1-)1.5(-2) µm, colorless, verrucose. Uredinia lacking. Telia mostly on abaxial surface, exposed, cinnamon brown, pulverulent; spores (22-)24-28(-30) x (17-)19-22(-24) µm, broadly ellipsoid or obovoid, wall (-1)1.5(-2) µm, cinnamon or golden brown except a paler poorly defined umbo, smooth except for a few indistinct ridges or linearly arranged verrucae; pedicel colorless, always broken near hilum.

Hosts and distribution: species of *Astragalus* and *Oxytropis:* northern New Mexico and Utah to Oregon and northward; circumboreal.

Type: on *Astragalus alpinus* L., Kvikkjokk in Lule Lappmark, Sweden, 1883, Lagerheim (S). Lagerheim cited more than one specimen but the text makes it clear that the above specimen was the basis of the description of telia, hence is the holotype. It is in the Stockholm Museum as "ex Herb. Lagerheim".

49. *UROMYCES PHACAE-FRIGIDAE* (Wahl.) Hariot, J. de Bot. 7: 376. 1893.
Aecidium phacae-frigidae Wahl. Fl. Lappon. 525. 1812.

Spermogonia, aecia and uredinia lacking. Telia systemic; spores as *U. lapponicus*, of which it doubtless is a microcyclic derivative.

Hosts and distribution: *Astragalus umbellatus* Bunge: Alaska; also in far northern Europe, in the U.S.S.R. and in northern Pakistan.

Type: on *Phaca frigida* L.(= *Astragalus frigidus* (L.) Gray), near Polmak in E. Finnmark, Norway (not seen).

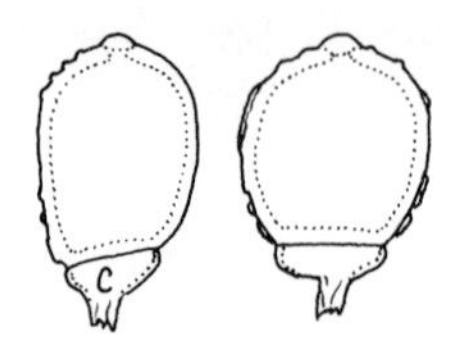

50. *UROMYCES ELEGANS* Lagerh. Tromsoe Mus. Aarsh. 17:34. 1895.

Spermogonia lacking. Aecia on abaxial leaf surface, scattered from a systemic mycelium, peridium short, whitish; spores (14-)17-19(-21) x (12-)14-18(-20) μm, globoid or more or less ellipsoid, wall 1(-1.5) μm thick, colorless, finely verrucose. Uredinia lacking; urediniospores rare with teliospores, wall echinulate, yellowish, pores 3 or 4, equatorial. Telia amphigenous and on stems, localized, exposed, pulverulent, dark cinnamon brown; spores (19-)21-24 (-26) x (15-)17-20(-21) μm, globoid, oblong ellipsoid or broadly ellipsoid, wall 1.5(-2) μm thick but with a small papilla over the pore, clear chestnut or golden brown, with discrete, rounded or flat warts, or these united in various degrees, or with ridges spaced 3-9 μm, usually with extensive smooth areas; pedicel colorless, 8-12 μm wide at hilum but collapsing abruptly below, usually broken short.

Hosts and distribution: *Trifolium carolinianum* Michx.: South Carolina to Arkansas and Texas.

Type: Aiken, South Carolina, Ravenel (S).

There is only one specimen in the Stockholm Museum as "ex Herb. Lagerheim" and I consider that it is the holotype. It consists of a single trifoliate leaf bearing telia. Date of collection is not recorded, nor was it when Lagerheim published the description of the telial stage. In fact, Lagerheim published the name as "*Uromyces elegans* (Berk. et Curt.) nob." adopting the epithet from *Aecidium elegans* Berk. & Curt.

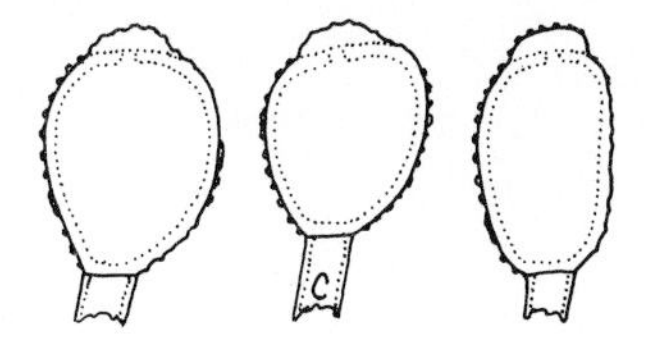

51. *UROMYCES COLORADENSIS* Ellis & Ever. Erythea 1:204. 1893.

Spermogonia amphigenous or mostly on abaxial leaf surface, distributed on a systemic mycelium. Aecia on abaxial surface, among the spermogonia, peridium short, whitish becoming yellowish, margin erose or dentate; spores (17-)21-26 (-28) x (15-)17-19(-20) µm, mostly broadly ellipsoid or globoid, wall 1-1.5 µm thick, colorless, finely verrucose. Uredinia lacking. Telia amphigenous but often mostly on one surface or the other, cinnamon brown or darker, mostly early exposed, pulverulent; spores variable, (20-)22-30(-38) x (-14)17-19(-21) µm, obovoid or oblong ellipsoid, wall 1.5 (-2) µm thick, thicker apically with a low, pale umbo, wall golden or pale chestnut brown, verrucose with scattered verrucae spaced about 1.5-2 µm or these in lines with same spacing or less commonly with complete or interrupted ridges; pedicels colorless, always broken near hilum.

Hosts and distribution: *Vicia* spp., especially *V. americana* Muhl.: Ontario to Alberta, New Mexico and California.

Type: on *Astragalus* or *Spiesia* (determined to be *Vicia truncata* Nutt. by Rydberg in 1908; this now treated as a synonym of *V. oregana* Nutt.), Ft. Collins, Colorado, 1893, Baker No. 118 (NY; isotype PUR 35379).

Arthur (2) separated the species into three doubtfully distinct varieties: *campester*, *montanus*, and *maritimus*.

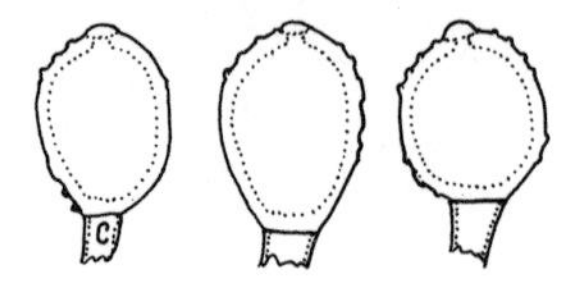

52. *UROMYCES MINOR* Schroet. in Cohn Krypt. Flora Schles. 3: 310. 1887.

Spermogonia lacking. Aecia mostly on abaxial leaf surface in groups or widely scattered, peridium usually short but may be cylindrical, margin erose; spores (14-)17-21(-23) x (13-)16-19(-21) µm, mostly globoid or broadly ellipsoid, wall 1-1.5 µm thick, finely verrucose. Uredinia lacking. Telia amphigenous or sometimes mostly on abaxial surface, tardily or often early exposed, pulverulent, blackish brown; spores (16-)18-22(-24) x (13-)15-19(-20) µm, broadly ellipsoid, obovoid or globoid, wall 1.5-2 µm thick, mostly clear chestnut brown, except a small pale, papilla over the pore, usually with a few longitudinal lines of verrucae or these fused to ridges, sometimes with cross ridges, occasionally with areas of randomly disposed verrucae, or rarely smooth; pedicels colorless, always short.

Hosts and distribution: *Trifolium* spp. mountainous regions of western Canada, the United States and Mexico; also in Europe, China and Japan.

Type: on *Trifolium montanum* L., Silesia, Germany (not seen).

The species doubtless is a derivative of *U. trifolii-repentis*.

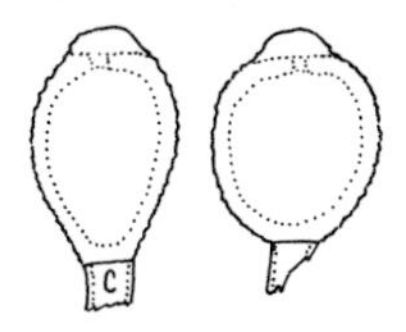

53. *UROMYCES HEDYSARI-OBSCURI* (DC.) Carc. & Picc. Erb. Critt. Ital. ed. 2, fasc. 9, No. 447. 1871.
Puccinia hedysari-obscuri DC. Syn. Pl. Flora Gall. Descr. 46. 1806.

Spermogonia mostly on adaxial leaf surface along veins and on petioles. Aecia in groups, mostly on abaxial surface and petioles, peridium whitish, cupulate or short cylindrical; spores (16-)18-21(-23) x (13-)15-18(-20) µm, globoid, or broadly ellipsoid, wall 1 µm thick, finely verrucose. Uredinia mostly on adaxial surface, with peridium like the aecia but without associated spermogonia, usually single and surrounded by telia; urediniospores like the aeciospores. Telia mostly on adaxial surface, exposed, pulverulent, blackish brown; spores (19-)21-27(-29) x (14-)16-18 µm, ellipsoid, broadly ellipsoid or obovoid; wall 1.5-2(-2.5) µm thick at sides, chestnut brown, 3-6(-7) µm at apex with an abrupt nearly colorless umbo, rugose or pseudoreticulate with variously anastomosed short or long ridges; pedicels colorless, always broken near the hilum.

Hosts and distribution: *Hedysarum* spp.: from Alaska to Utah and Colorado.

Type: on *Hedysarum obscurum* (= *H. hedysaroides* (L.) Sch. & Th.), France. Not seen.

This is one of the few species that produce aecidioid uredinia (repeating aecia). The true aecia are collected less often than the aecidioid uredinia and telia.

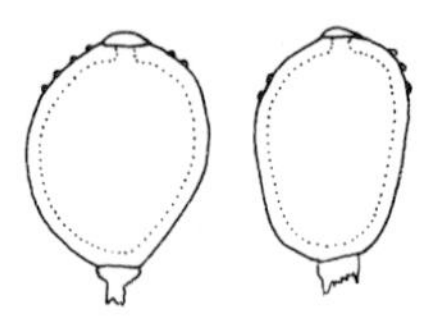

54. *UROMYCES TRIFOLII* (Hedw. ex DC.) Fckl. Symb. Mycol. p. 63: 1870.
    Puccinia trifolii Hedw. ex DC. Flore Fr. 2:225. 1805.
    Puccinia nerviphila Grog. Pl. Crypt. Saône-et-Loire p. 154. 1863.
    Uromyces flectens Lagerh. Sv. Bot. Tidskr. 3:36. 1909.
    Uromyces nerviphilus (Grog.) Hots. Publ. Puget Sound Biol. Sta. Univ. Wash. 4:368. 1925.

Spermogonia, aecia and uredinia lacking. Telia in groups along veins, midrib and petioles, often confluent, causing distortion and hypertrophy, exposed, pulverulent, chocolate brown; spores (18-)22-27(-30) x (15-)17-20(-22) µm, ellipsoid, obovoid or globoid, wall 1.5 µm thick, clear chestnut brown, with widely scattered or linearly arranged small verrucae, the pore with a low cap; pedicels colorless, broken near hilum.

Hosts and distribution: *Trifolium repens* L.: New York to Colorado and British Columbia; also in Europe, Asia, New Zealand and South America.

Lectotype: on *Trifolium repens*, Fontenai-aux-Roses, France (G-DC); lectotype designated here. The specimen is excellent and, together with the description by De Candolle ("...elle attaque les tiges, les pétioles, les nervures et les deux surfaces des feuilles; elle boursoufle, défigure, recroqueville souvent les organes sur lesquels elle croit et empeche le trefle de fleurir ..."), there is no reason to doubt the identity of the fungus. A detailed review of the nomenclature will be published soon by John Walker of the Biological and Chemical Research Institute, Rydalmere, Australia.

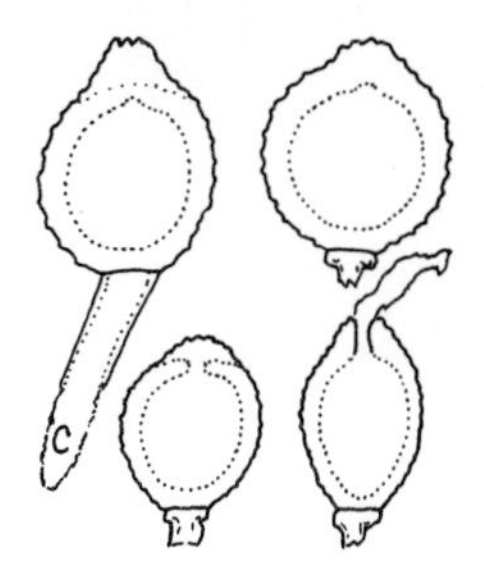

55. *UROMYCES BAUHINIICOLA* Arth. Bot. Gaz. 39:389. 1905.

Spermogonia few on adaxial leaf surface. Aecia and uredinia lacking. Telia amphigenous, exposed, the primary ones closely grouped, yellowish brown becoming gray from germination, secondary telia scattered, blackish brown, pulverulent; spores dimorphic, germinating form mostly 16-20 x 14-17 µm, wall golden or paler, indistinctly reticulate, resting form (18-)20-26(-30) x (15-)17-20(-22) µm, broadly ellipsoid, wall 2.5-3.5(-4) µm thick at sides, (4-)5-7(-8) µm at apex, chestnut brown except the apex progressively paler but not as a clearly defined umbo, irregularly reticulate, the meshes of various shapes but mostly 2-4 µm diam, tending longer than wide and merging apically to ridges with some cross connections; pedicels colorless, to 40 µm long, rugose basally but usually broken above the rugosity.

Hosts and distribution: *Bauhinia* spp.: south central Mexico.

Type: on *B. pringlei* S. Wats., Guadalajara, Jal., Holway No. 5060 (PUR 38650).

The primary telia and germinating teliospores have apparently been overlooked, perhaps because they are much less conspicuous than the resting telia, which must result from infection by the basidiospores produced by the primary teliospores.

EXCLUDED SPECIES

UROMYCES COLUTEAE Arth. Bull. Torrey Bot. Club 37:574. 1910.

This rust fungus undoubtedly was introduced with the host, *Colutea arborescens* L., and probably should be cited as *Uromyces laburni* (DC.) Otth. Because there is doubt that the fungus has persisted it is not described here.

3. UROPYXIS Schroeter

Hedwigia 14:165. 1875.

Spermogonia subcuticular, conical, type 7 (16). Aecia subepidermal in origin, erumpent, uredinoid; spores borne singly on pedicels. Uredinia subepidermal in origin, erumpent, with peripheral paraphyses; spores borne singly on pedicels. Telia subepidermal in origin, erumpent; spores borne singly on pedicels, 2 celled by horizontal septum, the wall usually conspicuously bilaminate with the outer layer pale or colorless and the inner wall pigmented, germ pores 2 in each cell; basidium external.

Type species: *Uropyxis amorphae* (Curt.) Schroet.

KEY TO SPECIES OF *UROPYXIS*

1. Teliospore wall conspicuously bilaminate, the outer hyaline layer mostly 6-10 μm thick .. *amorphae* (1)
1. Teliospore wall less conspicuously bilaminate, the outer layer less than 5 μm thick 2

    2. Teliospore pedicels hygroscopic, swelling 3
    2. Teliospore pedicels not hygroscopic, terete 6

3. Outer wall of teliospore smooth; species microcyclic .... *holwayi* (9)
3. Outer wall of teliospore verrucose echinulate 4

    4. Teliospores mostly 46-53 μm long; species microcyclic .... *diphysae* (7)
    4. Teliospores mostly less than 42 μm long 5

5. Teliospore pedicel swollen to 15-25 µm diam *daleae* (2)
5. Teliospore pedicel rugose and/or slightly swollen basally .... *nissoliae* (3)

6. Teliospore pedicel rugose and/or slightly swollen .... *nissoliae* (3)
6. Teliospore pedicel not swollen or rugose 7

7. Teliospores of 2 distinct size classes *heterospora* (6)
7. Teliospores of 1 size class only 8

8. Only spermogonia and telia produced .. *farlowii* (8)
8. Aecia and/or uredinia also produced 9

9. Teliospores mostly 36-42 x 23-27 µm *petalostemonis* (4)
9. Teliospores mostly 29-37 x 18-23 µm ..... *roseana* (5)

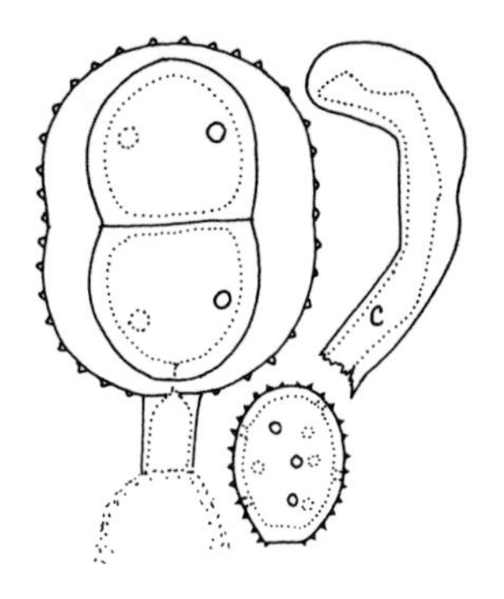

1. *UROPYXIS AMORPHAE* (M. A. Curt.) Schroet. Hedwigia 14: 165. 1875.
Puccinia amorphae M. A. Curt. Amer. J. Sci. II. 6:353. 1848.

Spermogonia mostly on adaxial leaf surface in small groups. Aecia mostly opposite the spermogonia, few in a group, yellowish brown, with peripheral, incurved, yellowish or colorless paraphyses, the inner wall 1-1.5 µm thick, the outer usually 2-3 µm; spores (19-)21-26(-28) x (14-)15-19 µm, obovoid or broadly ellipsoid, wall (1.5-)2-2.5 µm thick, yellowish, echinulate. Uredinia amphigenous or sometimes only on abaxial surface, yellowish brown, small, with peripheral, essentially colorless, strongly, often geniculately, incurved paraphyses, the inner wall 1-1.5 µm thick, the outer 4-7 µm thick; spores (16-)18-22(-24) x (12-)14-16(-18) µm, obovoid or ellipsoid, wall (1-)1.5(-2) µm thick, pale yellowish, finely echinulate, pores about 10, without caps, obscure. Telia amphigenous or often only on one surface, exposed, blackish brown, pulverulent, with paraphyses as in the uredinia; spores (35-)40-46(-50) x (28-)30-35(-38) µm, broadly ellipsoid, wall conspicuously bilaminate, the outer layer (4-)6-10(-12) µm thick but hygroscopic and commonly bursting, the outermost part thin, verrucose with small cones spaced 2-3 µm, separable from the inner hygroscopic material, inner wall 2-3 µm thick, chestnut brown, smooth; pedicels often swelling and disjoining near the hilum or sometimes terete throughout.

Hosts and distribution: *Amorpha* spp., *Parryella filifolia* Torr. & Gray: Manitoba to Florida, Arizona and California.

Type: on *Amorpha herbacea*, *"Car. Inf."* Ravenel (FH).

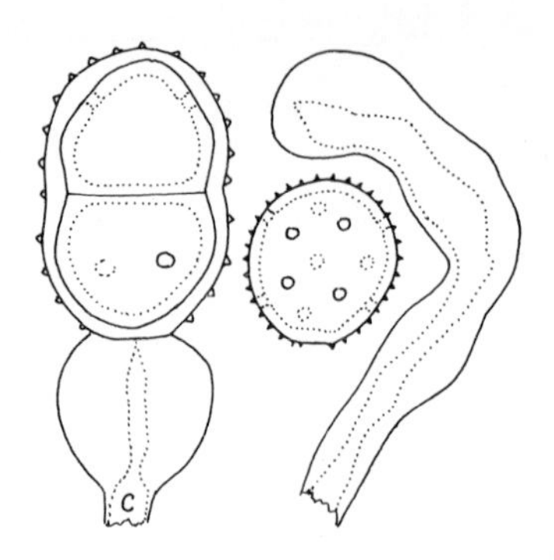

2a. *UROPYXIS DALEAE* (Diet. & Holw.) Magn. Ber. Dtsch. Bot. Ges. 17:115. 1899 var. *DALEAE*.
Puccinia daleae (Diet. & Holw.) in Holway, Bot. Gaz. 24: 27. 1897.

Spermogonia and aecia unknown. Uredinia on abaxial leaf surface or on small stem galls, yellowish brown, with colorless, mostly strongly incurved, peripheral paraphyses, the inner wall 1-1.5 µm thick, the outer wall 3-7 µm thick; spores (18-)20-24(-26) x (16-)18-20(-22) µm, broadly ellipsoid or broadly obovoid, wall (1-)1.5(-2) µm thick, pale cinnamon brown or golden, echinulate, pores 8-12, with no caps, obscure. Telia mostly on abaxial surface, rarely on stem galls, exposed, with paraphyses as in the uredinia, blackish brown, pulverulent; spores (33-)35-42(-46) x (23-)25-28(-30) µm, wall bilaminate, the outer layer nearly colorless, (1.5-)3-4(-5)µm thick, echinulate verrucose with small cones spaced (2-)3-4 µm, inner wall chestnut brown, (1.5-)2-2.5 µm thick, smooth; pedicel colorless, hygroscopic next to the spore and swelling to 15-25 µm diam, the lower portion thin walled and usually disjoining at the junction with the swollen part.

Hosts and distribution: *Dalea* spp.: western Texas and southern Arizona to Guatemala and El Salvador; also in South America.

Lectotype: on *Dalea mutabilis* Willd., near Tula, Mexico, 5 Oct. 1896, Holway (S; isotypes Barth. N. Amer. Ured. 1298). Arthur (1) and Baxter (4) cite this specimen as the type but two specimens were cited by Dietel and Holway, hence this lectotype designation.

The following variety has smaller urediniospores.

2b. *UROPYXIS DALEAE* var. *EYSENHARDTIAE* (Diet. & Holw.) J. W. Baxt. Mycologia 51:216. 1959.
Puccinia eysenhardtiae Diet. & Holw. in Holway, Bot. Gaz. 24:27. 1897.

Spermogonia in 2-3 mm diam slightly hypertrophied areas of leaves. Aecia mostly on abaxial leaf surface opposite the spermogonia, uredinoid, dark brown (teliospores common in aecia), with incurved peripheral paraphyses as in the uredinia; spores as the urediniospores. Uredinia and telia as in *U. daleae* var. *daleae* but the urediniospores 15-21 x 14-17 µm and with the wall 1 µm thick.

Hosts and distribution: *Dalea albiflora* Gray, *Eysenhardtia orthocarpa* (Gray) Wats., *E. polystachya* (Ort.) Sarg.: southern Arizona to south central Mexico.

Type: on *Eysenhardtia orthocarpa*, near Mexico City, Holway (S; isotype PUR 7217).

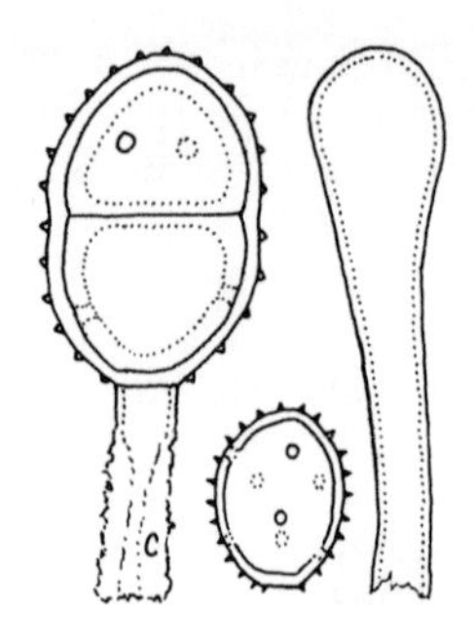

3. *UROPYXIS NISSOLIAE* (Diet. & Holw.) Magn. Ber. Dtsch. Bot. Ges. 17:115. 1899.
Puccinia nissoliae Diet. & Holw. in Holway, Bot. Gaz. 24: 27. 1897.

Spermogonia in small groups on adaxial leaf surface. Aecia on abaxial surface opposite the spermogonia, uredinoid, with colorless, clavate or capitate, uniformly thin walled paraphyses, yellowish brown; spores (14-)15-18(-20) x (12-) 14-16(-17) µm, broadly ellipsoid or globoid, wall 1-1.5 µm thick, pale yellowish, echinulate, pores 6-8, obscure. Uredinia and spores similar to the aecia except the sori scattered. Telia amphigenous, early exposed, blackish brown, pulverulent; spores (30-)33-38(-42) x (19-)21-24(26) µm, ellipsoid or oblong ellipsoid, wall bilaminate, the outer layer essentially colorless, 1.5-2(-3) µm thick, echinulate verrucose with small cones spaced (2-)2.5-4(-5) µm, the inner wall (2-) 2.5-3(-4) µm thick, chestnut brown, smooth; pedicels colorless, terete adjacent to the hilum and usually persisting, becoming rugose and slightly swollen basally and commonly dissolving, to 30-45 µm long.

Hosts and distribution: *Nissolia* spp.: Sonora, Mexico to Guatemala and El Salvador.

Type: on *N. confertifolia* Wats. (= *N. hirsuta* DC.), Guadalajara, Jal., Mexico, Holway (S; isotype PUR 7168).

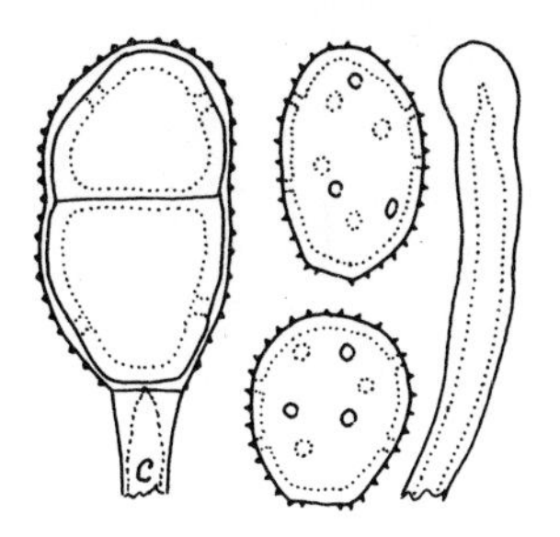

4. *UROPYXIS PETALOSTEMONIS* (Farl.) DeT. in Sacc. Syll. Fung. 7:735. 1888.
   Puccinia petalostemonis Farl. in Trelease, Trans. Wis. Acad. 6:129. 1884.
   Uropyxis affinis Arth. Manual Rusts U. S. & Canada p. 76. 1934.

Spermogonia on abaxial leaf surface, sometimes on stems, apparently from a systemic mycelium. Aecia scattered among the spermogonia, uredinoid, with colorless, thick walled, cylindrical paraphyses; spores (23-)26-32(-38) x (15-) 17-20 (-22) µm, mostly obovoid or ellipsoid, wall 1.5-2(-2.5) µm thick, yellowish, echinulate, pores 8-10, obscure. Uredinia similar to aecia but localized; spores as the aeciospores but in the shorter range. Telia on abaxial surface, sometimes with the spermogonia and aecia, exposed, chocolate brown or blackish, pulverulent; spores (33-)36-42(-45) x (20-)23-27 (-30) µm, ellipsoid or oblong ellipsoid, wall bilaminate, the outer layer colorless, 1-2 µm or to 3 µm over pores, finely echinulate verrucose with cones spaced 2-3 µm, inner wall (1.5-)2-2.5(-3) µm, chestnut brown, smooth; pedicels colorless, wall thick near hilum, thin below, not swelling, disjoining at junction of the two parts.

Hosts and distribution: *Petalostemon* spp.: Wisconsin to Saskatchewan south to Arizona and New Mexico.

Type: on *Petalostemon* sp.: La Crosse, Wis., Pammell (FH).

Baxter (4) considers that this fungus has an unstable life cycle, varying from microcyclic to macrocyclic.

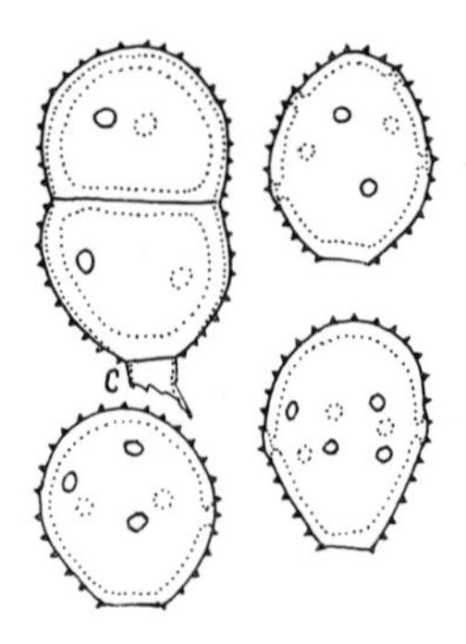

5. *UROPYXIS ROSEANA* Arth. N. Amer. Flora 7:157. 1907.

Spermogonia in small groups, amphigenous. Aecia amphigenous around the spermogonia, uredinoid, yellowish brown; spores (25-)28-33(-38) x (19-)20-23(-26) μm (orig. descr.: 23-27 x 17-22 μm), broadly ellipsoid or mostly obovoid, wall uniformly 1.5(-2) μm thick or slightly thicker apically, pale yellowish or the apex brownish, echinulate, pores (6)7-10 (orig.: 12 or more), scattered in shorter spores, mostly in the equatorial region in elongate spores, with slight or no caps. Telia not seen; teliospores 29-37 x 18-23 μm (from orig. descr.), ellipsoid, wall inconspicuously bilaminate, the inner layer 1.5-2.5 μm thick, about cinnamon brown, outer layer 0.5-1 μm thick, colorless, echinulate with small, fine echinulae spaced about 2-3 μm; pedicels colorless, thin walled, broken at the hilum.

Type: on *Cracca talpa* (Wats.) Rose (= *Tephrosia talpa* Wats., Hacienda Ciervo, Querétaro, Mexico, Rose No. 9694 (PUR 7200). Not otherwise known.

Baxter (4) treated this as a doubtful species because "No teliospores could be found in the scanty type material of this species." I found one spore, which is illustrated. The type packet has a slip with measurments of five teliospores, but the type is useless as regards teliospores.

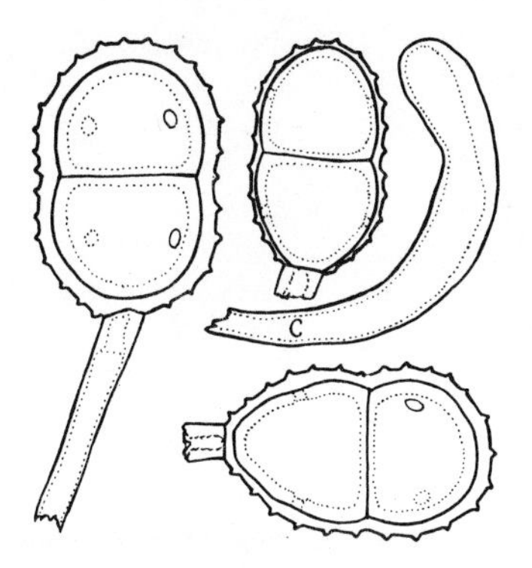

6. *UROPYXIS HETEROSPORA* Hennen & Cumm. Rept. Tottori Mycol. Inst. 10:182. 1973.

Spermogonia, aecia and uredinia unknown. Telia amphigenous, exposed, cinnamon brown or blackish brown, with peripheral, incurved, colorless paraphyses, the inner wall 1 μm thick, the outer wall 2-3.5 μm thick; spores variable, more or less of two classes, the small ones (22-)24-30 x (15-)17-20 μm, wall scarcely bilaminate, the outer part 0.5 μm thick or none, the inner part 1 μm thick, cinnamon brown or light chestnut brown, the larger spores (32-)34-40(-44) x (24-)25-28(-30) μm, broadly ellipsoid, wall conspicuously bilaminate, the outer colorless layer 2-3(-4) μm thick, verrucose echinulate with cones about 1 μm high spaced 2-3(-4) μm, the inner wall (1.5-)2-2.5(-3) μm thick, smooth, chestnut brown; pedicels colorless, to 60 μm long but usually broken near the hilum.

Hosts and distribution: *Apoplanesia paniculata* Presl: southern Jalisco and Michoacán, Mexico.

Type: north of Barra de Navidad, Jal., Cummins No. 71-361 (PUR 64753).

It is probable that the small pale spores germinate without dormancy but that the larger darker spores are resting spores.

7. *UROPYXIS DIPHYSAE* (Arth.) Cumm. Bull. Torrey Bot. Club 70:81. 1943.
Calliospora diphysae Arth. Bot. Gaz. 39:391 1905.

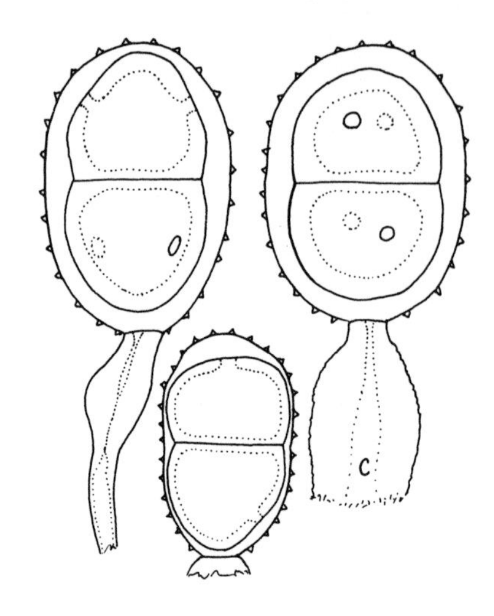

Spermogonia amphigenous in small groups. Aecia and uredinia lacking. Telia amphigenous and on petioles, closely associated with spermogonia, exposed, blackish brown, pulverulent; spores usually (42-)46-53(-56) x (29-)32-35(-37) µm, broadly ellipsoid, wall conspicuously bilaminate, the inner layer 3-4 µm thick, chestnut brown, the outer layer 3-4(-5) µm thick, yellowish, verrucose echinulate with broad based cones spaced 3-5 µm, pores 2 in each cell, equatorial; some collections have part or all spores (33-)36-44(-48) x (22-)24-27(-29) µm, wall bilaminate, the inner layer 1.5-2(-2.5) µm thick, cinnamon brown or golden, the outer wall (0.5-)1-1.5(-2) µm thick at sides, 3-4(-5) µm over the pores, pale yellowish, verrucose echinulate with cones spaced (2-)3-4(-5) µm, pore 1 in each cell, at or near the apex of the upper cell, midway or below in the lower cell; pedicels of 2 parts, the upper part swollen and usually bursting basally, the lower part terete, breaking away.

Hosts and distribution: *Diphysa robinioides* Benth. and *D. suberosa* Wats.: southern Mexico to Costa Rica.

Type: on *D. suberosa* Rio Blanco, Guadalajara, Jal., Mexico, Holway No. 5082 (PUR 7338; probable isotypes Barth. N. Amer. Ured. No. 208).

It is impossible to decide but probably the two kinds of spores belong to a single species. But when only the smaller, 1 pored spores are present the fungus has the characteristics of the genus *Prospodium*.

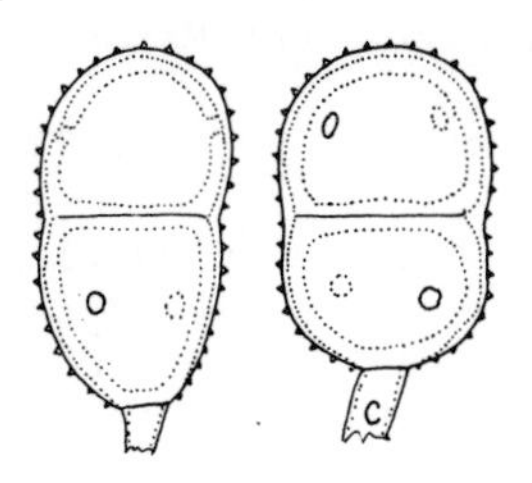

8. *UROPYXIS FARLOWII* (Arth.) J. W. Baxt. Mycologia 51:222. 1959.
Calliospora farlowii Arth. Bot. Gaz. 39:391 1905.

Spermogonia caulicolous, numerous. Aecia and uredinia lacking. Telia caulicolous on slightly enlarged areas, exposed, becoming more or less confluent, possibly causing small fasciations, dark cinnamon brown, pulverulent; spores (26-) 30-38(-42) x (18-)20-24 µm, mostly ellipsoid, wall rather inconspicuously bilaminate, the inner wall (1.5-)2(-2.5) µm thick, cinnamon brown, the outer wall 0.5-1.5(-2) µm thick, colorless, echinulate with cones spaced 1.5-3 µm; pedicels colorless, to 35 µm long but usually broken near the hilum.

Type: on *Dalea domingensis* DC., Orizaba, Mexico, Feb. 1885, Farlow (PUR 7353). Otherwise known from Cuba.

9. *UROPYXIS HOLWAYI* (Arth.) Arth. Manual Rusts U. S. and Canada. p. 77. 1934.
Calliospora holwayi Arth. Bot. Gaz. 39:390 1905.

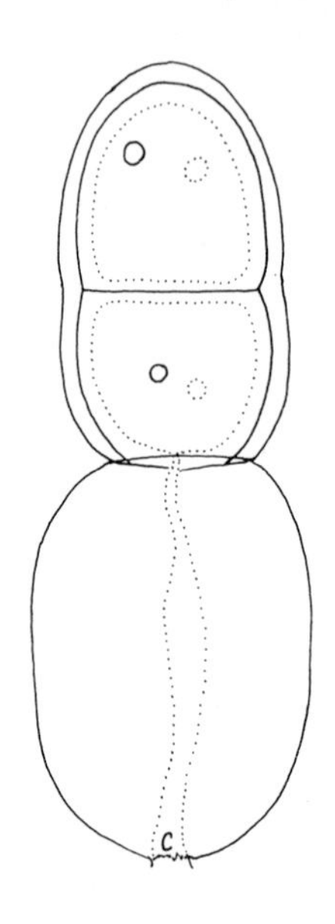

Spermogonia on adaxial surface of leaves in small or large groups. Aecia and uredinia lacking. Telia mostly on adaxial surface in confluent groups around the spermogonia, early exposed usually blackish brown but some collections also have cinnamon brown sori, pulverulent, without paraphyses; spores mostly (40-)46-56(-60) x (25-)28-34(-36) µm, ellipsoid or oblong ellipsoid, wall bilaminate, the outer wall 2-4 µm thick, essentially colorless, smooth, the inner wall 2-3 µm thick, chestnut brown, smooth, spores from cinnamon brown sori are paler and have thinner walls, the colorless layer often obvious only over the pores; pedicels colorless, the upper part hygroscopic, swelling to 30-60 µm wide, 40-60 µm long, the lower part thin walled, terete, usually disjoining at the junction with the upper part.

Hosts and distribution: *Diphysa floribunda* Peyr. and species of *Eysenhardtia*: southern Arizona to Guatemala.

Type: on *Eysenhardtia orthocarpa* (Gray) Wats., Etla, Oax., Mexico, Holway No. 5405 (PUR 7349).

4. PHRAGMOPYXIS Dietel in Engler & Prantl
Nat. Pflanzenfam. 1(1**):70. 1897.

Spermogonia subcuticular, conical, type 7 (16). Aecia subepidermal in origin, with or without paraphyses, caeomoid with catenulate spores, or uredinoid with spores borne singly on pedicels. Telia subepidermal in origin, erumpent; spores borne singly on pedicels which mostly are hygroscopic, mostly 3 celled by horizontal septa, germ pores 3 or 4 in each cell, the wall bilaminate, the outer part pale or colorless, the inner wall pigmented; basidium external

Type species: *Phragmopyxis deglubens* (Ber. & Curt.) Diet.

1. *PHRAGMOPYXIS DEGLUBENS* (Berk. & Curt.) Diet. in Engler & Prantl Nat. Pflanzenfam. 1(1**):70. 1897.
Triphragmium deglubens Berk. & Curt. Grevillea 3:55. 1874.

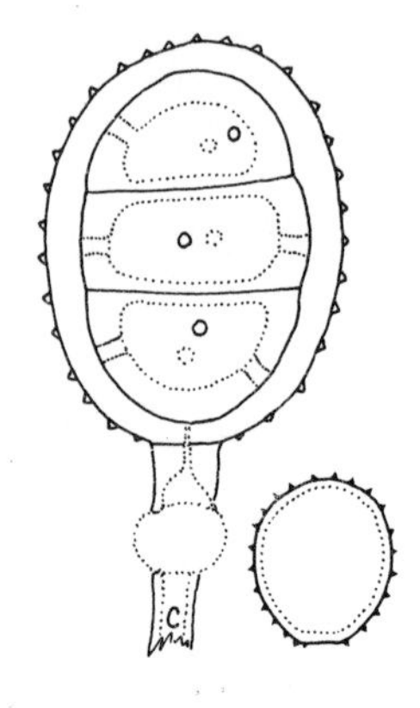

Spermogonia in small groups on adaxial leaf surface. Aecia grouped around the spermogonia, mostly on the abaxial surface, surrounded by upturned epidermis, yellowish brown, without peridium or paraphyses; spores catenulate, with intercallary cells, (15-)18-21(-23) x (13-) 15-18(-20) µm, mostly ellipsoid to globoid, wall (1.5-)2-2.5(-3) µm thick, finely verrucose rugose with small verrucae that usually merge in various patterns, pale yellow or colorless, pores several, scattered, obscure. Uredinia hypophyllous and on stems, yellowish brown; spores (17-)19-22(-24) x (15-)17-18(-20) µm, mostly obovoid or broadly ellipsoid, wall 1-1.5 µm thick, yellowish, echinulate, pores probably scattered, obscure. Telia amphigenous and on stems, early exposed, blackish brown, pulverulent; spores (42-)45-55(-60) x (32-)35-40(-44) µm, (2)3(4) celled, oblong ellipsoid or broadly ellipsoid, wall conspicuously bilaminate, the outer wall separable, yellowish or nearly colorless, 2-5 µm thick but hygroscopic and often swelling to more in mounts, echinulate verrucose with cones spaced 2-4 µm, inner wall uniformly (2.5-)3-4(-5) µm, chestnut brown, smooth, pores 3-5, commonly 4, about midway in each cell; pedicel of 2 parts, a short, thick walled, brownish part next to the hilum and a long, colorless, basal section, the parts separating at the junction, with or withslight swelling.

Hosts and distribution: *Benthamantha edwardsii* (Gray) Rose: southern Arizona, northern Sonora and Guatemala; also in Ecuador.

Type: on a leguminous plant, Texas (error for northern Sonora) (K).

Paraphyses have been ascribed to this species but I have been unable to verify this. There are compacted, colorless, pusher hyphae around the periphery of the aecia beneath the epidermis.

2. *PHRAGMOPYXIS NOELII* J. W. Baxt. Mycologia 56:287. 1964.

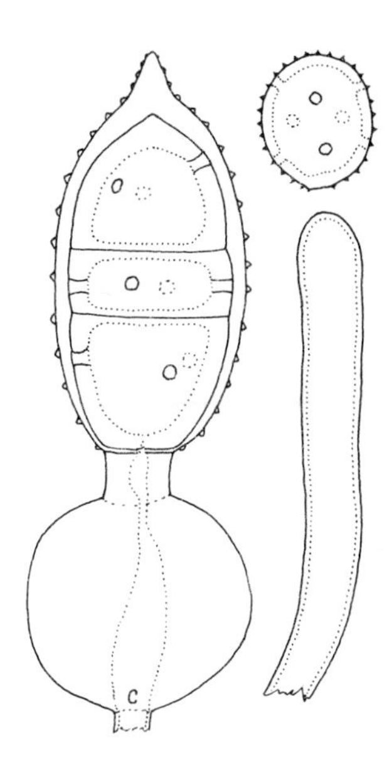

Spermogonia on both leaf surfaces. Aecia amphigenous, grouped around the spermogonia, uredinoid, pale brownish, with peripheral, thin walled, cylindrical paraphyses, 30-90 x 8-10 µm; spores (15-)18-22(-25) x (13-)15-18(-20) µm, obovoid or broadly ellipsoid, wall (1.5-)2-2.5 µm thick, pale golden brown, echinulate, pores scattered, 6-8, without caps, obscure. Uredinia similar to the aecia but not associated with spermogonia. Telia amphigenous, blackish brown, exposed, pulverulent; spores (2)3(4) celled, (50-)56-75(-95) x (30-)33-40 (-45) µm, mostly ellipsoid with the apex acuminately apiculate, wall conspicuously bilaminate, the inner wall wall uniformly (2.5-)3-4(-5) µm thick, chestnut brown, the outer wall separable, essentially colorless, 2-4(-6) µm thick at sides, prolonged apically to form an apiculus (sometimes) lacking) which extends beyond the inner spore 15-25 µm but occasionally more, verrucose with discrete, papillate verrucae spaced 3-6(-8) µm, pores 3 or 4 in each cell; pedicel of 3 parts, the upper part brownish and remaining terete, the central part colorless, hygroscopic and swelling to 30-45 µm and often bursting, and the basal part colorless, terete and usually breaking at the junction with the swollen part.

Hosts and distribution: *Coursetia glandulosa* Gray: southernmost Arizona south to Guadalajara, Mexico.

Type: Guaymas, Son., Mexico, Cummins No. 62-55 (PUR 59938).

This and *P. acuminata* differ in life cycle but have indistinguishable teliospores. *P. noelii* is common in Sonora at low elevations.

3. *PHRAGMOPYXIS ACUMINATA* (Long) P. Syd. & H. Syd. Monogr. Ured. 3:162. 1915.
Tricella acuminata Long, Mycologia 4:282. 1912.

Spermogonia in groups on both leaf surfaces. Aecia and uredinia lacking. Telia amphigenous, exposed, confluent around the spermogonia, pulverulent, blackish brown; spores indistinguishable from those of *P. noelii*.

Hosts and distribution: *Coursetia glandulosa* Gray; known only from Pima County, Arizona but doubtless occurs in northern Mexico.

Type: Sabino Canyon, Santa Catalina Mts., Arizona, Long & Hedgcock (BPI; isotype PUR 7058).

The species undoubtedly is a microcyclic derivative of *P. noelii*.

5. PILEOLARIA Castagne

Obs. Pl. Acotyl. Fam. Ured. 1:22. 1842.

Spermogonia subcuticular, conical, type 7 (16). Aecia subepidermal in origin, erumpent, uredinoid; spores borne singly on pedicels, similar to the urediniospores. Uredinia subepidermal in origin, spores borne singly on pedicels, the surface sculpture various but often in spiral or longitudinal patterns. Telia subepidermal in origin, erumpent; spores borne singly on pedicels, 1 celled, with 1 germ pore, wall pigmented; basidium external.

Type species: *Pileolaria terebinthi* Cast.

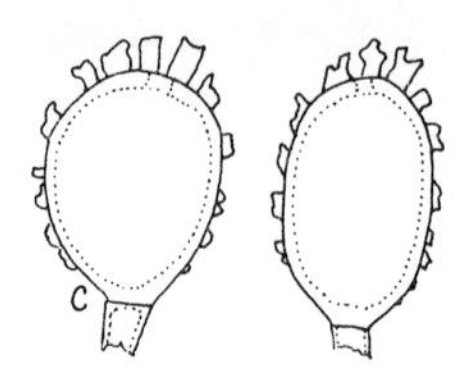

1. *PILEOLARIA INCRUSTANS* (Arth. & Cumm.) Thir. & Kern, Bull. Torrey Bot. Club 82:105. 1955.
Atelocauda incrustans Arth. & Cumm. Ann. Mycol. 31:41. 1955.

Spermogonia amphigenous in small groups. Aecia and uredinia lacking. Telia amphigenous around the spermogonia, mostly discrete, early exposed, chestnut brown, more or less pulverulent; spores (20-)22-28(-32) x (15-)17-20(-22) µm, mostly broadly ellipsoid or obovoid, wall uniformly 1-2 µm thick, dark cinnamon brown, with numerous cubical or apically divided processes, these larger and more numerous at the apex of the spore, often arranged in lines basally; pedicels colorless, fragile, seldom seen attached.

Type: on *Lonchocarpus* sp., Chinguinola, Dept. Bocas del Toro, Panama, Johnston (PUR 44631). One other collection from the same area is known.

This is one of the few species of *Pileolaria* that inhabit the Leguminosae and, although it has the principal features of *Pileolaria*, the surface sculpture is similar to that of species of *Dicheirinia*. The lack of an apical cell on the pedicel prevents placing the species in *Dicheirinia*. The genus *Atelocauda*, of which *A. incrustans* is the only species, is a synonym of *Pileolaria* under this treatment.

6. RAVENELIA Berkeley

Gard. Chron. 1853:132. 1853.

Spermogonia subcuticular, conical, type 7 (16), rarely type 5. Aecia mostly subepidermal but may be subcuticular, typically uredinoid with spores borne singly on pedicels, but sometimes aecidioid with catenulate spores. Uredinia mostly subepidermal but may be subcuticular, erumpent, often with paraphyses; spores borne singly on pedicles. Telia subepidermal or subcuticular in origin, erumpent; spores strongly adherent in pedicellate discs, spores 1 celled or less often 2 celled, with 1 germ pore in each cell, spore heads subtended by colorless, hygroscopic cysts, pedicel composed of several hyphae, i.e., fascicled, spore wall pigmented; basidium external.

Type species: *Ravenelia glandulosa* Berk. & Curt.

KEY TO SPECIES OF *RAVENELIA*

A. On the subfamily Mimosoideae

Teliospore heads smooth	Section I
Teliospore heads with surface sculpture	Section II

SECTION I

Uredinia lacking paraphyses or uredinia lacking	Section IA
Uredinia with paraphyses	Section IB

Section IA

1. Species microcyclic; spermogonia and telia only	2
1. Species with uredinia and telia	3

2. Infections localized on leaflets ... *versatilis* (1)
2. Infections systemic in witches' brooms *scopulata* (9)

3. Urediniospores striately sculptured ... *striatispora* (2)
3. Urediniospores echinulate or verrucose echinulate .... *entadae* (3)

Section IB

1. Urediniospore pores scattered 2
1. Urediniospore pores zonate 4

2. Paraphysis wall uniformly thin .... *texensis* var. *morongiae* (4b)
2. Paraphysis wall more or less thickened throughout 3

3. Urediniospore mostly 16-18 x 14-16 µm .... *texensis* var. *texensis* (4a)
3. Urediniospores mostly 18-21 x 14-16 µm .... *fragrans* var. *evernia* (31b)

4. Urediniospore pores bizonate 5
4. Urediniospore pores unizonate 9

5. Urediniospores densely verrucose echinulate .... *spegazziniana* (5)
5. Urediniospores with spaced echinulae 6

6. Uredinia and telia subepidermal .... *aurea* (6)
6. Uredinia and telia subcuticular 7

7. Cysts of teliospore head multiseriate *thornberiana* (7)
7. Cysts of teliospore heads uniseriate 8

8. Telia mostly on leaflets; uredinoid aecia on brooms .... *pringlei* (8)
8. Telia and uredinioid aecia on witches' brooms .... *scopulata* (9)

9. Central cells of teliospore head mostly less than 18 µm across 10
9. Central cells of teliospore head mostly more than 18 µm 12

10. Paraphyses mostly incurved .... *australis* (10)
10. Paraphyses straight, clavate or capitate 11

11. Urediniospores mostly 32-45 µm long .... *holwayi* (11)
11. Urediniospores mostly 27-33 µm long ..... *lysilomae* (12)

12. Urediniospores lemon shape, often apiculate .... *annulata* (13)
12. Urediniospores ellipsoid or obovoid 13

13. Paraphyses mostly incurved, dorsal wall thick .... *subtortuosae* (14)
13. Paraphyses straight, with cork-like apex .... *bifenestrata* (15)

SECTION II

Uredinia without paraphyses .... Section IIA
Uredinia with paraphyses .... Section IIB

Section IIA

1. Urediniospore pores scattered 2
1. Urediniospores pores zonate 4

2. Teliospore heads with low tubercles; urediniospores narrowly ellipsoid .... *cumminsii* (32)
2. Teliospore heads with conical spines; urediniospores broadly ellipsoid 3

3. Cells of teliospore heads with 3 or more spines each .... *echinata* var. *ectypa* (16)
3. Cells of teliospore heads with 1 spine each .... *bajacalensis* (17)

4. Urediniospore pores bizonate .... *cumminsii* (32)
4. Urediniospore pores unizonate 5

5. Urediniospores mostly less than 25 µm long 6
5. Urediniospores mostly more than 27 µm long 7

6. Urediniospores apiculate at apex ...... *distans* (18)
6. Urediniospores rounded at apex .... *alamosensis* (19)

7. Central cells of teliospore heads mostly 12-16 µm across .... *pithecolobii* (20)
7. Central cells of teliospore heads mostly 20 µm or more 8

8. Uredinia and telia subepidermal 9
8. Uredinia and telia subcuticular 10

9. Urediniospore wall 2.5-3.5 μm at sides, 5-7 μm at apex .... *floridana* (21)
9. Urediniospore wall 1-1.5 μm at sides, 2-3 μm at apex .... *multispinosa* (22)

10. Each cell of teliospore head with 6-20 spines .... *linda* (23)
10. Each cell of teliospore head with (0-)3-8(-10) spines 11

11. Teliospore heads often elliptical, 70-100 x 62-77 μm .... *havanensis* (24)
11. Teliospore heads isodiametric, 55-60 μm diam .... *siderocarpi* (25)

Section IIB

1. Urediniospore pores scattered 2
1. Urediniospore pores zonate 7

2. Teliospore heads with conical spines 3
2. Teliospore heads with rounded tubercles 4

3. Urediniospores with spaced echinulation *mainsiana* (26)
3. Urediniospores densely verrucose echinulate .... *mimosae-sensitivae* (27)

4. Urediniospores with spaced echinulation .... *verrucosa* (28)
4. Urediniospores densely verrucose echinulate 5

5. Each cell of teliospore head with 7-11 conspicuous tubercles .... *verrucata* (29)
5. Each cell of teliospore head with (0)2-7 bead-like tubercles 6

6. Paraphysis wall 5-8 μm apically, thin basally .... *expansa* (30)
6. Paraphysis wall more or less uniformly thick .... *fragrans* var. *fragrans* (31a)

7. Urediniospore pores bizonate 8
7. Urediniospore pores unizonate 11

8. Urediniospores pale yellow or colorless; paraphyses cylindrical or lacking .... *cumminsii* (32)
8. Urediniospores at least golden brown apically; paraphyses clavate or capitate 9

9. Urediniospores with sharp echinulae 10
9. Urediniospores echinulate verrucose with blunt cones .... *acaciae-pennatulae* (33)

10. Urediniospores mostly 22-27 x 16-19 µm .... *mexicana* (34)
10. Urediniospores mostly 30-38 x 11-16 µm .... *roemerianae* (35)

11. Each cell of teliospore head with forked processes .... *stevensii* (36)
11. Cells of teliospore head with other ornamentation .. 12

12. Each cell of teliospore head with 1 spine-like tubercle .... *arizonica* (37)
12. Each cell with more than 1 spine or tubercle ... 13

13. Urediniospore pores subequatorial, spores mostly 38-50 µm long .... *leucaenae* (38)
13. Urediniospore pores equatorial, spores mostly 35-42 µm long or less 14

14. Urediniospore pores 5 or 6, spores mostly 35-42 µm long .... *gracilis* (39)
14. Urediniospore pores 4 or 5, spores mostly less than 35 µm 15

15. Sori subepidermal; urediniospores 28-35 x 15-18 µm .... *hermosa* (40)
15. Sori subcuticular; urediniospores 27-33 x 19-20 µm .... *sololensis* (41)

B. On the subfamily Caesalpinioideae

1. Uredinia and telia subepidermal in origin 2
1. Uredinia and telia subcuticular in origin 3

2. Marginal cells of teliospore head with 1 tubercle each, cysts multiseriate ..... *bella* (42)
2. Marginal cells with 3-7 tubercles each, cysts uniseriate .... *antiguana* (43)

3. Uredinia without paraphyses; teliospore pedicel long, wide and conspicuous .... *cassiaecola* (44)
3. Uredinia with paraphyses; teliospore pedicels much less conspicuous 4

4. Urediniospores spirally striated ..... *corbula* (45)
4. Urediniospores echinulate or verrucose echinulate 5

5. Teliospore heads with rare or no papillae .... *mesillana* (46)
5. Teliospore heads with numerous papillae, tubercles or spines 6

6. Teliospore heads with 1 tubercle or spine per cell *spinulosa* (47)
6. Teliospore heads with 2-7 tubercles or bead-like warts per cell 7

7. Central cells of teliospore head mostly 13-18 μm across, with bead-like warts .... *humphreyana* (48)
7. Central cells of teliospore head mostly 18-22 μm across, with tubercles 4-7 μm long .... *inconspicua* (49)

C. On the subfamily Papilionoideae

1. Teliospore heads smooth 2
1. Teliospore heads with warts or tubercles 7

2. Only spermogonia and telia produced, microcyclic 3
2. Uredinia and telia produced, macrocyclic 4

3. Central cells of teliospore head 22-30 μm across, cysts multiseriate .... *opaca* (50)
3. Central cells of teliospore head 11-15 μm across, cysts uniseriate .... *lonchocarpicola* var. *mera* (51)

4. Uredinia with paraphyses 18-28 μm wide *laevis* (52)
4. Uredinia without paraphyses 5

5. Urediniospore pores mostly 6 and mostly equatorial .... *epiphylla* (53)
5. Urediniospore pores mostly 8 or more and scattered 6

6. Urediniospores mostly 23-27 x 19-23 μm, wall 2.5-3 μm thick .... *similis* (54)
6. Urediniospores mostly 19-23 x 14-16 μm, wall 1.5 μm thick .... *caulicola* (55)

7. Uredinia without paraphyses 8
7. Uredinia with paraphyses 11

8. Marginal cells of teliospore head with 1 large tubercle, other cells with warts .... *brogniartiae* (56)

8. Marginal cells of teliospore head without such a tubercle 9

9. Each cell of teliospore head with 3-15 warts 1.5-2 µm long .... *talpa* (57)

9. Each cell of teliospore head with tubercles 3 µm or longer 10

10. Cells of teliospore head with 5-12 tubercles each; urediniospore pores 3-5, approximately equatorial .... *rubra* (58)

10. Cells of teliospore head with 3-5 tubercles each; urediniospore pores about 8, scattered .... *irregularis* (59)

11. Teliospore head with low, inconspicuous warts .... *piscidiae* (60)

11. Teliospore head with tubercles 3 µm or longer .... *indigoferae* (61)

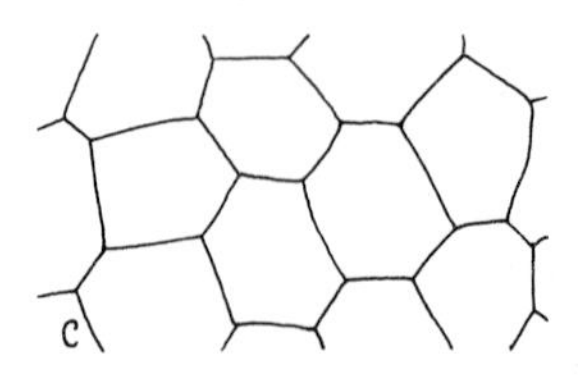

1. *RAVENELIA VERSATILIS* Diet. Hedwigia 33:64. 1894 (15 Apr.).
   Ravenelia farlowiana Diet. Hedwigia 33:369. 1894 (20 Dec.).
   Ravenelia acaciae-micranthae Diet. Bot. Centrlb. 20:371. 1906.

Spermogonia amphigenous, numerous in a close group, conspicuous, subcuticular. Aecia and uredinia lacking. Telia mostly on adaxial surface of leaflets, grouped, often circinately, about the spermogonia, blackish brown, subcuticular; spore heads (60-)70-100(-105) µm diam, light chestnut brown, (4)5-8 cells across, smooth or occasionally with a few inconspicuous, bead-like warts on the peripheral cells, central cells (13-)15-19(-22) µm across, cysts of same number as marginal cells, appressed but becoming semipendent and bursting.

Hosts and distribution: *Acacia* spp., Coahuila and Tamaulipas to San Luis Potosí, Mexico.

Lectotype: on *Acacia anisophylla* Wats., Jimulco, Coahuila, May 1885, Pringle (S; isotypes Reliq. Farl. No. 773).

The description of the telia and teliospores of *R. versatilis* obviously was drawn from *A. anisophylla* and not from *A. greggii* because Pringle's specimen on *A. greggii* is the broom forming uredinioid aecial stage of *R. pringlei* Cumm. (8).

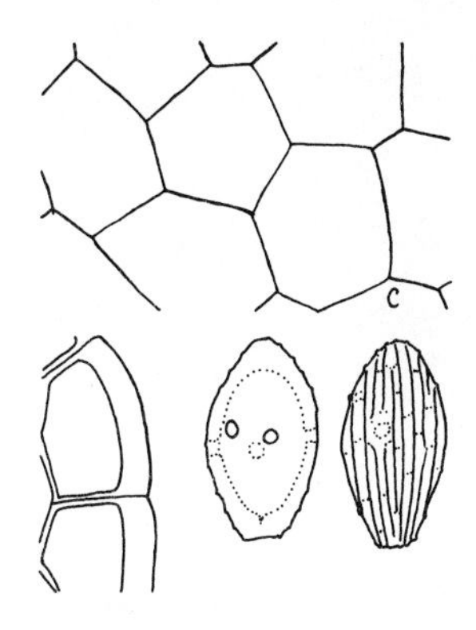

2. *RAVENELIA STRIATISPORA* Cumm. & J. W. Baxt. Mycotaxon 2: 202. 1975.

Spermogonia and aecia unknown. Uredinia amphigenous, subepidermal, cinnamon brown; spores (19-)21-25(-27) x (12-) 13-15(-16) µm, mostly ellipsoid, wall at base and apex (2.5-)3-3.5(-4) µm thick, (1.5-)2(-2.5) µm at sides, longitudinally striate, the striae spaced (1.5-)2(-2.5) µm and usually with some fine cross connections, pores (4)5 or 6, equatorial. Telia not seen; spore heads in the uredinia 80-110 µm diam, chestnut brown, smooth, 5 or 6 cells across, central cells (16-)18-22 µm across, cysts of same number as marginal cells, appressed.

Type: on *Pithecellobium mexicanum* Rose, Comanito, Sin., Mexico, 15 Mar. 1940, Gentry No. 5927 ex ARIZ 66892 (PUR 64929).

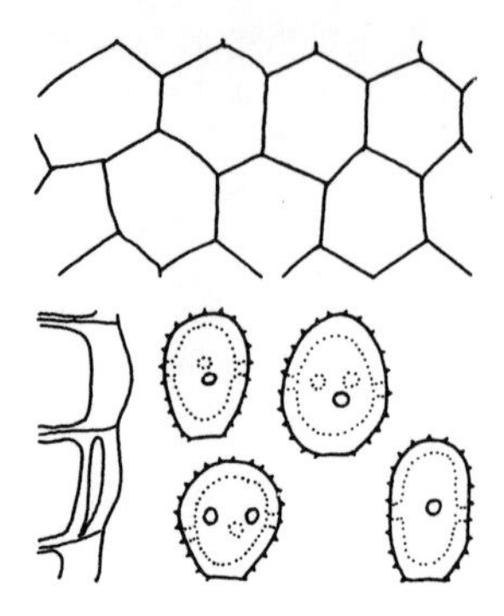

3. *RAVENELIA ENTADAE* Lager. & Diet. in Dietel, Hedwigia 33:62. 1894.

Spermogonia and aecia unknown. Uredinia mostly on adaxial surface of leaflets, subepidermal, about cinnamon brown; spores 14-19 x 10-14(-16) µm, mostly broadly ellipsoid or obovoid, wall (1.5-)2-3 µm thick, golden brown, echinulate, pores (3)4 or 5(6), equatorial, without caps. Telia mostly on adaxial surface, commonly in circles, blackish brown; spore heads (80-)90-120(-130) µm diam, smooth, 7-10(-12) cells across, central cells 10-16 µm diam, clear chestnut brown, cysts pendent.

Hosts and distribution: *Entada polystachia* (L.) DC.: Sinaloa, Mexico south to Guatemala and Panama.

Type: Panama, Oct. 1889, Lagerheim (S; isotype PUR 6119).

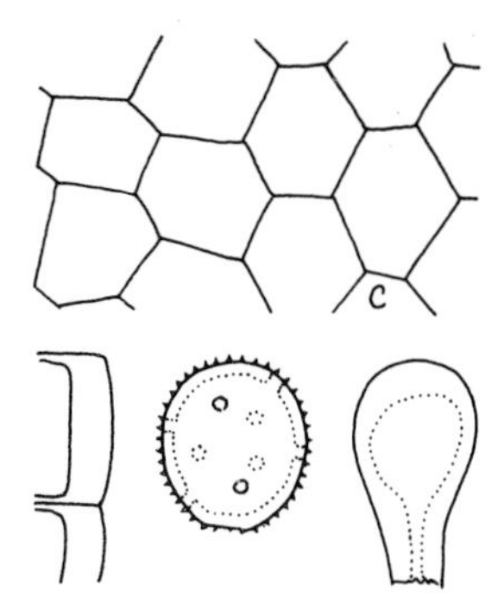

4a. *RAVENELIA TEXENSIS* Diet. Hedwigia 33:63. 1894 var. *TEXENSIS*.
Ravenelia texanus Ellis & Gall. in Jennings, Texas Agr. Exp. Sta. Bull. 9:28. 1890, *nom. nudum*.
Ravenelia reticulata Long, Bot. Gaz. 61:421. 1916.
Ravenelia gooddingii Long, Bot. Gaz. 72:41-42. 1921.

Spermogonia and aecia unknown. Uredinia amphigenous and on stems, subcuticular, yellowish brown, with abundant, mostly clavate or spatulate paraphyses to 15 µm wide, the stipe colorless and solid or nearly so, the head yellowish to golden brown with side wall 2.5-3.5 µm thick and apical wall to 8 µm thick; spores (14-)16-18(-20) x (13-)14-16(-18) µm, broadly ellipsoid, wall 2-2.5 µm thick, yellowish or pale brownish, densely echinulate or verrucose echinulate, pores 7-10, scattered, small, without caps. Telia amphigenous, subcuticular, blackish brown; spore heads 55-95(-110) µm diam, (4)5-8(9) cells across, golden to chestnut brown, smooth, central cells (11-)14-18(-20) µm across, cysts pendent, approximately or probably of the same number as the marginal cells.

Hosts and distribution: *Acacia angustissima* (Mill.) Kuntze, *Calliandra humilis* Benth., *C. reticulata* Gray, *Desmanthus cooleyi* (Eat.) Trel., Texas and southern Arizona to Durango, Mexico.

Type: on *Desmanthus* sp., Texas, 1889, Brunk (S; isotype PUR 6231).

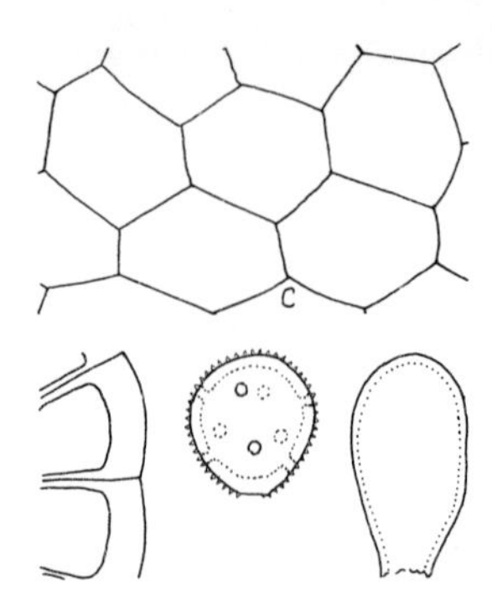

4b. *RAVENELIA TEXENSIS* Diet. var. *MORONGIAE* (Long) Cumm. Bol. Soc. Argent. Bot. 18:89. 1977.
Ravenelia morongiae Long, Bot. Gaz. 61:418. 1916.

Spermogonia and aecia unknown. Uredinia and urediniospores as in var. *texensis* except the paraphyses with thin walls. Telia mostly on abaxial surface of leaflets; spore heads 55-90(-100) μm diam, 4-6 cells across, chestnut brown, smooth, central cells (15-)16-20(-22) μm across; cysts pendent, of the same number as or a few more than the marginal cells.

Hosts and distribution: *Schrankia diffusa* Rose, *S. uncinata* Willd.: southeastern Texas and in Colima, Mexico.

Type: on *Morongia uncinata* (= *Schrankia uncinata*), Austin, Texas, Long No. 5474 (BPI; isotype PUR 6234).

5. *RAVENELIA SPEGAZZINIANA* Lindq. Bol. Soc. Argent. Bot. 1:300. 1946.
Ravenelia siliquae Long, Bot. Gaz. 35:118. 1903 (based on uredinia).

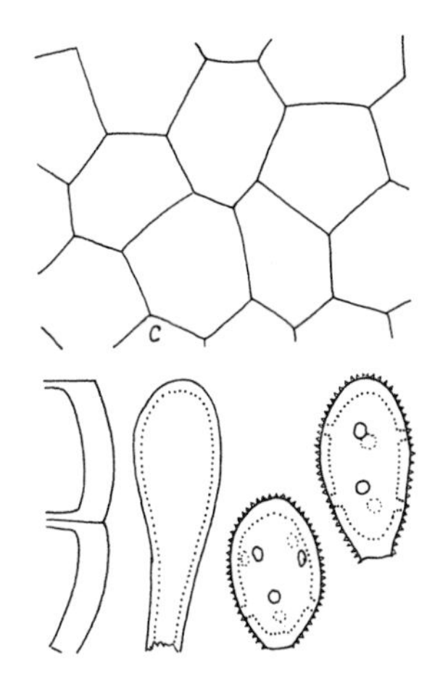

Aecia (according to Lindquist) in pods and deformed young branchlets, deep seated, without peridium or with scant peridial cells, opening by a rift in the epidermis; spores in chains, 30-36 x 12-18 µm, oblong fusoidal, pentagonal or irregularly polygonal, the base usually flat, the apex pointed, wall yellowish smoky, densely echinulate with fine spines, germ pores 3 or 4, equatorial. Uredinia on pods and leaflets, subcuticular, in large, confluent groups on pods or small and discrete on leaflets, rachis and petiole, cinnamon brown, with few or numerous mostly clavate, nearly colorless, uniformly thin walled paraphyses; spores (20-)23-28(-30) x (12-)14-16(-18) µm, mostly ellipsoid or narrowly obovoid, wall 1.5-2 µm at sides, 2-3 µm at apex, pale and rather dull cinnamon brown, densely verrucose echinulate with short mostly acute cones, pores in 2 bands of 4 each, above and below the equator, without caps. Teliospore heads (65-)70-95(-100) µm diam, (4)5-7 cells across, chestnut brown, smooth, central cells (13-)17-22(-24) µm across, cysts mostly of the same number as marginal cells, globoid, pendent.

Hosts and distribution: *Acacia smallii* Isley (*A. farnesiana* auth.): southern Texas to Guatemala; also in the islands of the Caribbean and in South America.

Type: on *Acacia aroma* Gill., La Plata, Argentina, Lindquist (LPS 12604; isotype PUR F11380).

Only the uredinial stage has been collected in North America.

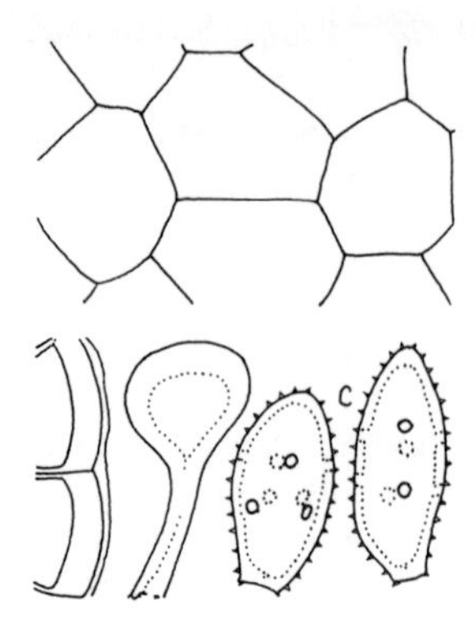

6. *RAVENELIA AUREA* Cumm. & J. W. Baxt. Mycotaxon 2:195-196. 1975.

Spermogonia amphigenous, conical, 55-70 μm diam, subcuticular, densely aggregated in slightly hypertrophied areas. Aecia amphigenous, subepidermal, circinately grouped around the spermogonia, brown, paraphyses numerous, capitate, to 22 μm diam in the head, the wall more or less uniformly 3.5-5 μm thick, chestnut brown apically to colorless basally; spores (22-)25-32(-35) x (11-)13-17(-18) μm, mostly ellipsoid or narrowly obovoid, wall 1.5-2 μm thick or slightly thicker apically, golden brown in the apex, paler below, echinulate, pores 8, bizonate, uredinia, if produced, similar to the aecia. Teliospore heads in the aecia (55-)60-70(-75) μm diam, 4 or 5 cells across, golden brown, the cells separating easily, smooth, central cells (14-)18-22(-24) μm across, cysts of the same number as the marginal cells, appressed or semipendent.

Type: on *Acacia pringlei* Rose, Mex hgw 190, Km 786 w of Tehuantepec, Oax., Mexico, 23 Feb. 1963, Barr No. 63-51 ex ARIZ 171780 (PUR 63746). Not otherwise known.

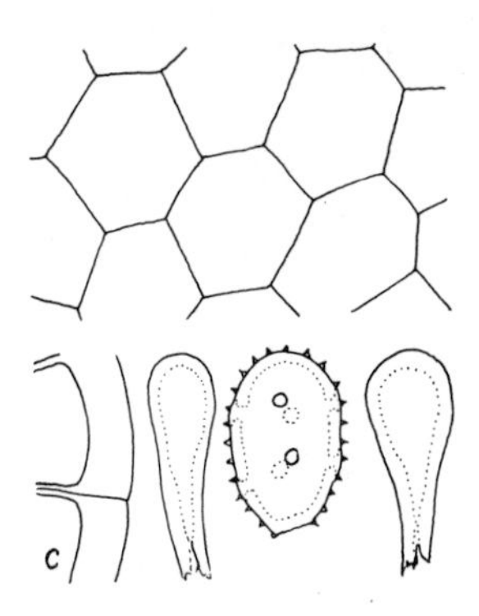

7. *RAVENELIA THORNBERIANA* Long, Bot. Gaz. 61:420. 1916.

Spermogonia amphigenous, subcuticular, on witches' brooms. Aecia on twigs and leaflets of the brooms, uredinoid, subcuticular, cinnamon brown, with abundant clavate or clavate capitate paraphyses 9-14 µm diam apically, wall (1.5-)2-3 µm thick and golden brown in the head, the stipe nearly solid and essentially colorless; spores (18-)22-27 (-30) x (12-)14-17(-19) µm, variable, mostly ellipsoid or oblong ellipsoid or the shorter ones obovoid or nearly globoid, wall 1.5(-2) µm thick, about cinnamon brown, closely verrucose echinulate, pores clearly bizonate in long spores to scattered in short robust spores, 5-9 but most often 8, with no or only slight caps. Uredinia following aecia on the brooms or separately on pods and leaflets, subcuticular, cinnamon brown, paraphysate as the aecia; spores like the aeciospores but tending to be more obovoid and a bit shorter. Telia amphigenous and on rachis, petioles and twigs, blackish brown, subcuticular; spore heads (65-)75-98(-105) µm diam, (4)5-7 cells across, chestnut brown, smooth, central cells (14-)17-20(-22) µm across, cysts in two series, pendent.

Hosts and distribution: *Acacia constricta* Benth.: western Texas and southeastern Arizona to Zacatecas, Mexico.

Type: on *A. constricta* var. *paucispina*, El Paso, Texas, Long No. 5506 (BPI).

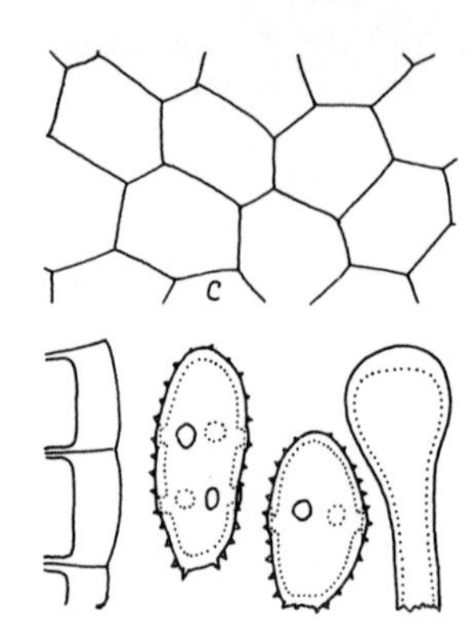

8. *RAVENELIA PRINGLEI* Cumm. Mycologia 67:1043. 1975.
Ravenelia versatilis auth. not Diet. 1894.

Spermogonia on twigs of witches' brooms, subcuticular. Aecia associated with the spermogonia, uredinoid, systemic in the brooms, about cinnamon brown, with clavate to capitate paraphyses, the wall at sides 1-1.5 μm thick, at apex to about 5 μm, colorless basally, golden brown apically; spores (20-)26-33(-35) x (10-)11-15(-17) μm, narrowly ellipsoid or oblong ellipsoid, wall (1-)1.5(-2) μm thick at sides, 2-3 μm at apex, pale yellowish basally to golden brown at apex, coarsely echinulate basally to nearly smooth at apex, pores bizonate with 4 pores in each zone, the upper zone equatorial or above, the lower zone near the base, small spores may have only 1 zone, without or with only slight caps. Uredinia amphigenous, not associated with spermogonia; spores and paraphyses as in the aecia. Telia amphigenous, blackish brown, subcuticular; spore heads (55-)70-95(-105) μm diam, (5)6-8 cells across, chestnut brown, smooth or the marginal cells with a few bead-like verrucae, central cells (12-)14-18(-20) μm across, cysts of the same number as the marginal cells, appressed or becoming semipendent, uniseriate.

Hosts and distribution: *Acacia greggii* Gray: southern Texas to California and in northern Mexico.

Type: on *Acacia greggii*, Bahia Kino, Son., Mexico, Cummins No. 71-685 (PUR 64070).

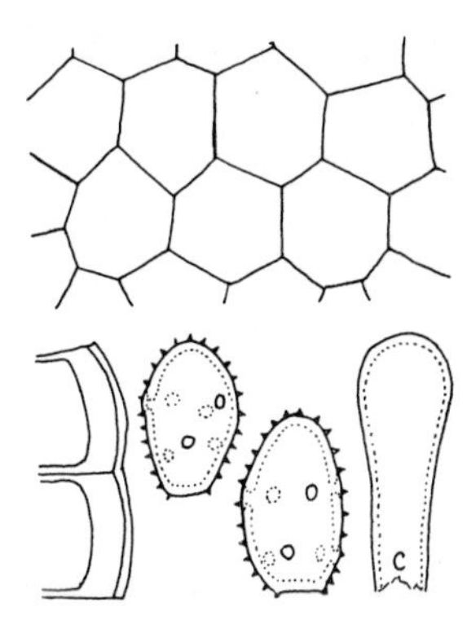

9. *RAVENELIA SCOPULATA* Cumm. & J. W. Baxt. Mem. N. Y. Bot. Gard. 28:40. 1976.

Spermogonia subcuticular, abundantly scattered over leaflets, rachis and branches of witches' brooms. Aecia subcuticular, uredinoid, associated with the spermogonia or perhaps sometimes lacking, with clavate or clavate capitate paraphyses to (13-)20-28(-35) μm long and 10-18 μm wide in the head, the wall uniformly 1-1.5 μm thick and hyaline or slightly thicker and pale brown apically; spores (17-)19-24 (-27) x (11-)12-14(-15) μm, mostly oblong ellipsoid or ellipsoid, wall (1-)1.5(-2) μm thick, cinnamon brown at the apex becoming paler or colorless basally, echinulate, pores 6 or 8, bizonate, the upper zone equatorial. Uredinia lacking. Telia subcuticular, associated with the aecia or directly with the spermogonia, black, more or less completely covering the branches of the witches' brooms; spore heads (55-)65-100(-110) μm diam, 5-8 cells across, chestnut brown, smooth, central cells (13-)16-19(-21) μm across, cysts of same number as marginal cells, semipendent or more or less appressed, pedicel persistent, colorless, to 150 μm long and 30 μm wide, composed of several hyphae.

Hosts and distribution: *Acacia greggii* Gray, *A. occidentalis* Rose: central Sonora, Mexico.

Type: on *Acacia greggii*, Nuri, Son., Cummins No. 75-12 (PUR 65068).

The species probably is a reduced cycle derivative of *R. pringlei*. The black witches' brooms are conspicuous in the trees.

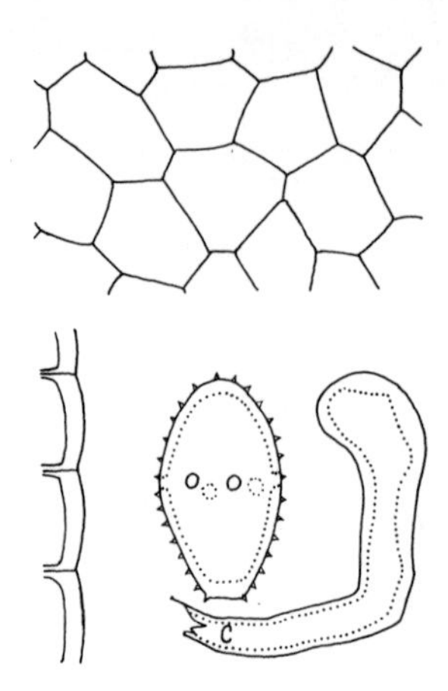

10. *RAVENELIA AUSTRALIS* Diet. & Neger, Bot. Jahrb. 24:161. 1897.

Spermogonia and aecia (according to Linquist, Rev. Fac. Agron. B. Aires 33:111. 1954) occur on witches brooms. Aecia (*Uredo hieronymii* Speg.) subepidermal, light brown, with rudimentary or no peridium; spores catenulate, (22-)26-33 (-38) x (13-)16-20 µm, variable in size and shape, obovoid, ellipsoid or oblong fusiform, often apiculate, wall 1.5-2 µm thick at sides, 3-5 µm at one or both ends, dull golden brown, densely verrucose echinulate, pores mostly 5 or 6, equatorial, sometimes 6-8 and more or less bizonate. Uredinia amphigenous, small, subepidermal, about cinnamon brown, with abundant, incurved, often geniculate, brown, dorsally thick walled paraphyses; spores (14-)16-19(-21) x (23-)26-31 (-35) µm, ellipsoid or obovoid, wall uniformly 1.5(-2) µm thick or 1.5-2.5 µm at apex, golden brown, echinulate, pores 4-6, equatorial, without obvious caps. Telia similar to the uredinia but blackish brown; spore heads (70-)85-115(-130) µm, 7-11 cells across, chestnut brown, smooth, central cells (10-)12-16(-20) µm across; cysts numerous, globoid, pendent, in 2 rows.

Hosts and distribution: *Acacia smallii* Isley (*farnesiana* auth.) *A. pennatula* (Schlecht. & Cham.) Benth. (vel aff.): southern Texas and Nayarit, Mexico; also in South America.

Type: on *Acacia cavenia* Hook. & Arn., Concepcion, Chile, Neger (S).

11. *RAVENELIA HOLWAYI* Diet. Hedwigia 33:61. 1894.

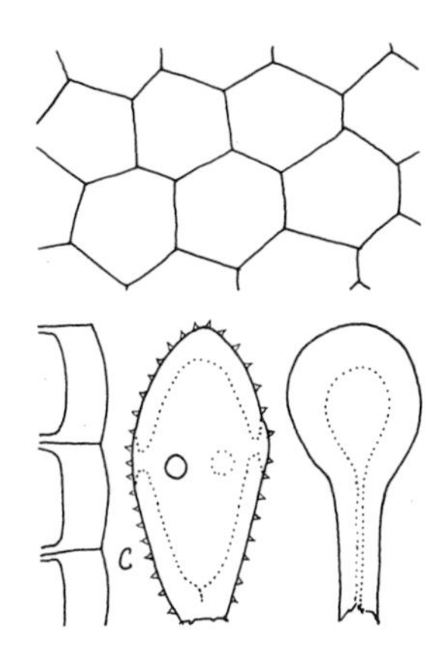

Spermogonia amphigenous on hypertrophied area of leaflets, rachis and petioles, subcuticular. Aecia associated with the spermogonia, amphigenous subepidermal but soon exposed, dull brown, without peridium but the spores catenulate; spores (24-)30-38(-42) x (13-)15-19 (-20) µm, more or less oblong ellipsoid, often irregular and angular, wall 2-3.5 µm thick, golden brown, closely verrucose, pores 4, equatorial. Uredinia amphigenous, subepidermal, about cinnamon brown, with capitate paraphyses to 27 µm diam in the head, wall 5-10 µm thick in the head and mostly chestnut brown, the stipe pale and solid or nearly so; spores (26-)30-40(-44) x (15-)17-19(-20) µm, mostly elongately obovoid, wall 2.5-3(-4) µm thick at the sides, usually thickened to 4-7 µm apically and often basally, echinulate, pores 4(5), equatorial, with slight caps. Telia amphigenous, often in circles, subepidermal, blackish, without paraphyses; spore heads (65-)85-125(-150) µm diam, (6)7-12 cells across, dark chestnut brown, smooth, central cells (10-)13-18(-20) µm across, cysts pendent, multiseriate.

Hosts and distribution: *Prosopis glandulosa* Torr. var. *torreyana* (L. Bens.) M. C. John.: Texas to southern California.

Type: on *Prosopis juliflora* (now considered to be as above), San Bernardino, California, 1893, Parish (holotype not in S; isotype MIN 317835).

Dietel (loc. cit.) described and illustrated (Tab. V, Fig. 26) teliospores but indicated that they were rare. After learning that there is no material in the Dietel Herbarium (in S), I examined the isotype leaflet by leaflet without finding teliospore heads. But Texas specimens have telia and the teliospore heads agree with those described by Dietel.

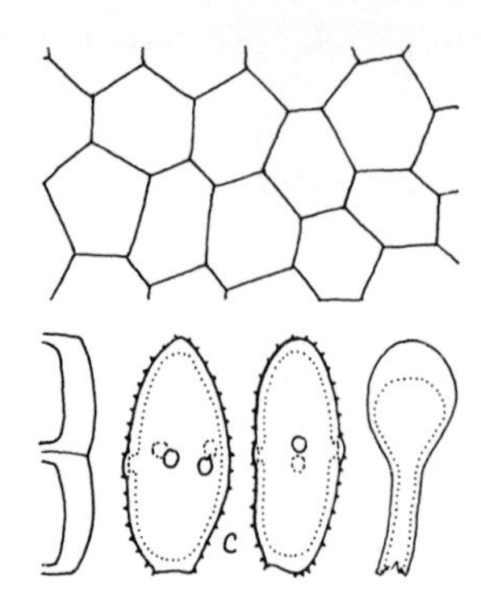

12. *RAVENELIA LYSILOMAE* Arth. Bot. Gaz. 39:392. 1905.
Ravenelia leucaenae-microphylli Diet. Bot. Centralbl. (Beih.) 20:375. 1906.

Spermogonia in small groups on adaxial surface of leaflets (described by Arthur). Uredinia mostly on abaxial surface, cinnamon brown, subepidermal, with clavate or capitate, yellowish or golden paraphyses to about 15 µm diam, the wall 4-6 µm apically; spores (24-)27-33(-37) x (11-)13-17(-19) µm, mostly oblong ellipsoid or ellipsoid, about cinnamon brown, echinulate, pores in an almost colorless equatorial band, large but difficult to count, 4-6, with slight or no caps. Telia amphigenous or mostly on adaxial surface, subepidermal, blackish brown; spore heads (75-)80-110(-120) µm diam, (6)7-9 cells across, chestnut brown, smooth, central cells (9-)12-18(-21) µm across, cysts of same number as marginal cells, appressed.

Hosts and distribution: *Acacia angustissima* (Mill.) Kuntze, *Lysiloma* spp.: southern Sonora to southern Tamaulipas, Guerrero and in Baja California Sur, Mexico, and in Guatemala.

Type: *Lysiloma tergemina* Benth., Iguala, Gro., Mexico, Holway No. 5317 (PUR 6140; probable isotypes Barth. F. Columb. No. 4626; Barth. N. Amer. Ured. No. 1112).

The host of *R. leucanae-microphylli* has been determined to be *Acacia angustissima*.

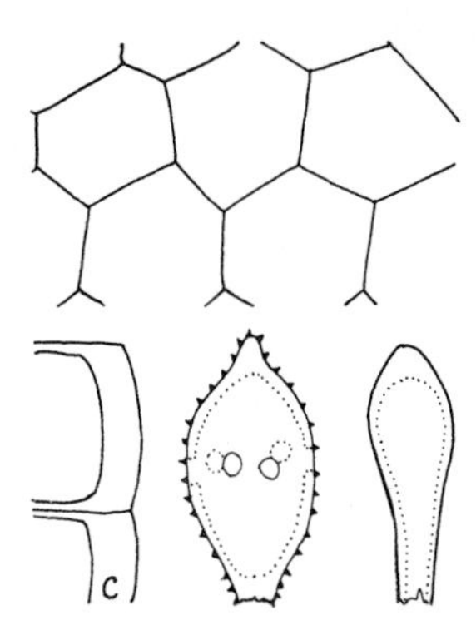

13. *RAVENELIA ANNULATA* Long, Bot. Gaz. 61:423. 1916.

Spermogonia and aecia unknown. Uredinia on adaxial surface of leaflets, subepidermal, paraphyses variable, from cylindrical to clavate, apical wall brown and from 2-7 µm thick; spores (25-)28-33(-37) x (17-)18-20(-22) µm, mostly lemon shape, the ends acuminate or the apex often apiculate, wall 1.5 µm thick at sides, 2-5 µm at apex, echinulate, cinnamon brown, pores in a paler equatorial band, difficult to count but probably always 6, with slight or no caps. Telia on adaxial surface, subepidermal; spore heads (50-)60-95 (-115) µm diam, (3)4-7 cells across, smooth, chestnut brown, central cells (14-)17-22(-26) µm diam; cysts appressed to underside of spore head, of same number as marginal cells.

Type: on *Lysiloma bahamensis* Benth. (= *L. latisiliqua* Benth.), Miami, Florida, Long No. 4623 (BPI; isotypes Barth. N. Amer. Ured. No. 1882). Not known elsewhere.

This species differs from *R. lysilomae* because of the lemon shape urediniospores and the larger cells of the teliospore head.

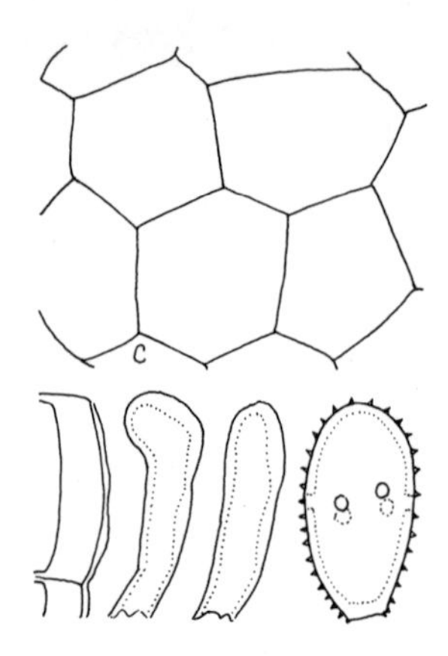

14. *RAVENELIA SUBTORTUOSAE* Long, Bot. Gaz. 72:40. 1921.

Spermogonia scattered over twigs of witches' brooms, becoming obscured by the aecia. Aecia densely distributed in the witches' brooms, peridium prominent at first then breaking at surface of the host; spores (15-)18-23(-28) x (13-)15-18(-20) µm, variable but mostly angularly broadly ellipsoid or globoid, wall 2-2.5(-3) µm thick, yellowish or pale golden brown, finely and densely verrucose, pores scattered, 8-10, obscure. Uredinia amphigenous, small, subepidermal, with abundant peripheral, mostly incurved paraphyses, mostly with dorsal wall much thickened, chestnut brown; spores (24-)26-32(-36) x (14-)16-20(-22) µm, mostly obovoid or ellipsoid, wall uniformly 1.5(-2) µm thick, dark cinnamon brown, echinulate, pores 5 or usually 6, equatorial, without caps. Telia similar to the uredinia but blackish brown; spore heads (40-)55-85(-100) µm diam, variable in shape and size, (3)4 or 5(6) cells across, clear chestnut brown, smooth, central cells (16-)18-28 µm across, usually or perhaps always 2 celled by a usually oblique septum, cysts of same number as marginal cells, subappressed, bursting rapidly.

Type: on *Acacia subtortuosa* Shafer (probably now treated as *A. schaffneri* Herm. var. *bravoensis* Isley) Corpus Christi, Texas, Long No. 6891 (BPI; isotype PUR 6076). Known only in this locality.

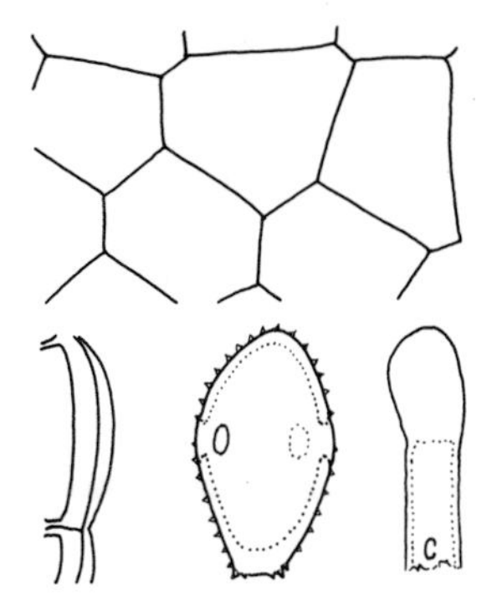

15. *RAVENELIA BIFENESTRATA* Mains, Carnegie Inst. Wash. Publ. 461:97. 1935.

Spermogonia and aecia unknown. Uredinia mostly on adaxial surface of leaflets, subepidermal, cinnamon brown, with abundant, peripheral, long, cylindrical, colorless or pale golden paraphyses, the wall thin except the apex which typically is solid for as much as 20 µm; spores (24-)27-32 (-35) x (15-)17-19(-21) µm, mostly ellipsoid, wall 1-1.5 µm thick at sides, 1.5-2(-2.5) µm at apex, cinnamon brown except a pale equatorial band around the pores, echinulate, pores 4, equatorial, large, without caps. Telia similar to the uredinia except blackish brown; spore heads 74-110 µm diam, 5 or 6 cells across, chestnut brown, smooth, central cells more variable than in most species, (16-)18-30(-33) µm across; cysts of same number as and appressed to the marginal cells but becoming semipendulous and bursting.

Type: on *Pithecellobium platylobum* (Spreng.) Urban, Tuxpeña, Camp., Mexico, Lundell No. 1296A (MICH; isotype PUR 47916). One other collection from the same locality is known.

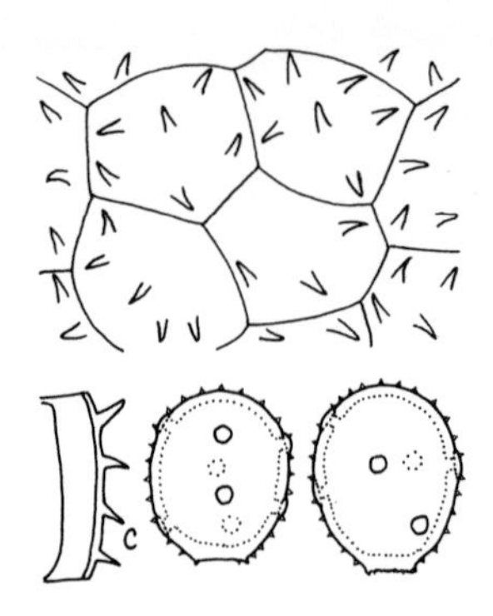

16. *RAVENELIA ECHINATA* Lager. & Diet. in Dietel, Hedwigia 33:65. 1894 var. *ECTYPA* (Arth. & Holw.) Cumm. Bol. Soc. Argent. Bot. 18:85. 1977.
Ravenelia ectypa Arth. & Holw. in Arthur, Mycologia 10: 120. 1918.

Spermogonia amphigenous, in a small group, subcuticular. Aecia uredinoid, amphigenous, grouped around the spermogonia, cinnamon brown, with very few peripheral, thin walled, colorless, cylindrical or clavate paraphyses; spores as the urediniospores. Uredinia amphigenous, subcuticular, pale cinnamon brown; spores (18-)20-25(-28) x (16-)17-19(-20) µm, mostly broadly ellipsoid, wall 1-1.5 µm thick, golden or pale brownish, echinulate, pores scattered (5)6-8(-10?), with slight caps. Telia amphigenous, subcuticular, blackish brown; spore heads (46-)52-62(-66) µm diam, chestnut brown, (3)4(5) cells across, each cell with 3-9(-12) sharply conical spines 2-2.5 µm wide at base and 3-5 µm long, central cells almost always 4, (16-)18-24(-28) µm across, cysts appressed to the underside of the spore head, of same number as marginal cells.

Hosts and distribution: *Calliandra* spp.: Sinaloa and Tamaulipas, Mexico to Costa Rica; also in South America.

Type: on *Calliandra gracilis* Klotsch, San José, Costa Rica, Holway No. 296 (PUR 6160).

Ravenelia echinata var. *echinata* has larger urediniospores, larger teliospore heads typically with 6 central cells and 8 marginal cells. It has not been collected in North America.

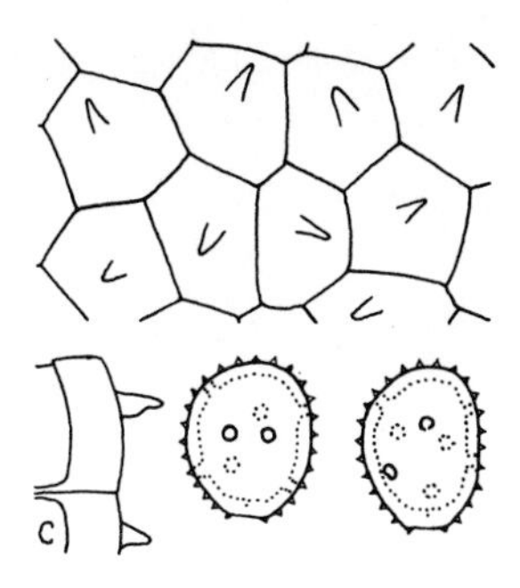

17. *RAVENELIA BAJACALENSIS* Cumm. & J. W. Baxt. Mycotaxon 2: 199. 1975.

Spermogonia and aecia unknown. Uredinia subcuticular, mostly on abaxial surface of leaflets, associated with the midrib, yellowish brown; spores (15-)16-19(-21) x (11-)13-16(-17) µm, oblong ellipsoid, broadly ellipsoid or obovoid, wall (1.5-)2(-2.5) µm thick, echinulate or verrucose echinulate, pale golden brown or yellowish, pores obscure, scattered or more or less bizonate, apparently 8. Telia similar to the uredinia but blackish brown; spore heads (74-)80-110 (-120) µm diam, (6)7-9(10) cells across, chestnut brown, each cell with 1 tubercle 2-3.5 µm wide and (2-)3-5(-6) µm long, central cells (11-)13-18(-22) µm across, cysts numerous, multiseriate, pendent.

Type: on *Lysiloma candida* Brand., Los Encinos, Sierra Giganta, Baja California Sur, Mexico, Gentry No. 4263 ex ARIZ 66230 (PUR 64928). Not otherwise known.

18. *RAVENELIA DISTANS* Arth. & Holw. in Arthur, Amer. J. Bot. 5:424. 1918.

Spermogonia and aecia unknown. Uredinia on abaxial surface, subepidermal, yellowish brown; spores (18-)22-26 (-28) x (12-)13-16 µm, mostly ellipsoid or ovoid with an abruptly narrowed, apiculate apex, wall 1.5 µm at sides, the apiculus solid, 3-7 µm thick, golden brown or paler, echinulate with broad-based echinulae, pores equatorial, 4 or 5, without obvious caps. Telia similar to the uredinia but blackish brown, subepidermal; spore heads (45-)55-66 (-75) µm diam, (3)4-6 cells across, clear chestnut brown, each cell with (3-)5-8 spines 1.5-2 µm wide as base and 3-4 µm long, central cells (12-)16-22(-24) µm across, cysts of same number as marginal cells, appressed to cells but becoming semipendent.

Type: on undetermined Mimosoideae, Retalhuleu, Guatemala, Holway No. 535 (PUR 6118). Not otherwise known. The host perhaps is a species of *Calliandra*.

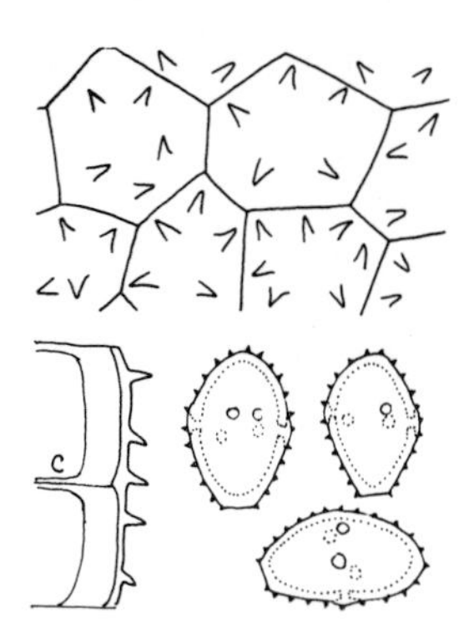

19. *RAVENELIA ALAMOSENSIS* Cumm. & J. W. Baxt. Mycotaxon 2: 200. 1975.

Spermogonia and aecia unknown. Uredinia amphigenous, subcuticular, brown; spores (16-)18-21(-23) x (11-)13-14 (-15) µm, mostly ellipsoid, wall 1.5(-2) µm thick, cinnamon brown or golden brown, echinulate, pores 4-6, frequently 5 or 6, equatorial. Telia similar to the uredinia but blackish brown; spore heads (48-)60-80(-85) µm diam, 5 or 6 cells across, chestnut brown, each cell with 3-6 spines 2.5-4(-5) µm long, central cells 16-22 µm across, cysts of same number as marginal cells, appressed.

Hosts and distribution: on *Pithecellobium tortum* Mart.: southern Sonora and Sinaloa, Mexico.

Type: Mt. Alamos, Alamos, Son., Cummins No. 70-106 (PUR 63539).

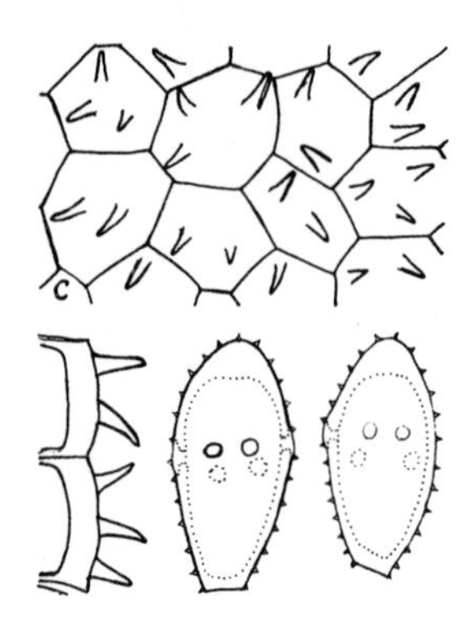

20. *RAVENELIA PITHECOLOBII* Arth. Bot. Gaz. 39:394. 1905.

Spermogonia and aecia unknown. Uredinia amphigenous, often in circles, subepidermal, cinnamon brown; spores (24-) 27-33(-38) x (15-)16-20(-22) μm, mostly ellipsoid or elongately obovoid, wall (1.5-)2(-2.5) μm thick at sides (3-)4-6 (-7) μm at apex, uniformly echinulate, golden brown, pores 4-6, commonly 5, equatorial, with small or no caps. Telia as the uredinia but blackish brown; spore heads (65-)70-85 (-100) μm diam, (4)5-8 cells across, central cells 12-16(-18) μm diam, each cell with (1)2 or 3(4) narrowly conical, essentially colorless tubercles 5-7 μm long or the central cells rarely without, chestnut brown; cysts of the same number as marginal cells, appressed to the spore head from pedicel to periphery; pedicel rather stout but usually detached.

Hosts and distribution: *Pithecellobium dulce* (Roxb.) Benth.: central Mexico; also in the islands of the Caribbean.

Type: on *P. dulce*, Guadalajara, Jal., Mexico, Holway No. 505 (PUR 6112).

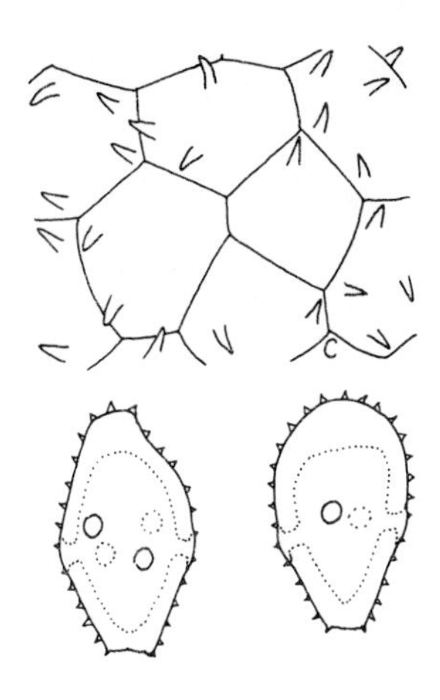

21. *RAVENELIA FLORIDANA* Cumm. & J. W. Baxt. Mycotaxon 2:199. 1975.

Spermogonia and aecia unknown. Uredinia on adaxial surface of leaflets, subepidermal, long covered partially by epidermis, brown; spores (24-)26-35(-40) x (15-)18-22(-24) µm, variable but mostly obovoid, wall (2-)2.5-3.5(-4) µm thick at sides, (3.5-)5-7(-9) µm at apex, golden brown or nearly colorless, echinulate, pores (4)5 or 6, equatorial. Telia similar to the uredinia but blackish brown; spore heads (55-)60-75(-83) µm diam, chestnut brown, (3)4-6 cells across, each cell with (0-)2-5(6) spines 3-6 µm long, 2-3 µm wide at base, central cells (17-)19-24(-26) µm across, cysts of same number as marginal cells, semipendent.

Type: on *Pithecellobium unguis-cati* (L.) Mart., Matheson's Hammock, Dade County, Florida, Stevenson No. 1817 (PUR 6115). One other Florida specimen has uredinia only. Not known elsewhere.

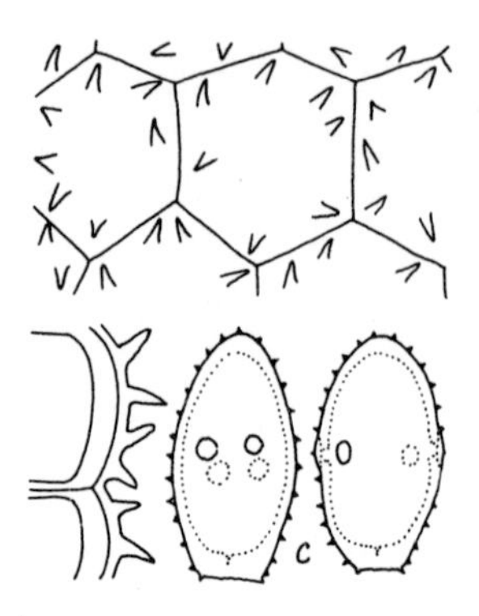

22. *RAVENELIA MULTISPINOSA* Cumm. & J. W. Baxt. Mycotaxon 2: 200-201. 1975.

Spermogonia and aecia unknown. Uredinia mostly on adaxial surface of leaflets, subepidermal, cinnamon brown; spores (26-)29-35(-38) x (16-)18-21(-23) µm, ellipsoid, wall 2.5-3.5 µm at base, 2-3 µm thick at apex, 1-1.5 µm at sides, dark cinnamon brown, pores 4, equatorial. Telia mostly on adaxial surface, subepidermal, blackish brown; spore heads (60-)65-85(-90) µm diam, chestnut brown, 4(5) cells across, each cell with 5-10 spines, central cells 22-30 µm across, cysts of the same number as marginal cells, appressed.

Hosts and distribution: *Pithecellobium tortum* Mart.: Baja California Sur, Sinaloa and Sonora, Mexico.

Type: east of El Fuerte, Sin., Cummins No. 71-612 (PUR 64165).

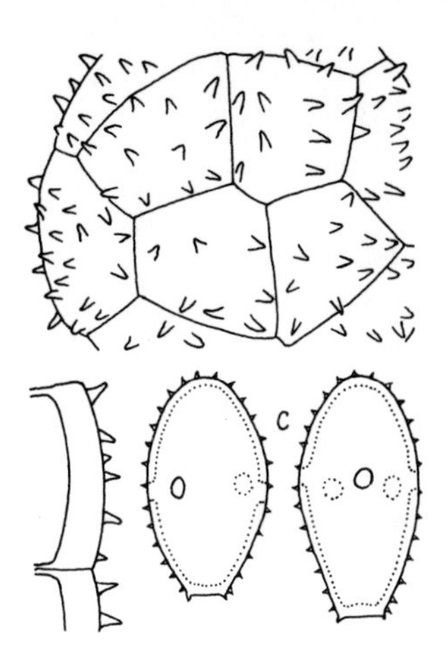

23. *RAVENELIA LINDA* Cumm. & J. W. Baxt. Mycotaxon 2:196. 1975.

Spermogonia and aecia unknown. Uredinia amphigenous, scattered, subcuticular, brown; spores (26-)30-36(-40) x (15-)17-20(-24) µm, ellipsoid or narrowly obovoid, wall uniformly 1-1.5 µm thick or to 2.5 µm thick at apex, pale cinnamon brown, echinulate, pores 4 or 5, equatorial. Telia similar to the uredinia but blackish brown; spore heads 55-70(-80) µm diam, mostly 4 cells across, typically with 4 central and 6 peripheral cells, chestnut brown, each cell with (6-)10-20 spines or spine like tubercles, central cells (16-)19-24(-28) µm across, cysts of the same number as the marginal cells, globose.

Type: on *Calliandra tapirorum* Standl. 5 Km s of Ojo de Agua, Region of Quebrada de Dantas, Dept. El Paraiso, Honduras, Standley, Williams & Molina No. 1254 (PUR 51702). Known also from one collection on *C.* sp. from Chiapas, Mexico.

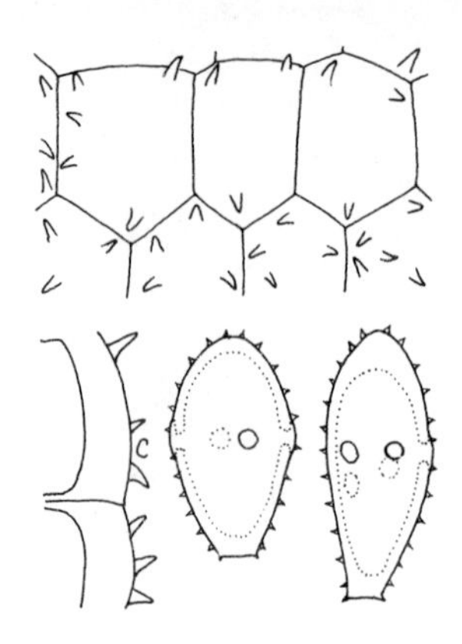

24. *RAVENELIA HAVANENSIS* Arth. Bull. Torrey Bot. Club 48:35. 1921.

Spermogonia on leaflets, petioles and swollen young stems, abundant. Aecia uredinoid, associated with the spermogonia, otherwise and the spores as the uredinia. Uredinia mostly on adaxial side of leaflets, scattered, yellowish brown, subcuticular; spores (25-)28-38(-42) x (12-)14-18 (-20) µm, ellipsoid or mostly elongately obovoid, wall 1.5-2 µm thick at sides, 2-5(-7) µm thick at apex, golden brown apically to nearly colorless basally, echinulate, pores (3)4 or 5, equatorial, with slight caps. Telia on adaxial surface, subcuticular, blackish brown; spore heads circular or often broadly elliptical in surface view 70-100 x 62-77 µm, (3)4 or 5(6) cells across, chestnut brown, each marginal cell with 5-10 and each central cell with 0-5 narrowly conical spines 3-5 µm long and 2-2.5 µm wide at base, central cells (18-)22-28 µm across, cysts of same number as marginal cells, appressed or becoming semipendent.

Hosts and distribution: *Enterolobium cyclocarpum* (Jacq.) Griseb.: Costa Rica; also in Cuba.

Type: Capdevilla, Havana, Cuba, Johnston (PUR 6110).

The record of *Ravenelia oligothelis* Speg. for Costa Rica (Cummins & Stevenson, 10) doubtless is not correct. The host is now considered to be *Enterolobium cyclocarpum* and the fungus *R. havanensis*.

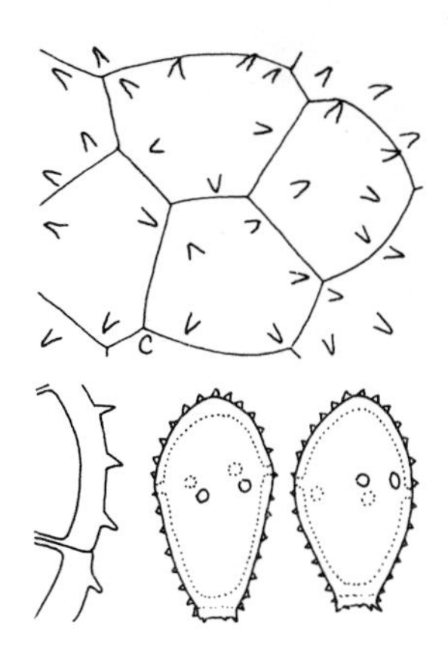

25. *RAVENELIA SIDEROCARPI* Long Bot. Gaz. 64:57-58. 1917.

Spermogonia and aecia unknown. Uredinia mostly on adaxial surface of leaflets, dark cinnamon brown, subcuticular; spores (23-)27-35(-38) x (14-)15-19(-21) µm, variable in both size and shape but mostly elongately obovoid, wall 1.5-2(-3) µm thick at sides, 2-3(-3.5) µm at apex, cinnamon or deep golden brown, echinulate with broad based, mostly sharp cones, pores 6, small but distinct, without caps, equatorial. Telia mostly on adaxial surface, subcuticular, blackish brown; spore heads (45-)55-60(-66) µm, (2)3 or 4 cells across, chestnut brown, each cell with (0-)3-6(-8) conical tubercles or dull spines, 1.5-2.5 µm wide at base and 2-3.5(-5) µm long, central cells (commonly 4), (16-)20-27(-29) µm across, cysts pendent from and of same number as the marginal cells.

Hosts and distribution: *Pithecellobium flexicaule* (Benth.) Coult.: southern Texas.

Type: on *Siderocarpus flexicaulis* (= *P. flexicaule*, near Brownsville, Texas, Long No. 6174 (BPI; isotype PUR 6117).

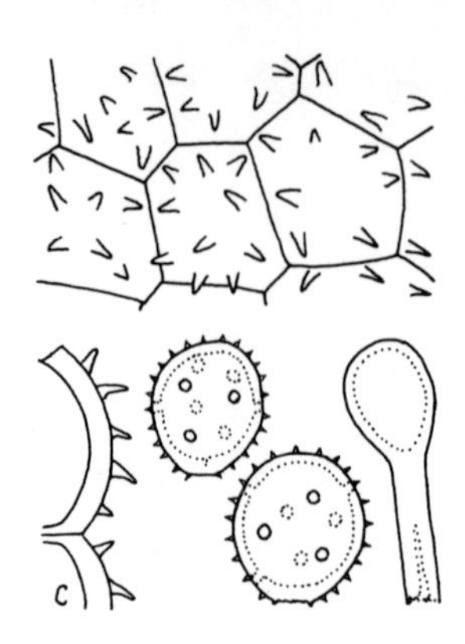

26. *RAVENELIA MAINSIANA* Arth. & Holw. in Arthur, Amer. J. Bot. 5:426. 1918.

Spermogonia and aecia unknown. Uredinia amphigenous, subepidermal, cinnamon brown, with numerous clavate or capitate, golden or paler paraphyses, the wall uniformly 1-2 µm thick in the head, the stipe commonly solid; spores (17-)19-23(-24) x (16-)17-19 µm, mostly broadly ellipsoid or broadly obovoid, wall 1.5-2 µm thick, pale cinnamon brown or golden, echinulate, pores scattered, 8-10, without caps, obscure. Telia as the uredinia except blackish brown; spore heads (46-)55-93 µm diam, irregular in outline, the marginal cells usually protruding conspicuously, (3)4-5(6) cells across, chestnut brown, each with 5-11 usually narrowly rounded, spine-like tubercles about 2-2.5 µm wide at base and up to 4(-5) µm long, central cells (16-)18-22(-24) µm across, cysts of same number as marginal cells, pendent, crowded.

Hosts and distribution: *Mimosa albida* H. & B., *M. manzanilloana* Rose: Sinaloa and Jalisco, Mexico south to Guatemala and El Salvador.

Type: on *Mimosa albida*, Guatemala City, Holway No. 13 (PUR 6291).

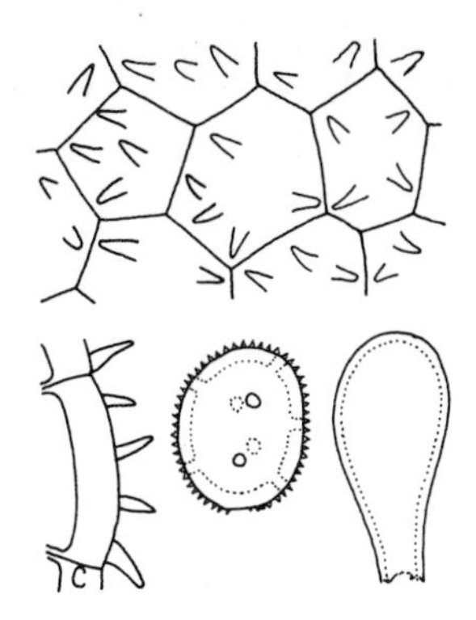

27. *RAVENELIA MIMOSAE-SENSITIVAE* P. Henn. Hedwigia 35:246. 1896.
Ravenelia mimosae-albidae Diet. Bot. Centralb. (Beih.) 20:378. 1906.
Ravenelia mimosae-caeruleae Diet. *ibid.* 20:378. 1906.
Ravenelia mimosicola Arth. N. Amer. Flora 7:137. 1907.

Spermogonia and aecia unknown. Uredinia amphigenous, subcuticular, yellowish brown, with abundant, mostly spatulate or clavately capitate, golden paraphyses to as much as 20 µm wide in the head but usually only 8-12 µm wide, the wall to 8 µm thick in the head, the stipe usually solid; spores (17-)18-21(-25) x (13-)15-17 µm, mostly ellipsoid or broadly so, wall 1.5-2(-2.5) µm thick, golden or pale cinnamon brown, closely and conspicuously verrucose echinulate, pores scattered, difficult to count, 8-10, without caps. Telia as the uredinia except blackish brown and without paraphyses; spore heads (55-)65-90(-100) µm diam, (3)4-6(7) cells across, chestnut brown, each cell with (3)4-8(-10) usually more or less cylindrical or apically narrowed tubercles (2-)2.5-3 µm wide at the base and (2-)3-7 µm long, central cells (16-)19-25(-28) µm across, cysts of same number as marginal cells, pendent.

Hosts and distribution: *Mimosa* spp. and *Schrankia distachya* Moc. & Sess.: central Mexico to Guatemala and Costa Rica; also in South America.

Type: on *Mimosa sensitiva* L. (perhaps better considered to be *M. argentinensis* Burk.?), Tucumán, Argentina, Lorentz (B).

This treatment follows that of Baxter (5).

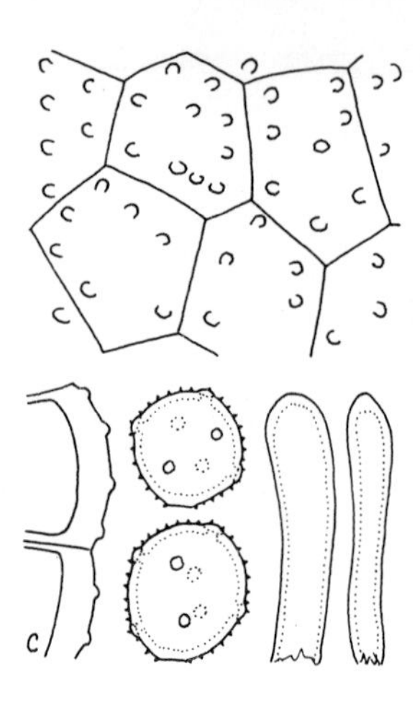

28. *RAVENELIA VERRUCOSA* Cooke & Ellis, Grevillea 15:112. 1887.

Spermogonia and aecia unknown. Uredinia mostly on adaxial surface of leaflets, subepidermal, dark cinnamon brown with abundant peripheral, straight or incurved, mostly cylindrical paraphyses to 90 µm long and 15 µm wide, the wall about 1 µm thick and yellowish basally gradually thickening to 2-3 µm apically and becoming chestnut brown; spores (16-) 17-19(-21) x (14-)16-18 µm, essentially globoid, wall 1.5-2 µm thick, dull brown, echinulate, pores 6-8, scattered, with small caps. Telia amphigenous, subepidermal, blackish brown, with paraphyses as in uredinia (if formed anew?); spore heads (60-)75-95(-110) µm diam, (3)4 or 5(-7) cells across, chestnut brown, central cells (18-)20-27(-30) µm diam, each cell with 4-10 low, rounded tubercles 2-2.5 µm diam and 1.5-2 µm high, the central cells usually with fewer tubercles than the peripheral cells; cysts appressed to the underside of the head, of the same number as the marginal cells.

Type: on *Lecania* sp.? (now known to be *Leucaena lanceolata* Wats.) Mexico, Palmer (holotype in K; isotype NY). Not otherwise known.

The type locality undoubtedly is Hacienda San Miguel, about one mile from Batopilas in southewestern Chihuahua where Edward Palmer collected it in 1885. Rusted material apparently was separated from plant specimens that Watson (31) described as *Leucaena lanceolata*.

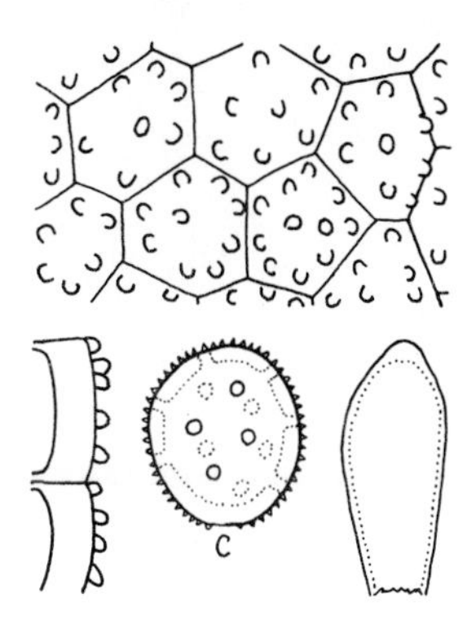

29. *RAVENELIA VERRUCATA* Cumm. & J. W. Baxt. Mycotaxon 2: 202-203. 1975.

Spermogonia and aecia unknown. Uredinia on adaxial surface of leaflets, subcuticular, yellowish brown, with mostly clavate, colorless or yellowish paraphyses, the wall uniformly 1 µm thick or the apex to 6 µm; spores (17-)19-22 (-23) x (16-)17-19(-20) µm, broadly ellipsoid or globoid, wall 2-2.5(-3) µm thick, yellowish brown, densely verrucose echinulate, pores 10-15, scattered. Telia similar to the uredinia but blackish brown and without paraphyses; spores heads 60-90 µm diam, 4-7 cells across, chestnut brown, each cell with (4-)7-11 bead-like tubercles, central cells (15-) 17-21 µm across, cysts of the same number as the marginal cells, appressed or semipendent.

Hosts and distribution: *Mimosa* spp., Sinaloa and Nayarit, Mexico.

Type: on *M. spirocarpa* Rose, Mex hgw 15, Km 41 n of Mazatlán, Sin., Cummins No. 71-591 (PUR 64152).

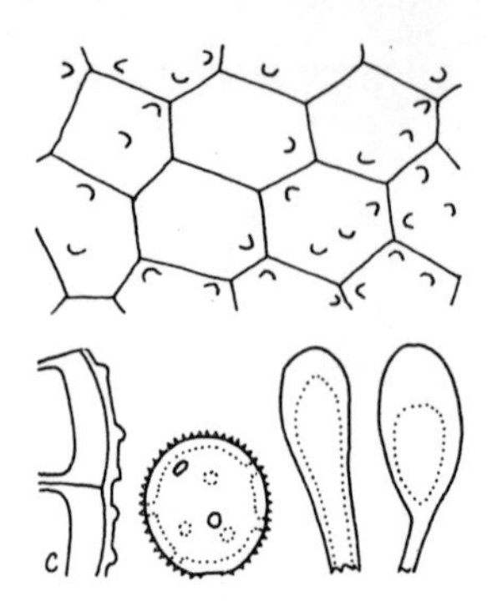

30. *RAVENELIA EXPANSA* Diet. & Holw. in Holway, Bot. Gaz. 24: 35. 1897.
Ravenelia igualica Arth. N. Amer. Fl. 7:136. 1907.

Spermogonia and aecia unknown. Uredinia amphigenous or often mostly on adaxial surface of leaflets, subcuticular, about cinnamon brown, with clavate, capitate or spatulately capitate paraphyses 8-14 µm diam and golden brown apically, the wall at apex (3-)5-8(-11) µm thick; spores (13-)15-18 (-20) x (11-)13-16(-18) µm, essentially globoid, wall (1-) 1.5(-2) µm thick, about pale cinnamon brown or golden brown, closely verrucose echinulate, the echinulate layer tending to swell, pores 6-10, scattered, without caps. Telia amphigenous or mostly on adaxial surface, subcuticular, blackish brown, without paraphyses when formed anew; spore heads (55-)60-90(-100) µm diam, (4)5 or 6(7) cells across, chestnut brown, central cells (13-)15-18(-20) µm diam, each cell with 2-7(9) tubercles 2-3(-3.5) µm wide and 2-3(-4) µm high or the central cells occasionally smooth; cysts pendent, of same number as marginal cells.

Hosts and distribution: on *Acacia* spp., *Leucaena glauca* (L.) Benth.: southern Texas to Baja California Sur and Guatemala.

Type: on *Acacia tequilana* Wats., Guadalajara, Jal., Mexico, 13 Oct. 1896, Holway (S; isotype PUR 6213; probable isotypes Barth. N. Amer. Ured. 179).

The species is similar to *R. verrucosa* but differs markedly in the type of paraphyses and the size of the cells of the teliospores heads.

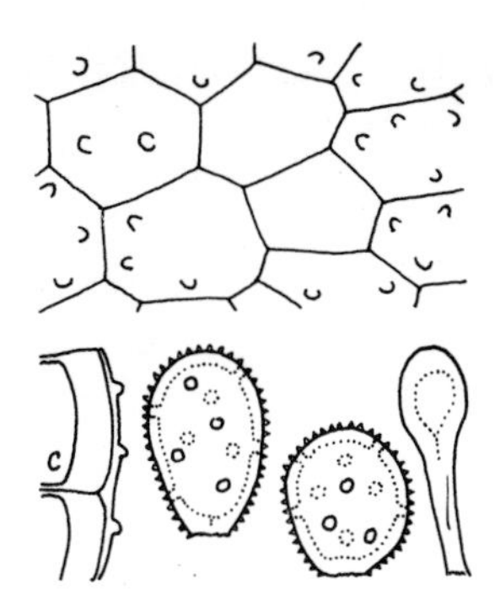

31a. *RAVENELIA FRAGRANS* Long, Bot. Gaz. 35:123. 1903 var. *FRAGRANS*.

Spermogonia and aecia unknown. Uredinia amphigenous and on petioles, stems and pods, subcuticular, yellowish brown, with abundant mostly clavate, spatulate or clavately capitate, thick walled, pale golden or occasionally colorless, thin walled paraphyses, 6-10(-12) µm wide apically; spores (15-)18-23(-26) x (13-)15-19(-21) µm, mostly ellipsoid or broadly ellipsoid, wall 1.5-2 µm thick, closely verrucose echinulate, dull golden brown or paler, pores scattered, 8-12, without caps. Telia located as the uredinia, blackish brown; spore heads (55-)60-100(-110) µm diam, (4)5-7 (8) cells across, chestnut brown, each cell with 0-6 low, bead-like warts or tubercles commoner on peripheral than on central cells, central cells, (13-)17-21(-24) µm across, cysts of same number as marginal cells, pendent, crowded.

Hosts and distribution: *Mimosa* spp. (especially *M. biuncifera* Benth.): southern Texas and southern Arizona to San Luis Potosí and Sinaloa, Mexico.

Type: on *Mimosa fragrans* (now = *M. borealis* Gray, Austin, Texas, Long No. 142 (BPI).

This and the following variety intergrade morphologically and geographically and also parasitize many of the same host plants.

31b. *RAVENELIA FRAGRANS* Long var *EVERNIA* (Syd.) J. W. Baxt. Mycologia 57:79. 1965.
Ravenelia evernia Syd. Ann. Mycol. 31:87. 1939.
Ravenelia dysocarpae Long & Good. Mycologia 31:670. 1939.

Uredinia, paraphyses and urediniospores as in var. *fragrans*; teliospore heads as in var. *fragrans* except all cells smooth.

Hosts and distribution: *Mimosa* spp.: southern New Mexico and southern Arizona to Sinaloa, Mexico; also in India.

Type: on *Mimosa rubicaulis* Lam., Majhgawan, India, 1930, Tandon (HC10; isotype PUR F2103).

This treatment follows that of Baxter (loc. cit.), who recognized that, while extremes are distinctive, numerous collections can be assigned only more or less arbitrarily. In fact, this variety is scarcely distinguishable from *R. texensis* var. *texensis*.

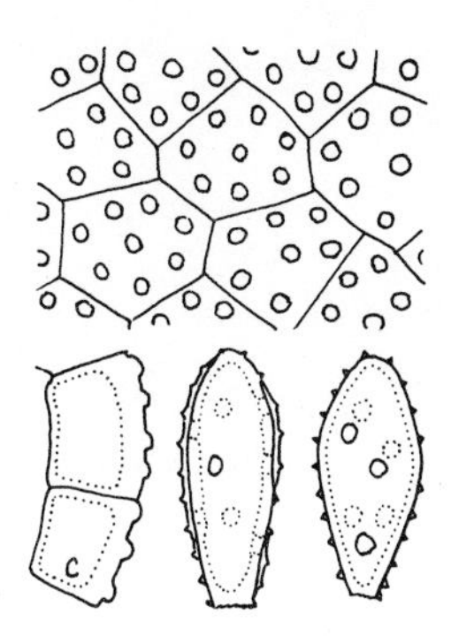

32. *RAVENELIA CUMMINSII* J. W. Baxt. Mycologia 56:285. 1964.

Spermogonia and aecia unknown. Uredinia mostly on abaxial surface of leaflets, subcuticular, in groups, often along midvein, dull brown, compact, with (or sometimes without) inconspicuous, cylindrical, peripheral paraphyses, the wall uniformly 1-1.5 µm thick, colorless; spores (20-)24-30 (-33) x (8-)10-14(-15) µm, narrowly ellipsoid, somewhat asymmetrical, slightly wider with pores in face view than when seen with pores lateral, wall 2-2.5 µm thick and nearly bilaminate on the two principal pore bearing sides, 1-1.5 µm thick on other areas, verrucose echinulate, yellowish or nearly colorless, pores scattered or tending to be bizonate, mostly 6-8. Telia similar to the uredinia but blackish brown, subcuticular, early exposed; spore heads (70-)80-105 (-120) µm diam, chestnut brown, mostly 6-8 cells wide, central cells 10-18(-20) µm diam, each cell has 6-10(-12) discrete tubercles 2-3(-4) µm diam, cysts, colorless, of the same number as marginal cells, extending from center to margin, appressed to under side of spore head.

Hosts and distribution: *Acacia willardiana* Rose: west central Sonora, Mexico.

Type: near Bocachibampo Bay, Guaymas, Cummins No. 62-57 (PUR 59932).

The host is limited to the same region of Sonora and to Baja California.

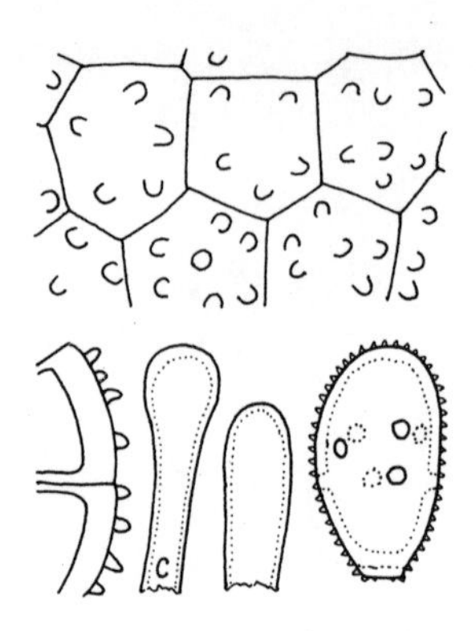

33. *RAVENELIA ACACIAE-PENNATULAE* Diet. Bot. Centralbl. (Beih.) 20:373. 1906.

Spermogonia and aecia unknown. Uredinia mostly on the adaxial surface of the leaflets, subcuticular, cinnamon brown, with abundant clavate or clavately capitate, pale golden paraphyses, the wall uniformly thin or to 3 µm apically; spores (20-)24-30(-33) x (14-)17-20(-22) µm, mostly ellipsoid or elongately obovoid, wall uniformly 1.5-2 µm thick or to 3 µm apically, about dark cinnamon brown, echinulate verrucose with narrowly to bluntly rounded cones, pores usually in 2 bands of 4 each in the equatorial area or less commonly 5-8 without precise arrangement. Telia similar to the uredinia but blackish brown and without paraphyses (when formed de novo); spore heads 4-6(7) cells across, chestnut brown, each cell with (4)5-8(10) rounded tubercles 2-2.5 µm wide and 2-3 µm high, central cells (15-)18-24(-26) µm across, cysts pendent, more numerous (?) than the marginal cells, bursting quickly.

Hosts and distribution: *Acacia* spp., perhaps mainly *A. pennatula* Benth.: Sinaloa, Mexico to Guatemala.

Type: on *A. pennatula*, Etla, Oax., Mexico, Holway, Nov. 1903 (S; probable isotype PUR 6191).

1a

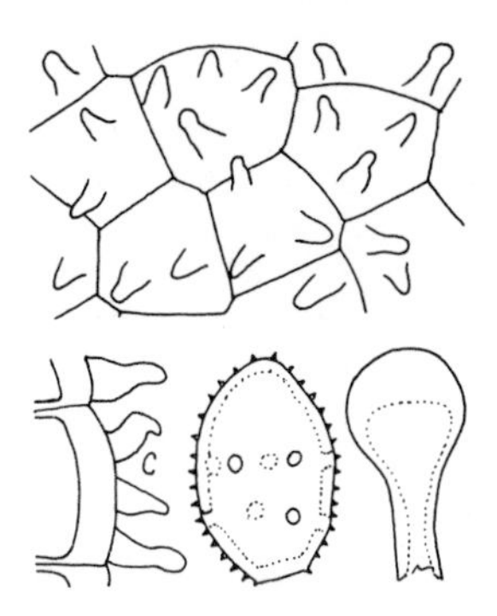

34. *RAVENELIA MEXICANA* Tranz. in Dietel, Hedwigia 33:370. 1894.
Ravenelia bizonata Arth. & Holw. in Arthur, Amer. J. Bot. 5:424. 1918.

Spermogonia and aecia unknown. Uredinia on abaxial surface of leaflets, subcuticular, yellowish brown to near chocolate brown, with abundant, mostly capitate paraphyses 10-16(-19) µm wide in head, wall golden brown apically and to 10 µm thick, paler and thin below, some clavate, uniformly thin walled paraphyses occur; spores (20-)22-27(-29) x (15-)16-19 µm mostly ellipsoid or obovoid, wall (1-)1.5(-2) µm thick at sides, 2-2.5(-3) µm at apex, echinulate with short, fine echinulae, these more prominent basally, golden or cinnamon brown above, usually paler below, pores bizonate with one band equatorial, the other below, usually 4 or 5 in each band, with slight or no caps. Telia as the uredinia but chestnut brown and without paraphyses; spore heads (50-)55-75(-80) µm diam, (3)4-6 cells across, clear chestnut brown, each cell with 2-4 more or less cylindrical or bottle shape tubercles 3-5 µm wide at the base and up to 10 µm long, those on the peripheral cells longer than those on the central cells, central cells variable in size, 15-21 (-23) µm across, cysts of same number as marginal cells, appressed or becoming semipendent.

Hosts and distribution: *Calliandra anomala* (Kunth) Macbr., *C. houstoniana* (Will.) Standl.: west central Mexico and Guatemala.

Type: on *Calliandra grandiflora* (L'Her.) Benth. (now considered to be *C. anomala*), Mexico, State of Jalisco. Mountains near Chapala, Pringle (S).

Long (23) gives the date of collection as Sept. 12, 1889, but Davis (11) shows that Pringle was in Laredo, Texas on this date but did collect *Calliandra grandiflora* in "the mountains on the north shore of Lake Chapala, Dec. 9, 1889" and this is the date on the isotype in Stockholm. This isotype is no longer of value, but by the courtesy of Prof. D. S. Barrington, curator of the Pringle Hervarium, I was able to examine the Pringle specimen (Pl. Mex. No. 2426). A few infected leaflets bore sori. The uredinia have abundant paraphyses, the urediniospores have bizonate pores and the teliospore heads are as in *R. bizonata*. Hence, *R. bizonata* is synonymous with *R. mexicana*.

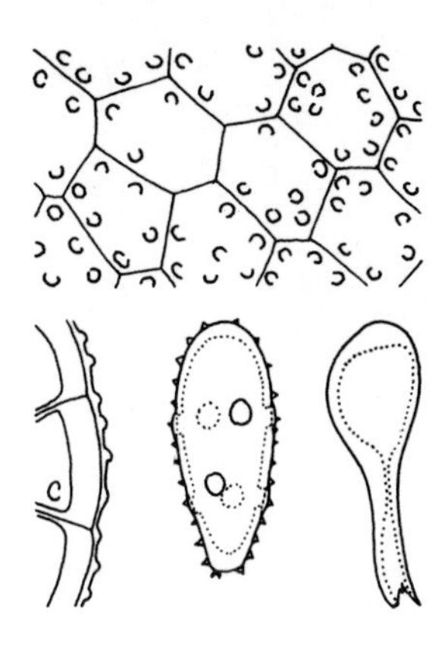

35. *RAVENELIA ROEMERIANAE* Long, Bot. Gaz. 64:59. 1917.

Spermogonia on pods, twigs of witches' brooms and leaflets, subcuticular. Aecia associated with spermogonia, often concentrically so, uredinoid, with abundant clavate or more or less capitate, golden brown paraphyses, the apical wall to 5 µm thick; spores variable, (24-)30-38(-42) x (10-)11-16(-18) µm, narrowly oblong ellipsoid or elongately obovoid, wall 1-1.5 µm thick at sides, 2-3.5 µm and deep golden brown at apex, becoming progressively paler below, strongly echinulate basally but less so apically, pores bizonate (in small spores sometimes unizonate), the upper zone equatorial or slightly above, the lower zone slightly to considerably below the equator, 3 or 4 pores in each zone. Uredinia not grouped about spermogonia, paraphyses as in the aecia; spores similar to the aeciospores but less variable. Telia amphigenous or mostly on abaxial surface of leaflets, also on pods, blackish brown; spore heads (50-)65-90(-100) µm diam, (4)5-7 cells across, chestnut brown, each cell with 3-10 tubercles 1-2 µm high and about 2.5 µm wide, often indistinct in surface view, central cells (12-)14-18 µm diam, cysts appressed to underside of head, of same number as marginal cells.

Hosts and distribution: *Acacia roemeriana* Scheele: southern Texas and Nuevo Leon, Mexico.

Type: San Marcos, Texas, Long No. 5498 (BPI; isotype PUR 6172).

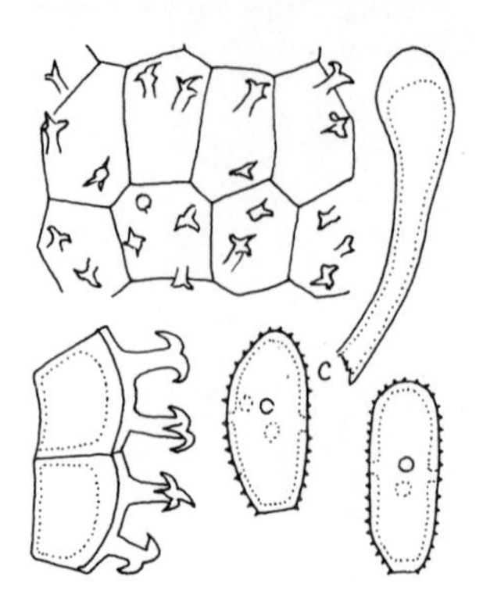

36. *RAVENELIA STEVENSII* Arth. Mycologia 7:178. 1915.

Spermogonia and aecia unknown. Uredinia on abaxial surface of leaflets and on pods, scattered, dull brown, subcuticular, with abundant peripheral, capitate or clavate, brownish paraphyses, the apex 9-12 μm wide, the apical wall mostly 2-6 μm thick; spores (18-)22-28(-30) x (8-)9-13 μm, oblong or very narrowly ellipsoid, wall 1 μm thick, pale brownish to pale yellowish, finely echinulate, pores equatorial, difficult to count, 4 or perhaps also 5 or 6, without caps. Telia on abaxial surface, subcuticular, blackish brown; spore heads (40-)50-70(-75) μm diam, (3)4 or 5(6) cells across, chestnut brown, the central cells (9-)12-17 (-20) μm diam, each cell with 1-4 nearly colorless, elongated, apically 2-4-forked tubercles, (6-)10-13(-18) μm long, 2-4 μm wide, cysts colorless, globoid, in a single ring, swelling greatly.

Hosts and distribution: *Acacia* spp.: San Luis Potosí, Sinaloa and Jalisco, Mexico.

Type: on *Acacia ripari* H.B.K., Guayanilla, Puerto Rico, Stevens No. 5881 (PUR 6216).

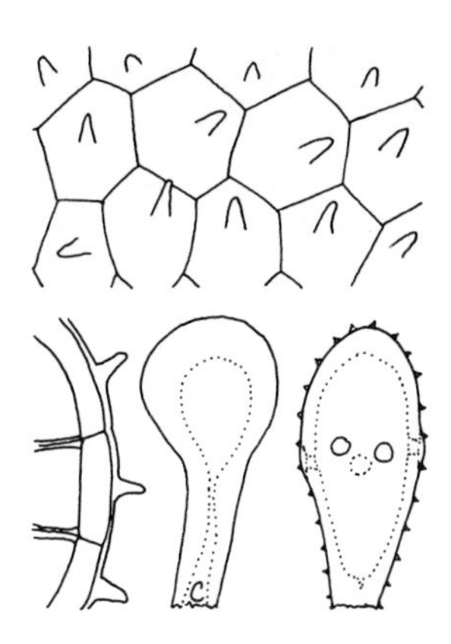

37. *RAVENELIA ARIZONICA* Ellis & Ever. Bull. Torrey Bot. Club 22:363. 1895.

Spermogonia not seen. Aecia (presumably) on woody galls on stems, uredinoid, dark brown, with occasional paraphyses as in the uredinia; spores (25-)28-42(-44) x (15-)18-22 µm, more variable than but similar to the urediniospores. Uredinia amphigenous, subepidermal, but mostly on abaxial surface of leaflets, dark cinnamon brown, paraphyses abundant, mostly capitate, to 30 µm diam, the head chestnut brown with wall 4 µm or thicker, colorless below; spores (30-)33-46(-50) x (16-)17-22(-24) µm, mostly elongately obovoid, wall 2-2.5 µm thick at sides, 2.5-4(-6) µm at apex, about golden brown basally to chestnut brown apically, echinulate, pores 4 or 5(6), equatorial with slight or no caps. Telia similar to uredinia but darker brown; spore heads (60-)75-100(-110) µm diam, (5)6-8(9) cells across, chestnut brown, central cells 12-18 µm diam, each cell with a single apically rounded spine 2-3 µm wide at base and 4-7 µm long; cysts pendent, multiseriate, not united.

Hosts and distribution: *Prosopis glandulosa* Torr. (*P. juliflora* auth.): Texas to southern California and south to Jalisco, Mexico.

Type: Tucson, Arizona, Aug. 1894, Toumey No. 37 (NY).

The woody galls are considered here to be the site of the uredinoid aecial stage but this has not been proved.

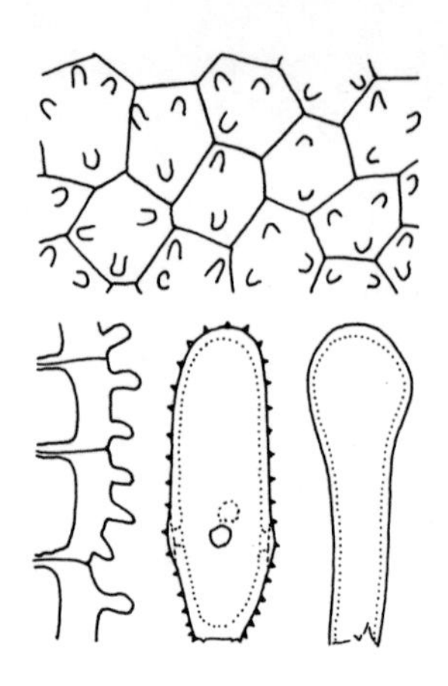

38. *RAVENELIA LEUCAENAE* Long, Bot. Gaz. 35:126. 1903.

Spermogonia and aecia unknown. Uredinia mostly on adaxial surface of leaflets, cinnamon brown, with clavate to capitate paraphyses, the wall uniformly thin or only slightly thicker apically, colorless or usually brownish apically, to 20 μm diam; spores (32-)38-50(-52) x (12-)14-18(-20) μm, mostly oblong ellipsoid, wall (1-)1.5 μm thick at sides, 1.5-2(-2.5) μm thick at apex, nearly colorless basally to pale chestnut brown apically, echinulate, pores 4-6, in lower 1/3 of spore, with slight caps. Telia on adaxial surface, subcuticular, blackish brown; spore heads (55-)65-90(-100) μm diam, mostly 5 or 6 cells across, each cell with 2-5 cylindrical, apically rounded projections 2-3 μm wide and 3-6 μm long, central cells 11-17(-20) μm diam, chestnut brown, cysts adherent to under side of head, of same number as marginal cells.

Hosts and distribution: *Leucaena* spp.: Jalisco to Oaxaca, Mexico, also in southern Texas.

Type: on *Leucaena diversifolia* Benth., Etla, Oax., Holway 3826 (BPI; isotype PUR 57675).

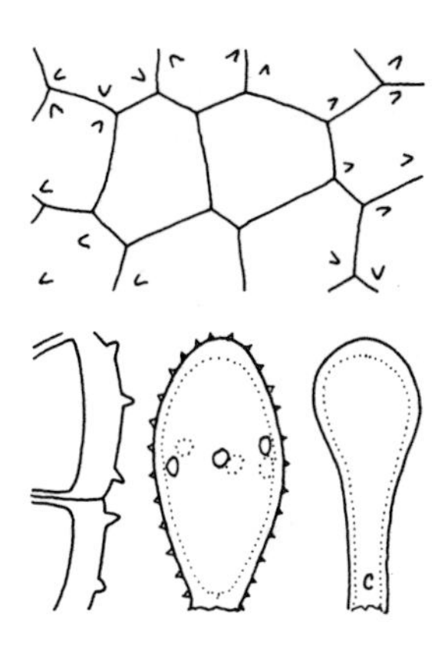

39. *RAVENELIA GRACILIS* Arth. Bot. Gaz. 39:393. 1905.

Spermogonia amphigenous, subcuticular. Aecia on adaxial surface of leaflets, uredinoid, about cinnamon brown, with cylindrical or more or less capitate, apically brown, mostly periperal paraphyses; spores (30-)35-42(-46) x (16-)18-20(-22) µm, ellipsoid or elongately obovoid, wall at sides 1.5-2 µm thick, at apex (2-)2.5-3.5(-4) µm, uniformly deep golden brown, echinulate, pores (4)5 or 6, equatorial, with slight or no caps. Uredinia and spores similar to aecia and aeciospores except not associated with spermogonia. Telia on pods and on adaxial surface, subepidermal, blackish brown; spore heads (70-)80-95(-100) µm diam, 5 or 6(7) cell across, chestnut brown, each cell with (0-)2-5(-7) small, rounded conical tubercles 2-3(-4) µm high, the central cells without or with fewer tubercles than the periperal cells, central cells (15-)17-21(-24) µm diam, cysts adpressed to under side of spore head, of same number as marginal cells.

Hosts and distribution: *Pithecellobium pallens* (Benth.) Standl.: southern Texas to Oaxaca, Mexico.

Type: on undetermined Mimosaceae (later identified as *Pithecellobium brevifolium* Benth. now = *P. pallens*), Cardenas, San Luis Potosí, Holway No. 31441/2 (PUR 6171).

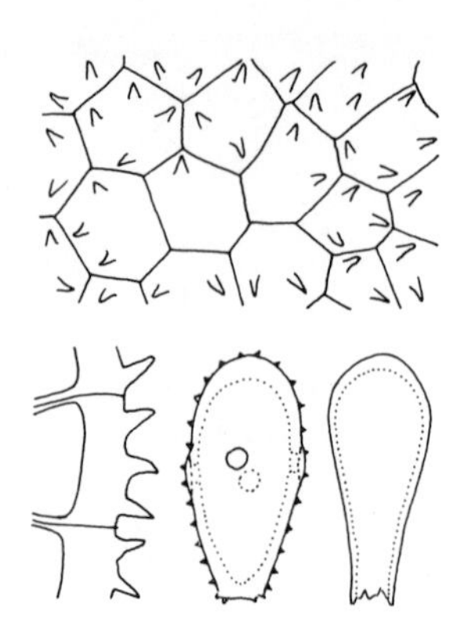

40. *RAVENELIA HERMOSA* Cumm. & J. W. Baxt. Mycotaxon 2:198. 1975.

Spermogonia subcuticular. Aecia subepidermal, uredinoid, cinnamon brown, peripheral paraphyses not abundant; spores (25-)28-35(-38) x 15-18(-20) µm, mostly ellipsoid or obovoid, wall 1.5-2 µm thick at sides, 3-5 µm at apex, pale chestnut brown, echinulate, pores (3)4 or 5, equatorial. Uredinia amphigenous, with numerous mostly clavate, brownish paraphyses, the wall uniformly thin or slightly thicker apically; spores as the aeciospores. Telia mostly on the abaxial surface of leaflets, subepidermal, blackish brown, without paraphyses; spore heads 70-105 µm diam, (5)6 or 7(8) cells across, chestnut brown, each cell with (0-)2-4(-5) spines, 3-4.5 µm long, central cells (10-)12-17(-20) µm across, cysts of same number as marginal cells, appressed.

Hosts and distribution: *Leucaena lanceolata* Wats., *L. palmeri* Britt. & Rose: Nayarit, Sinaloa and Sonora, Mexico.

Type: on *L. palmeri*, Mt. Alamos, Alamos, Son., Cummins No. 70-142 (PUR 63574).

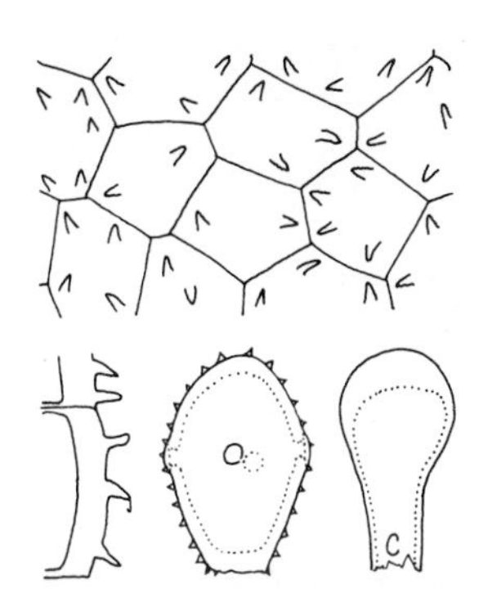

41. *RAVENELIA SOLOLENSIS* Arth. & Holw. in Arthur, Amer. J. Bot. 5:425. 1918.

Spermogonia and aecia unknown. Uredinia amphigenous and on fruits, subcuticular, cinnamon brown, with clavate or capitate paraphyses to 20 µm wide apically, the wall to 6 µm thick and chestnut brown at apex, thin and pale below; spores (25-)27-33(-38) x (16-)17-20(-22) µm, ellipsoid or obovoid, wall 1.5-2 µm thick at sides and pale golden, 2-3.5 µm thick at apex and deep golden brown, echinulate, pores equatorial, 4 or 5, with only slight caps. Telia similar to the uredinia but blackish brown and apparently without paraphyses when formed anew; spore heads (60-)70-100(-110) µm diam, (5)6-8(9) cells across, chestnut brown, each cell with (1-)3-5(6) narrowly conical spines, 2-3 µm wide at base and 2.5-4 µm long, the apex usually narrowly rounded, central cells (14-)17-20(-24) µm diam, cysts pendent, one from each marginal cell.

Type: on *Lysiloma acapulcensis* (Kunth) Benth., Solola, Guatemala, Holway No. 147 (PUR 6141). Not otherwise known.

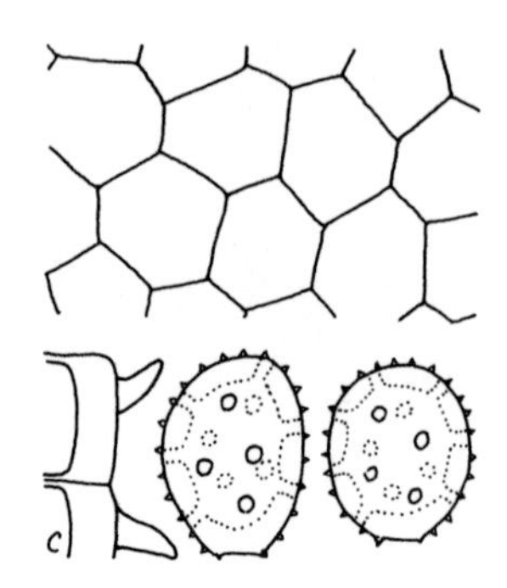

42. *RAVENELIA BELLA* Cumm. & J. W. Baxt. N. Y. Bot. Gard. Mem. 28:38. 1976.

Spermogonia and aecia unknown. Uredinia mostly on adaxial surface of leaflets, subepidermal, mostly in concentric rings, yellowish brown, without paraphyses; spores (19-)21-25(-28) x (16-)18-21(-22) µm, globoid, broadly ellipsoid or broadly obovoid, wall (2-)2.5-3(-3.5) µm thick, yellowish or golden, echinulate, pores scattered, 11-15, without caps. Telia as the uredinia but blackish brown; spore heads (75-)90-120(-130) µm diam, 7-11 cells across, chestnut brown, with 1 apically rounded tubercle 2.5-4 µm wide at base and 4-8 µm long on 1-3 ranks of the marginal cells but lacking on the central cells, central cells (10-) 12-17(-19) µm across, cysts multiseriate, pendent.

Hosts and distribution: *Cassia atomaria* L., *C. emarginata* L.: southern Sonora to San Luis Potosí and Oaxaca, Mexico.

Type: on *Cassia emarginata*, Alamos, Son., Cummins No. 70-110 (PUR 63578).

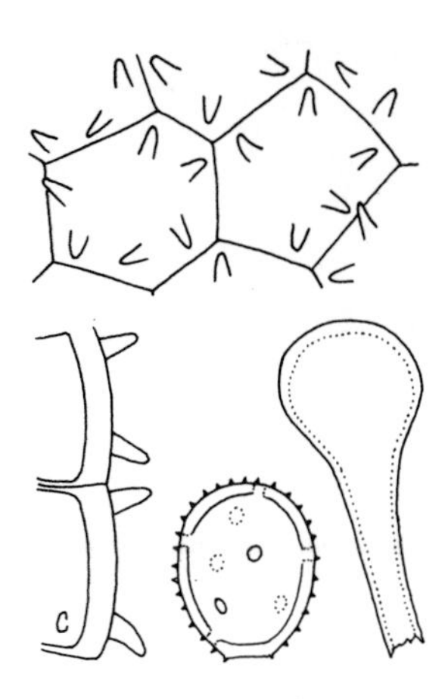

43. *RAVENELIA ANTIGUANA* Cumm. Bull. Torrey Bot. Club 67: 608-609. 1940.

Spermogonia and aecia unknown. Uredinia on abaxial surface of leaflets, subepidermal, cinnamon brown, with abundant brownish, capitate paraphyses to 25 µm wide in the head, the wall 2-3 µm thick apically, thinner below; spores 21-25 x 16-21 µm, ellipsoid or broadly so, wall 1.5 µm thick, golden, uniformly echinulate, pores 8-10, scattered, without caps. Telia on abaxial surface, subepidermal, without paraphyses, blackish brown; spore heads 65-115 µm diam, 4-6 cells across, chestnut brown, each cell with 3-7 cylindrical tubercles 2-3 µm wide at base and 4-8 µm long, central cells 18-27 µm across, cysts equal in number to the marginal cells, appressed to the cells from margin to pedicel.

Type: on *Cassia biflora* L., near Antigua, Dept. Sacatepequez, Guatemala, Standley No. 63356 (PUR 49070). Not otherwise known.

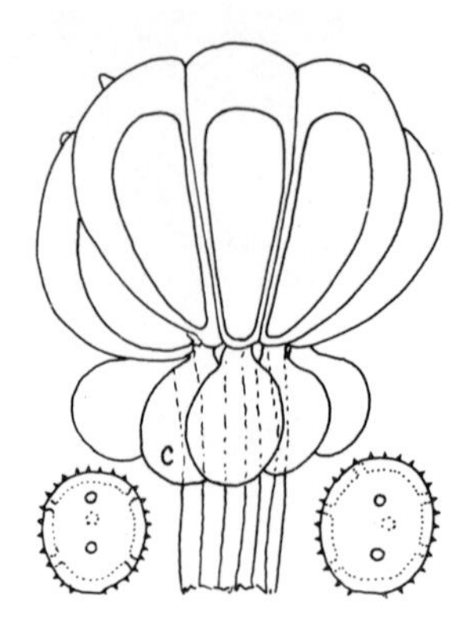

44a. *RAVENELIA CASSIAECOLA* Atk. Bot. Gaz. 16:313. 1891 var. *CASSIAECOLA*.

Spermogonia and aecia unknown. Uredinia amphigenous and on stems and pods, subcuticular, yellowish brown; spores (13-)15-18(-20) x (12-)13-16(-17) µm, globoid or broadly ellipsoid, wall 2-2.5(-3) µm thick, golden to near cinnamon brown, echinulate, pores 7 or 8, scattered, without caps. Telia amphigenous and most often confluent on stems, subcuticular, blackish brown; spore heads (44-)55-84(-90) µm diam, (3)4-6 cells across, chestnut brown, smooth or some or all cells with a low bead-like papilla, central cells (14-)17-20(-22) µm across, cysts multiseriate, with a short stipe-like base, pendent, persistent, pedicel to 100 µm long and 25 µm wide, persistent and conspicuous.

Hosts and distribution: *Cassia* spp. (*Chamaecrista* spp.): Florida to southern Ohio and Texas; also in Puerto Rico and Barbados.

Lectotype: on *Cassia nictitans* L., Auburn, Alabama, 4 Sept. 1890, Atkinson (CUP-A).

In the original publication, Atkinson cited, without dates or numbers, his own collection from Auburn, Alabama and a collection from Starkville, Mississippi by S. M. Tracy. Arthur (1) cited the type as from Auburn, but the specimen marked (but not in Atkinson's script) as type in the Atkinson Herbarium is that of Tracy. But this does not obviate the need to designate a lectotype and I follow Arthur in selecting an Atkinson collection. Moreover, it is obvious from the original diagnosis that Atkinson was describing the species from his own collections.

44b. *RAVENELIA CASSIAECOLA* Atk. var. *BERKELEYI* (Mundk. & Thirum.) Cumm. & J. W. Baxt. Mem. N. Y. Bot. Gard. 28: 38-39. 1976.
Ravenelia berkeleyi Mundk. & Thirum. Imp. Mycol. Inst. Kew Mycol. Papers 16:19-20. 1946.
Ravenelia indica auth., not Berkeley 1853.

Uredinia and telia and teliospore heads as in var. *cassiaecola* except the heads all smooth and less deeply pigmented, the compound pedicel, although equally long and persistent, usually is golden or paler, and the stipe-like base of the cysts usually is more conspicuous.

Hosts and distribution: *Cassia abusus* L., *C. hispidula* Vahl: Oaxaca and Jalisco, Mexico; also in Cuba, Venezuela, Ceylon and India.

Type: on *Cassia absus*, Damboul (Ceylon), Mar. 1868, Thwaites (PDA 517).

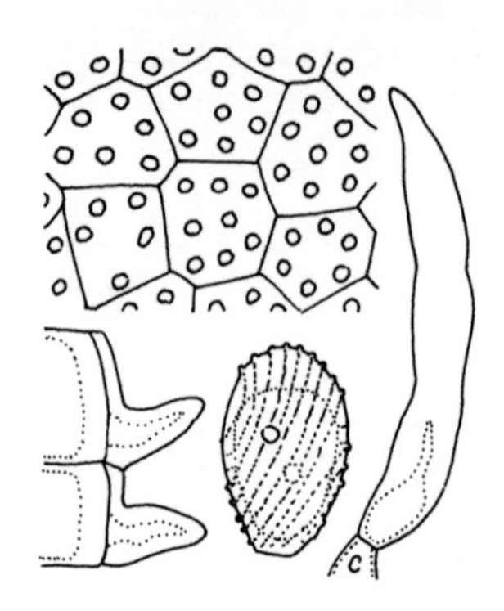

45. *RAVENELIA CORBULA* J. W. Baxt. Mycologia 58:336. 1966.

Spermogonia amphigenous, subcuticular. Aecia amphigenous, subcuticular, uredinoid, few in a group, cinnamon brown, with paraphyses similar to those in the telia but less conspicuous; spores (20-)24-28(-32) x (13-)15-18(-20) µm, obovoid or ellipsoid, wall (1-)1.5(-2) µm thick at sides, (3-)4-6(-8) µm at apex, golden or cinnamon brown, verrucose in conspicuous spiral lines, the verrucae discrete or merged, the spirals spaced (1.5-)2-2.5(-3) µm, pores 4 or 5(6), equatorial. Uredinia on abaxial surface of leaflets, minute, corbiculoid like the telia; spores similar to the aeciospores but usually shorter and with the apical wall less thickened and sometimes not thickened. Telia on abaxial surface, small, with only 1 or few spore heads, subcuticular, with abundant peripheral, colorless to golden, thick walled or mostly solid, basally united, incurved paraphyses that form a basket-like sorus; teliospore heads (55-)70-95(-120) µm diam, mostly 6-8 cells wide, central cells (12-)14-18(-21) µm diam, each marginal cell with 1 colorless or yellowish, mammiform protuberance to 16 µm long and 6-10 µm wide at base, the other cells each with 4-8 low, hemispherical tubercles 2-3 µm diam, the ornamentation part of a differentiated, pale, outer layer of the apical wall of the cells, cysts globoid, pendent, swelling and bursting.

Hosts and distribution: *Caesalpinia eriostachya* Benth: Sinaloa, Mexico.

Type: Pueblo Viejo, south of Culiacán, Cummins No. 63-740 (PUR 59937).

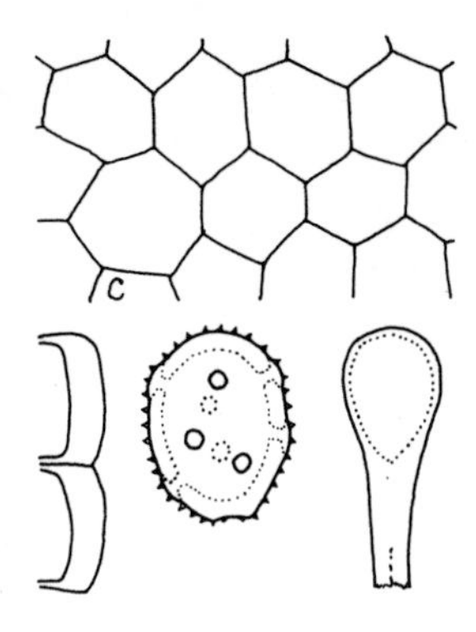

46. *RAVENELIA MESILLANA* Ellis & Barth. in Ellis & Everhart, Bull. Torrey Bot. Club 25:508. 1898.
Ravenelia longiana H. Syd. & P. Syd. Hedwigia 40 (Beibl.):128. 1901.
Ravenelia cassiae-covesii Long & Good. in Long, Bot. Gaz. 72:42. 1921.

Spermogonia amphigenous, mostly along the veins of leaflets or on petioles and stems, subcuticular. Aecia with the spermogonia, often extensively confluent, subcuticular, cinnamon brown; spores and paraphyses as in the uredinia. Uredinia amphigenous, subcuticular, cinnamon brown, with variable numbers of paraphyses that vary from cylindrical to more or less capitate, the cylindrical ones mostly with uniform wall, the capitate ones with solid stipe and thin walled head; spores (19-)22-26(-29) x (15-)17-19(-21) µm, mostly ellipsoid or oblong ellipsoid, wall (2-)2.5-3 µm thick, cinnamon brown or dark golden brown, echinulate, pores (6)7-10, irregularly bizonate in long spores, scattered in short spores, with slight caps. Telia amphigenous, subcuticular, blackish brown; spore heads (60-)80-115(-150) µm diam, (5)6-9(10) cells across, chestnut brown, smooth or rarely a few peripheral cells with 1 papilla each, central cells (11-)13-17(-19) µm across, cysts multiseriate, pendent.

Hosts and distribution: *Cassia* spp. (herbaceous species): Texas to southern Arizona and adjacent northern Mexico.

Type: on *Cassia bauhinioides* Gray, near Mesilla, New Mexico, Oct. 1897, Wooton (NY; isotype PUR 6307).

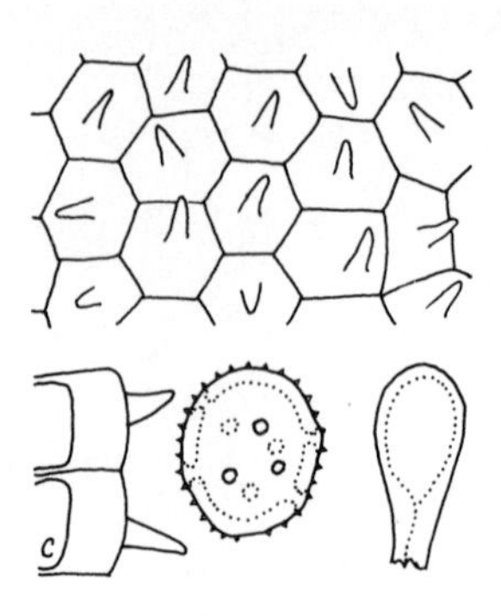

47a. *RAVENELIA SPINULOSA* Diet. & Holw. in Holway, Bot. Gaz. 31:336. 1901 var. *SPINULOSA*.

Spermogonia and aecia unknown. Uredinia amphigenous, subcuticular, yellowish brown, with numerous mostly clavate or clavate capitate, nearly colorless paraphyses, the wall 1.5-2 µm thick in the head and nearly solid in the stipe or sometimes uniform throughout; spores (18-)19-22(-24) x (14-) 16-19(-20) µm, mostly broadly ellipsoid, wall 1.5-2 µm thick, golden to near cinnamon brown, echinulate, pores 9-11, scattered, with slight or no caps. Telia amphigenous, subcuticular, blackish brown, without paraphyses; spore heads (60-)75-110(-120) µm diam, 6-9 cells across, chestnut brown, each cell with 1 (rarely 0, 2 or 3) nearly cylindrical, apically rounded tubercles 2.5-3.5 µm wide at base and 4-7 µm long, central cells (10-)12-17(-18) µm across; cysts multiseriate, pendent.

Hosts and distribution: *Cassia* spp., especially *C. biflora* L.: southern Sonora, Mexico to Honduras; also in South America and the West Indies.

Type: on *Cassia multiflora* Mart. & Gall. not Vog. (= *C. holwayana* Rose), Oaxaca, Oax., Mexico, Holway No. 3675 (S; probable isotypes Barth. N. Amer. Ured. 1484).

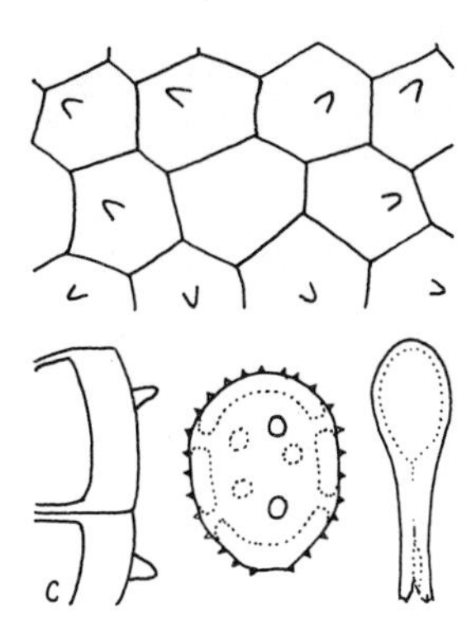

47b. *RAVENELIA SPINULOSA* Diet. & Holw. var. *PAPILLIFERA* (H. Syd. & P. Syd.) Cumm. & J. W. Baxt. N. Y. Bot. Gard. Mem. 28:39. 1976.
Ravenelia papillifera H. Syd. & P. Syd. Ann. Mycol. 1: 330. 1903.

Spermogonia amphigenous, subcuticular, in a close group. Aecia amphigenous, in a ring around the spermogonia, often confluent, uredinoid, subcuticular, cinnamon brown, with varying numbers of paraphyses but usually few; spores as the urediniospores. Uredinia amphigenous, subcuticular, cinnamon brown, with paraphyses as in var. *spinulosa*; spores (20-)22-27(-30) x (15-)16-19(-20) µm, mostly ellipsoid or broadly ellipsoid, wall 2-2.5(-3) µm thick, cinnamon brown or golden, echinulate, pores 8-10, scattered, with slight or no caps. Teliospore heads (60-)75-110(-125) µm diam, (5)6-9 (10) cells across, chestnut brown, each cell with 1 papilla 2-3 µm wide at the base and 1-3(4) µm long, or the central cells smooth, central cells (11-)13-18(-20) µm across, cysts multiseriate, pendent.

Hosts and distribution: *Cassia* spp. (perhaps only herbaceous species): Texas and Arizona south to Zacatecas, Mexico; also in the Bahamas.

Type: on *Cassia lindheimeriana* Scheele, Austin, Texas; collector not stated but probably W. H. Long (S; isotypes Sydow Ured. No. 1737).

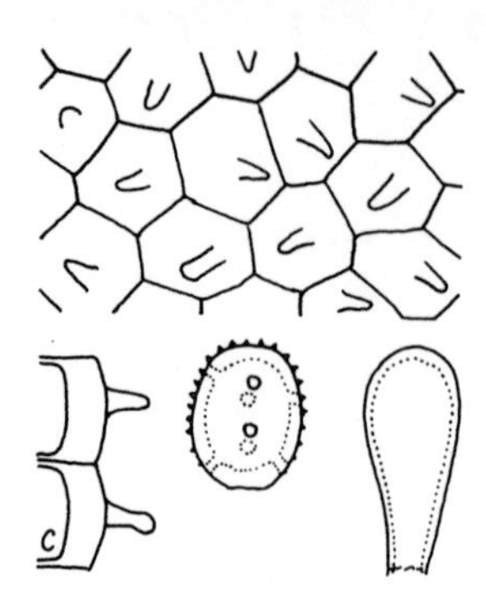

47c. *RAVENELIA SPINULOSA* Diet. & Holw. var. *MICROSPORA* Cumm. & J. W. Baxt. N. Y. Bot. Gard. Mem. 28:39. 1976.

Spermogonia and aecia unknown. Uredinia amphigenous, subcuticular, with abundant clavate or clavate capitate, pale golden paraphyses to 17 µm wide apically, the wall mostly uniformly 1.5-2 µm thick, brownish sporogenous basal cells conspicuous; spores 15-18 x 13-16 µm, mostly broadly ellipsoid or oblong ellipsoid, wall (1.5-)2-2.5 µm thick, pale golden, echinulate apically becoming smooth or nearly so at base, pores 7-9, scattered or tending to be bizonate. Telia subcuticular, without paraphyses, spore heads (70-)80-115 µm diam, (6)7-10 cells across, chestnut brown, each cell with 1 cylindrical tubercle 2.5-3 µm wide at base, 5-8(-9) µm long, apex rounded, central cells (11-)13-17 µm across; cysts multiseriate, pendent.

Type: on *Cassia nicaraguensis* Benth., San Juan Sacapulas, Guatemala, J. R. Johnson No. 1447 (PUR 49425). Known from one other Guatemalan collection on the same host species.

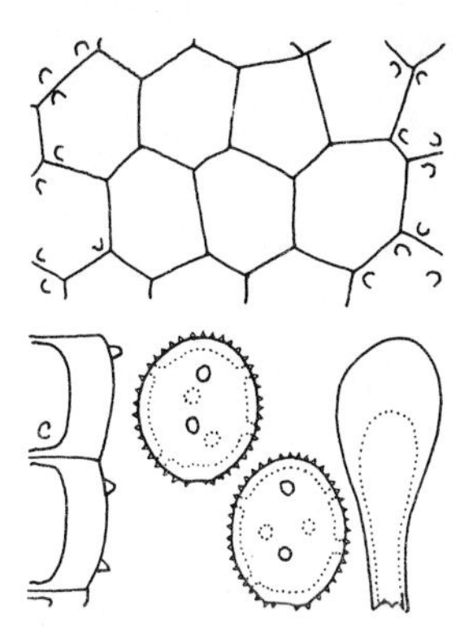

48. *RAVENELIA HUMPHREYANA* P. Henn. Hedwigia 37:278. 1898.
Ravenelia pulcherrima Arth. Bot. Gaz. 39:395. 1905.

Spermogonia and aecia unknown. Uredinia amphigenous, subcuticular, cinnamon brown, with abundant, apically chestnut brown or rarely only yellowish, clavate, spatulate or more or less capitate paraphyses to 15 µm wide apically, spores (15-)17-19(-21) x (13-)14-17(-18) µm, broadly ellipsoid or globoid, wall 1-1.5(-2) µm thick, golden or dull cinnamon brown, closely verrucose echinulate with narrowly rounded or acute cones, pores 6-9, scattered, with slight or no caps. Telia similar to the uredinia but without paraphyses (when formed de novo), blackish brown; spore heads (60-)70-100(-120) µm diam, 5-8 cells across, chestnut brown, smooth or each with 3-5 low, bead-like tubercles, these commoner toward the periphery, central cells (11-)13-18(-20) µm across, cysts of same number as marginal cells, adherent but swelling to appear semipendent.

Hosts and distribution: *Caesalpinia* (especially *pulcherrima* (L.) Sw.): southern Florida, southern Texas and central Sonora, Mexico south to Guatemala and Costa Rica; also in the Caribbean Islands.

Type: on *Cassia* (now considered to be *Caesalpinia pulcherrima*), Kingston, Jamaica, Humphrey (B; isotype PUR 6362).

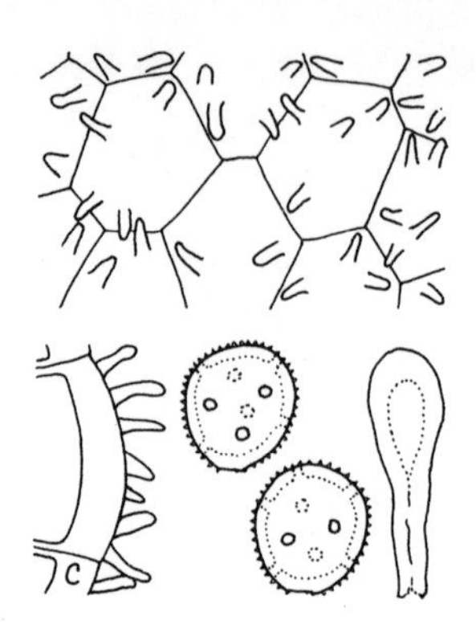

49. *RAVENELIA INCONSPICUA* Arth. Bot. Gaz. 39:395. 1905.

Spermogonia in close groups on the abaxial side of leaflets, subcuticular. Aecia on abaxial surface, subcuticular, uredinoid, in a ring around the spermogonia, paraphyses and spores as in uredinia. Uredinia on abaxial surface, subcuticular, with abundant variable but mostly clavate, yellowish or golden paraphyses to 15 µm wide, the wall in upper part more or less uniformly 3-6 µm thick, the stalk solid; spores 14-17 µm diam, essentially globoid, wall (1.5-)2(-2.5) µm thick, golden brown, closely verrucose echinulate, pores 7-10, scattered, with slight or no caps. Telia on abaxial surface, subcuticular, blackish brown; spore heads (60-)65-85(-100) µm diam, chestnut brown, 4-7 (8) cells across, each cell with (2-)4-7(-10) cylindrical tubercles 2-3 µm wide and 4-7 µm long, the central cells usually with fewer tubercles than the peripheral cells, central cells (13-)18-22(-24) µm across, cysts of same number as marginal cells, appressed or semipendent.

Type: on *Cassia* (or *Caesalpinia*) sp. (considered here to be *Caesalpinia* sp.), Zapotlán, Jal., Mexico, Holway No. 5135 (PUR 6352). Not otherwise known.

This species differs from *R. humphreyana* because of the more numerous and longer tubercles on the spore heads and the larger central cells.

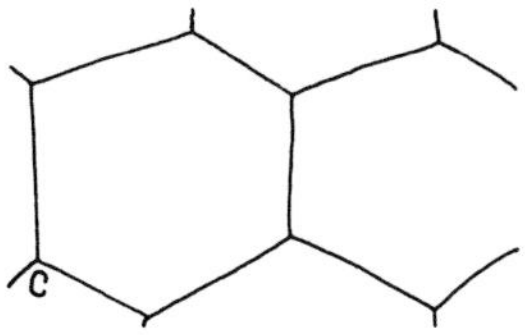

50. *RAVENELIA OPACA* Diet. Hedwigia 34:291. 1895.

Spermogonia in small groups, subcuticular, amphigenous. Aecia and uredinia wanting. Telia amphigenous or mostly on the adaxial surface of leaflets, subepidermal, blackish; spore heads (75-)90-120(-130) µm diam, (4)5-7 cells across, dark chestnut brown, relatively opaque, smooth, central cells (18-)22-30(-32) µm across; cysts multiseriate, pendent.

Type: on *Gleditsia triacanthos* L., Clear Creek, Union County, Illinois, Earle (S; isotypes Seymour & Earle Econ. F. No. 203). Known only from this locality.

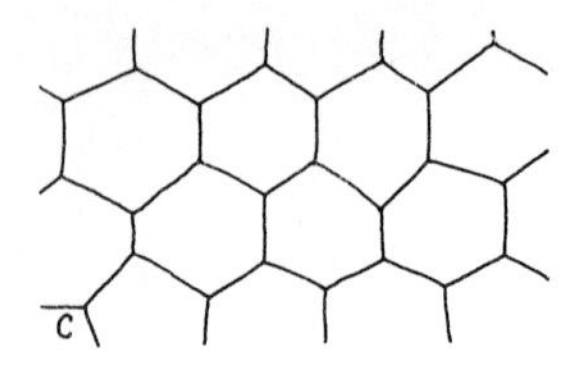

51. *RAVENELIA LONCHOCARPICOLA* Speg. var. *MERA* (Cumm.) J. W. Baxt. Mycologia 60:43. 1968.
Ravenelia mera Cumm. Bull. Torrey Bot. Club 70:78. 1943.
Ravenelia lonchocarpicola Speg. var. *ROBUSTA* J. W. Baxt. Mycologia 60:45. 1968.

Spermogonia in close groups, subcuticular, amphigenous. Aecia and uredinia wanting. Telia amphigenous or usually more abundant on the adaxial surface, usually closely grouped around the spermogonia, subepidermal, blackish; spore heads (50-)70-110(-122) µm diam, (5)6-10(11) cells across, chestnut brown, smooth, central cells (9-)11-15(-17) µm across, cysts of same number as marginal cells, appressed to the cells from margin to pedicel, occasionally slightly decurrent into the pedicel.

Hosts and distribution: *Lonchocarpus* spp.: southern Sonora and Tamaulipas, Mexico south to Costa Rica.

Type: on *Lonchocarpus michelianus* Pitt., Jutiapa, Guatemala, Standley No. 75109 (PUR 50340).

R. lonchocarpicola var. *lonchocarpicola* has not been recorded in North America.

52. *RAVENELIA LAEVIS* Diet. & Holw. in Holway, Bot. Gaz. 24: 35. 1897.

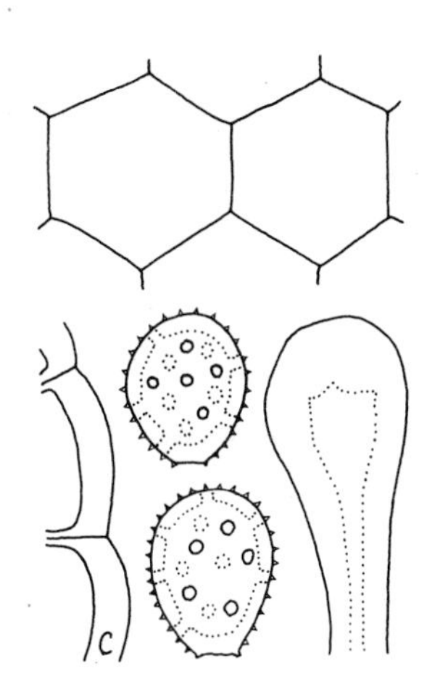

Spermogonia in small groups on the adaxial surface of leaflets, subcuticular. Aecia mostly in close groups opposite the spermogonia, uredinoid, often along the nerves, otherwise as the uredinia. Uredinia mostly on the abaxial surface, subepidermal, dull cinnamon brown, with abundant capitate or clavately capitate paraphyses 18-28(-30) μm wide apically, wall more or less uniformly 5-8 μm thick or to 12 μm at apex, golden to chestnut brown apically, paler below; spores (20-)22-25(-27) x (17-)18-20(-22) μm, mostly broadly ellipsoid, wall (2-)2.5(-3) μm thick, dull golden or cinnamon brown, echinulate, pores many, 12-16 but difficult to count, scattered, without caps. Telia amphigenous or mostly on the abaxial surface, probably without paraphyses when formed de novo, chestnut brown or blackish; spore heads (60-)75-120(-130) μm diam, (3)4 or 5(6) cells across, chestnut brown, smooth, central cells (20-)22-28(-30) μm across, cysts of same number as marginal cells, appressed to undersides of cells from margin to pedicel.

Hosts and distribution: *Indigofera* spp.: Mexico from southern Sonora and San Luis Potosí south to Oaxaca.

Lectotype: on *Indigofera* sp., Eslava, near Mexico City, 3 Oct. 1896, Holway (S). Arthur (1) designated this collection as the type and it is here formalized as the lectotype because two collections were cited in the original. Isotype PUR 6457; probable isotypes Barth. N. Amer. Ured. No. 180.

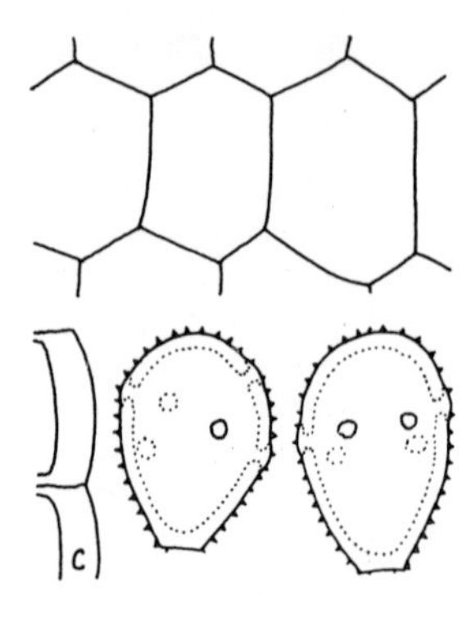

53. *RAVENELIA EPIPHYLLA* (Schw.) Diet. Hedwigia 33:27. 1894.
Sphaeria epiphylla Schw. Schr. Nat. Ges. Leipzig 1:40. 1822.

Spermogonia in small groups on adaxial side of the leaflets, subepidermal. Aecia amphigenous, closely associated with the spermogonia, otherwise as the uredinia. Uredinia amphigenous, subepidermal, yellowish brown; spores (23-)27-33(-38) x (17-)19-22(-24) μm, mostly obovoid or ellipsoid, wall 1.5-2 μm thick, golden brown or paler, echinulate with short thin aculeae, pores (5)6(7) usually equatorial but may be scattered in short broad spores, with slight or no caps. Telia amphigenous and on rachis and stems, subepidermal, blackish; spore heads (75-)80-125 (-140) μm diam, often irregular in outline, 5-8(9) cells across, chestnut brown, smooth, central cells (16-)18-24 (-30) μm across, cysts of same number as marginal cells, appressed to the cells and decurrent into the persistent pedicels.

Hosts and distribution: *Tephrosia* spp., especially *T. virginiana* (L.) Pers.: the United States mostly east of the Mississippi River and in Mexico and El Salvador.

Type: on *Galega virginiana* (= *Tephrosia v.*), North Carolina (PH).

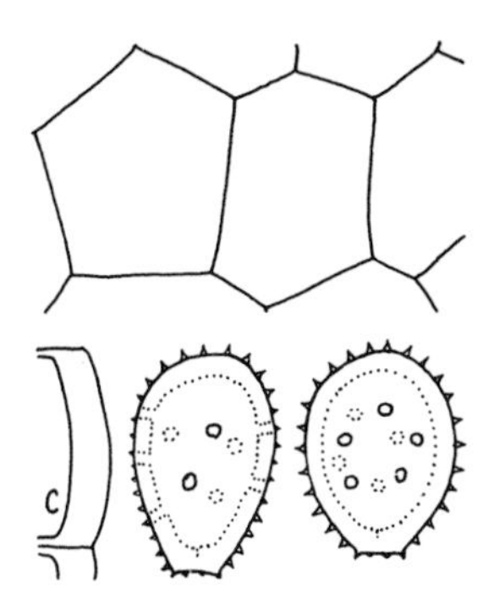

54. *RAVENELIA SIMILIS* (Long) Arth. Bot. Gaz. 39:396. 1905.
Pleoravenelia similis Long, Bot. Gaz. 35:128. 1903.

Spermogonia in small groups on the adaxial side of leaves, subcuticular. Aecia opposite the spermogonia in small groups, otherwise as the uredinia. Uredinia amphigenous, subepidermal, about cinnamon brown; spores (20-)23-27(-30) x (17-)19-23(-25) µm, mostly broadly ellipsoid or obovoid, wall (2-)2.5-3 µm thick, cinnamon brown or dark golden brown, echinulate, pores mostly grouped in opposite sides of the spore, 9-12, scattered but tending to be in the equatorial region. Telia as the uredinia except blackish brown; spore heads (65-)80-140(-150) µm diam, (4)5-7(8) cells across, chestnut brown, smooth, central cells (17-)20-26(-30) µm across, cysts of same number as marginal cells, adherent to the cells or separating easily, decurrent into the pedicel.

Hosts and distribution: *Brogniartia* spp.: Jalisco, Mexico.

Type: on *Brogniartia* sp., Guadalajara, 14 Sept. 1899, Holway No. 3145 (error for 3415) (BPI).

It is obvious that the published collection number 3145 is an error. Other Holway numbers in the range of 3145 were taken in 1898, whereas No. 3400 was collected 12 Sept. 1899 and No. 3424 was collected 15 Sept. 1899.

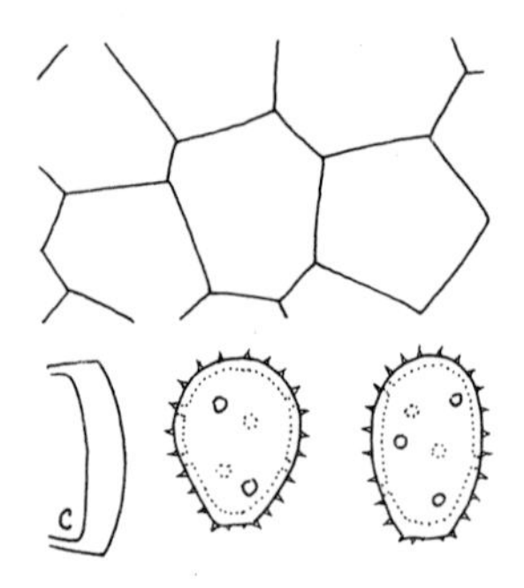

55. *RAVENELIA CAULICOLA* Arth. N. Amer. Flora 7:143. 1907.

Spermogonia and aecia unknown. Uredinia amphigenous and caulicolous, often grouped along the veins, subepidermal, yellowish brown; spores (17-)19-23(-28) x (12-)14-16 (-18) µm, mostly ellipsoid or obovoid, wall (1-)1.5(-2) µm thick, golden brown or paler, echinulate with narrow spines, pores scattered, about 8, small, obscure, with slight or no caps. Teliospores in the uredinia or the telia in separate, usually confluent, caulicolous groups; spore heads (65-)75-105(-120) µm diam, often irregular, (4)5-7(8) cells across, chestnut brown, smooth, central cells (14-)17-22 µm across, cysts of same number as marginal cells, appressed to the cells and decurrent into the pedicel.

Hosts and distribution: *Tephrosia leicocarpa* Gray: central Sonora, Mexico. Also in the Bahamas, Puerto Rico and South America.

Type: on *Cracca cinerea* (L.) Morong (= *Tephrosia cinerea*), Bahamas, Britton & Millspaugh No. 2807 (PUR 6408).

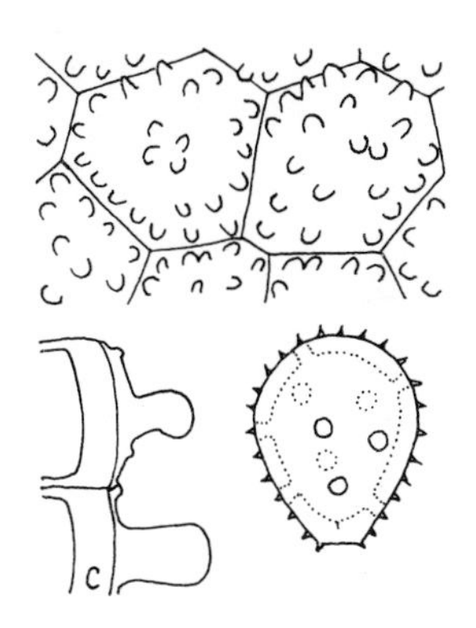

56. *RAVENELIA BROGNIARTIAE* Diet. & Holw. in Holway, Bot. Gaz. 24:35. 1897.

Spermogonia on adaxial side of leaves, subcuticular. Aecia grouped around the spermogonia, uredinoid, otherwise as the uredinia. Uredinia amphigenous, subepidermal, about cinnamon brown; spores (22-)24-28(-32) x (18-)19-22(-24) µm, mostly obovoid, wall 2-3 µm thick, golden or dull cinnamon brown, echinulate, pores 9-12, scattered, mostly in opposite slightly flattened sides, without caps. Telia as the uredinia except blackish brown; spore heads (70-)85-140(-150) µm diam, (4)5-7(8) cells across, chestnut brown, each cell with few or usually many (to 30) bead-like verrucae 2-4 µm high and 2-3 µm wide (occasional cells are smooth), each or most marginal cells in addition have a single cylindrical or slightly capitate tubercle 4-14 µm long and 3-8 µm wide, central cells (17-)20-25(-28) µm across, cysts of same number as marginal cells, appressed to the cells and decurrent into the pedicel.

Hosts and distribution: *Brogniartia* spp.: central and southern Mexico.

Type: on *Brogniartia* sp., Cuernavaca, Mor., Holway (S).

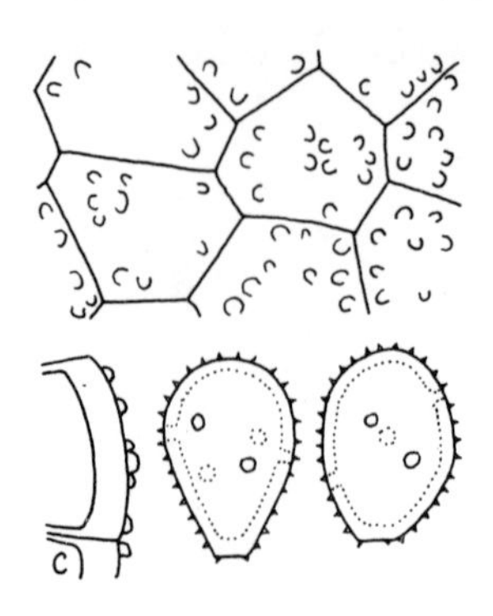

57. *RAVENELIA TALPA* (Long) Arth. Bot. Gaz. 39:396. 1905.
Pleoravenelia talpa Long, Bot. Gaz. 35:130-131. 1903.

Spermogonia and aecia unknown. Uredinia amphigenous but commonly on the adaxial surface, subepidermal, yellowish brown; spores (19-)21-25(-27) x (15-)17-20(-22) µm, mostly obovoid or broadly ellipsoid, wall 1.5(-2) µm thick, echinulate, pores 4-6, commonly 5, mostly in the equatorial region but rarely strictly zonate. Telia as the uredinia but blackish brown; spore heads (55-)65-95(-110) µm diam, often irregular in shape, (3)4-6(-8) cells across, chestnut brown, each cell with 3-15 low, bead-like verrucae 1.5-2 µm wide and high, central cells (18-)21-27(-33) µm across, cysts of same number as marginal cells, appressed to underside of spore head but decurrent on the pedicel.

Hosts and distribution: *Tephrosia* spp.: Jalisco and Oaxaca, Mexico and Honduras.

Type: on *Tephrosia talpa* Wats., Oaxaca, Oax., Holway No. 3679 (BPI; isotype PUR 6365).

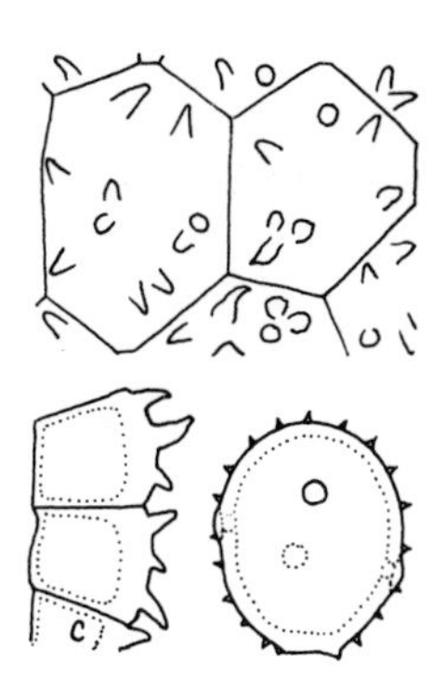

58. *RAVENELIA RUBRA* J. W. Baxt. Mycologia 57:83. 1965.

Spermogonia and aecia unknown. Uredinia amphigenous, often in circles, subepidermal, cinnamon brown; spores (19-) 21-25(-27) x (16-)18-21(-23) µm, mostly broadly ellipsoid or globoid, wall 1.5-2(-2.5) µm thick, dull brown, echinulate, pores mostly 3-5, approximately equatorial, with slight caps. Telia like the uredinia but blackish brown; spore heads (65-)80-115(-125) µm diam, mostly 4-6 cells across, the central cells 20-28(-30) µm diam, each cell with 5-12 irregularly arranged, colorless to brownish tubercles (2-) 3-5(-7) µm long, spore heads dark reddish chestnut, rather opaque, cysts of the same number as marginal cell, appressed to underside of head; pedicel colorless, obviously compound, broad, to at least 35 µm wide, persistent.

Hosts and distribution: *Brogniartia glabrata* Hook. & Arn.: Nayarit and Sinaloa, Mexico.

Type: on *Brogniartia* sp., Mex. hgw 40, 11 mile E of hgw 15, Sinaloa, Cummins No. 63-647 (PUR 59931).

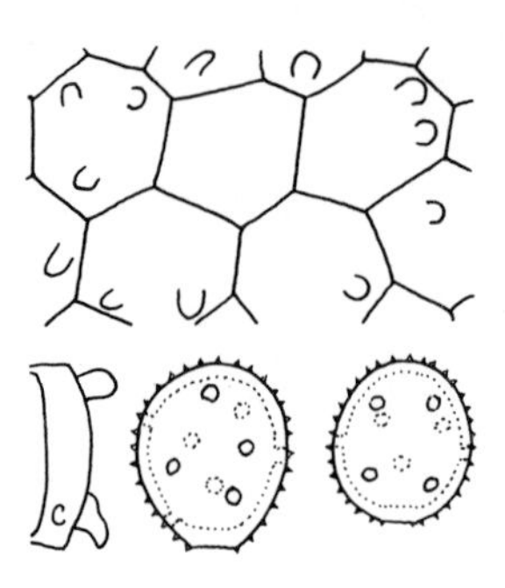

59. *RAVENELIA IRREGULARIS* Arth. N. Amer. Flora 7:142-143. 1907.

Spermogonia on adaxial surface of leaflets. Aecia opposite the spermogonia, subepidermal, uredinoid, yellowish brown; spores (19-)20-22(-24) x (16-)17-19(-20) µm, mostly broadly ellipsoid, wall (1-)1.5 µm thick, pale yellowish to essentially colorless, finely echinulate, pores scattered, about 8, without caps. Uredinia and spores similar to the aecia and spores except occurring on both leaf surfaces. Telia amphigenous or mostly on adaxial surface, subepidermal, blackish brown; spore heads 60-90 µm diam, 4-6(7) cells across, chestnut brown, each cell with 3-5 broadly rounded tubercles 2.5-4 µm wide at base and 3-5(-7) µm long, or these often lacking on the central cells, central cells (15-)16-20(-22) µm across, cysts of same number as marginal cells, decurrent into the pedicel.

Type: on *Cracca macrantha* (= *Tephrosia macrantha* B. L. Rob. & Greenm.), Etzatlán, Jal., Mexico, Holway No. 5100 (PUR 6407). Not otherwise known.

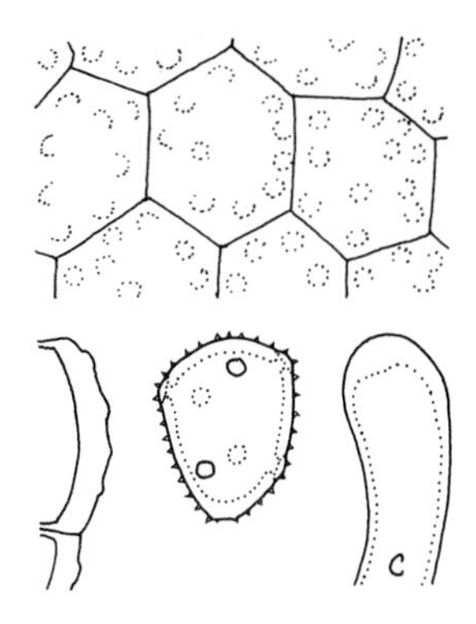

60. *RAVENELIA PISCIDIAE* Long, J. Mycol. 12:234. 1906.
Ravenelia piscidiae ssp. *rugosa* Hennen & Cumm. Rept. Tottori Mycol. Inst. 10:179. 1973.

Spermogonia and aecia unknown. Uredinia mostly on the abaxial surface of leaves, subepidermal, dark cinnamon brown, with abundant peripheral, incurved, cylindrical or clavate paraphyses 9-17 µm wide apically, gradually narrowing basally, the wall thin below but usually becoming 2.5-7 µm thick dorsally and apically, chestnut brown or paler apically, pale basally; spores (18-)20-22(-24) x (14-)16-20 µm, mostly broadly ellipsoid or obovoid, commonly angular and frequently more or less square in end view, wall 1-1.5(-2) µm thick, cinnamon brown or slightly darker, echinulate, pores 6-8, bizonate or tending to be so, near the ends, with small caps. Telia commonest on the adaxial surface, blackish, without paraphyses; spore heads (60-)75-110(-120) µm, (3)4-6 cells across, chestnut brown, smooth or each cell with several low, scarcely visible warts 3-4 µm wide, these more obvious on immature than mature heads, central cells (15-)18-23(-25) µm across, cysts of same number as marginal cells, appressed from margin to pedicel.

Hosts and distribution: *Piscidia* spp.: southern Florida and the east and west coasts of Mexico; also in Cuba.

Type: on *Piscidia erythrina* L., Miami, Florida, 25 Mar. 1903, Holway (BPI; isotype PUR 6442; probable isotypes Barth. N. Amer. Ured. Nos. 677, 780).

The teliospore heads of ssp. *rugosa* are more obviously warted, both when mature and immature, but there is no other distinction.

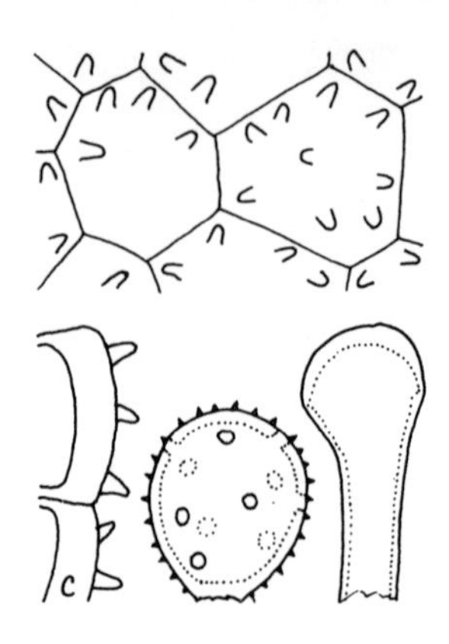

61. *RAVENELIA INDIGOFERAE* Tranz. in Dietel Hedwigia 33:369. 1894.

Spermogonia few in a close group, amphigenous, subcuticular. Aecia mostly on abaxial surface of leaflets or on the rachis, closely grouped, usually circinately, with the spermogonia, uredinoid, otherwise as the uredinia. Uredinia mostly on abaxial surface, subepidermal, yellowish brown, with abundant, mostly capitate paraphyses, the head 15-30 µm wide, the wall cinnamon brown apically and usually 2-3 µm thick, colorless and thinner below, occasionally thick throughout, the head commonly collapsing downward; spores (21-)23-25(-27) x (18-)20-22(-24) µm, mostly broadly ellipsoid or obovoid, wall (1-)1.5(-2) µm thick, dark golden or cinnamon brown, echinulate, pores scattered, 9-12, without caps. Telia mostly on abaxial surface but may be on stems, subepidermal, blackish brown; spore heads (65-)85-120(-127) µm diam, (3)4-6 cells across, chestnut brown, each cell with (0-)3-7(-10) cylindrical tubercles 2-3 µm wide and 3-7 µm long, central cells (18-)21-26(-30) µm across, cysts of same number as marginal cells, appressed to the cells from margin to pedicel.

Hosts and distribution: *Indigofera* spp.: southern Arizona to Guatemala; also in the West Indies, South America, Africa, China and the Philippines.

Type: on *Indigofera palmeri* Wats., rocky hills near Guadalajara, Jal., Mexico, Pringle (S).

7. SPUMULA Mains

Mycologia 27:638. 1935.

Spermogonia subcuticular, conical, type 7 (16). Aecia subepidermal in origin, erumpent, with peridium; spores catenulate. Uredinia subepidermal in origin, erumpent; spores borne singly on pedicels. Telia subepidermal in origin, erumpent; spores 1 celled, adherent in discoid heads subtended by colorless cysts, with 1 germ pore in each cell, pedicel simple; basidium external.

Type species: *Spumula quadrifida* Mains.

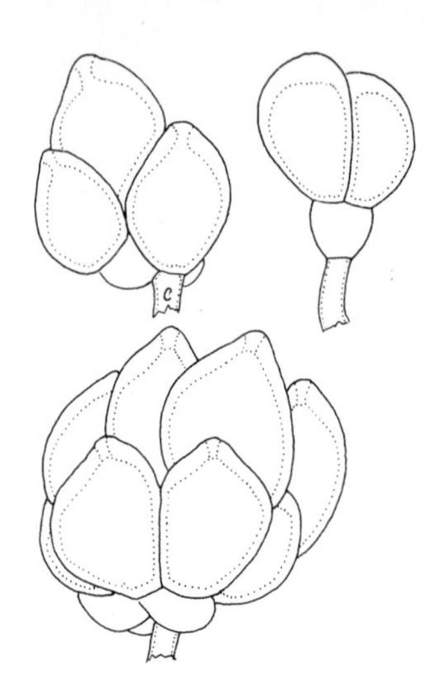

1. *SPUMULA HETEROMORPHA* J. W. Baxt. Mycologia 58:337. 1966.

Spermogonia and aecia systemic in small witches' brooms. Aecia with cylindrical peridium, brownish; spores (16-)19-27 (-30) x (13-)16-19(-20) µm, ellipsoid, oblong ellipsoid or globoid, often apiculate or tailed, wall (1.5-)2-2.5(-3) µm thick, yellowish or brownish, finely verrucose, pores often obvious, 7-9, scattered. Uredinia unknown. Telia on either or both leaf surfaces, chocolate brown; spore heads variable, (1-)3-10(-20) celled, commonly 3 celled with 2 basal and 1 apical, wall at sides 1.5-2.5 µm thick, (3-)4-5(-6) µm thick, chestnut brown or golden brown, smooth, pore apical in each cell, often in a paler area, cysts or cyst like cells present or absent; main pedicel simple.

Hosts and distribution: *Acacia farnesiana* (L.) Willd.: central Mexico.

Type: near Durango, Dgo., Cummins No. 63-617 (PUR 59934).

Baxter described this fungus in *Spumula* because of the simple pedicel but it has some features of *Ravenelia*, some of *Hapalophragmium*, and some of *Dicheirinia*. The aecial stage has been mistaken for that of *Ravenelia hieronymii* Speg., which does not occur in North America.

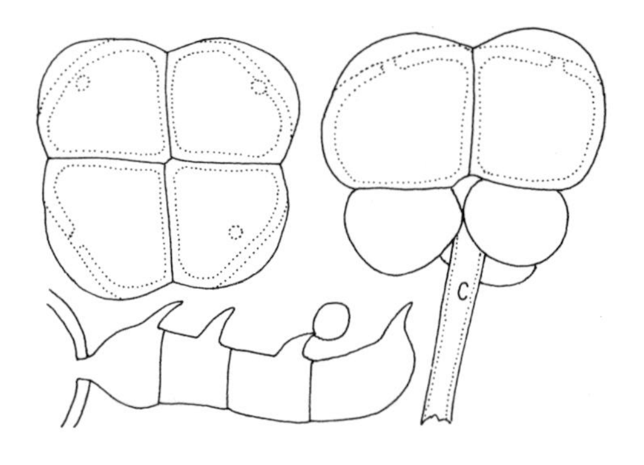

2. *SPUMULA QUADRIFIDA* Mains, Mycologia 27:638. 1935.

Spermogonia unknown. Aecia on abaxial surface of leaves, scattered or few in a group, peridium cupulate to cylindrical; spores 16-22 x 12-16 μm, broadly ellipsoid, wall colorless, 1 μm thick, finely verrucose. Uredinia apparently lacking. Telia mostly on abaxial surface, exposed, pale cinnamon brown, rather loose but not pulverulent; spores (38-)42-53 (-55) μm wide when seen from the top or bottom and square or nearly so, mostly about 35 μm high, composed of 4 cells, the wall 1.5-2 μm thick, golden brown except a broad, pale, umbo 3-5(-6) μm over each pore, pore 1 in each cell somewhat lateral to the apex, cysts colorless, pendent, 1 per cell; pedicel colorless.

Type: on *Calliandra bijuga* Rose, Trail San Sebastian to Real Alto, Jalisco, Mexico, Ynes Mexia No. 1638 (MICH; isotype PUR 63675). Not otherwise known.

8. CYSTOMYCES H. Sydow

Ann. Mycol. 24:290. 1926.

Spermogonia subepidermal, conical, type 5 (16). Aecia and uredinia unknown. Telia subepidermal in origin, erumpent; spores composed of 3, rarely 4 or 5, radially adherent cells, probably 1 germ pore in each cell, each cell subtended by a colorless, hygroscopic cyst, the spore wall pigmented, pedicel simple, attached to the cysts; basidium external.

Type species: *Cystomyces costaricensis* Syd.

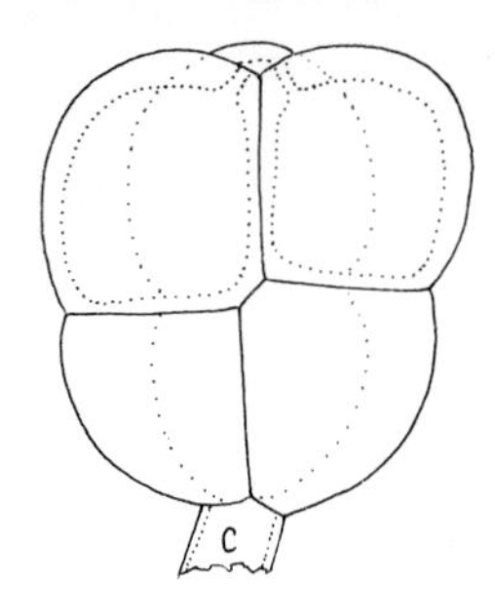

1. *CYSTOMYCES COSTARICENSIS* H. Syd. Ann. Mycol. 24:290-91. 1926.

Spermogonia in small groups on the adaxial leaf surface. Aecia and uredinia lacking. Telia on the adaxial surface surrounding the spermogonia, early exposed, conspicuous, blackish; spores (21-)24-29(-31) µm high, (40-)42-54(-62) µm wide, usually composed of 3 equal cells, occasionally of 4 or 5, wall uniformly 3-4 µm thick or slightly thicker apically, dark chestnut brown, smooth, subtended by colorless cysts of the same number and approximately the same size as the cells, pedicel attached to the cysts, yellowish, collapsing, to 200 µm long but often broken at or near the cysts.

Type: on undetermined Leguminosae, San Pedro de San Ramon, Costa Rica, 5 Feb. 1925 (holotype probably lost; isotypes Sydow F. exot. exsic. No. 595). Known from this and one other collection from the same area.

The host probably is a species of *Lonchocarpus*, perhaps *L. guatemalensis* Benth.

9. DICHEIRINIA Arthur

N. Amer. Flora 7:147. 1907.

Spermogonia subcuticular, conical, type 7 (16). Aecia subepidermal in origin, erumpent, uredinoid; spores borne singly on pedicels. Uredinia subepidermal in origin, erumpent, mostly with elaborate paraphyses; spores borne singly on pedicels. Telia subepidermal in origin, erumpent; spores borne singly on pedicels, 2 or 3 cells, each cell subtended by 1 apical cell of the pedicel, pedicel simple below the apical cells, spore wall pigmented and mostly adorned with block like warts; basidium external.

Type species: *Dicheirinia binata* (Berk. & Curt.) Arth.

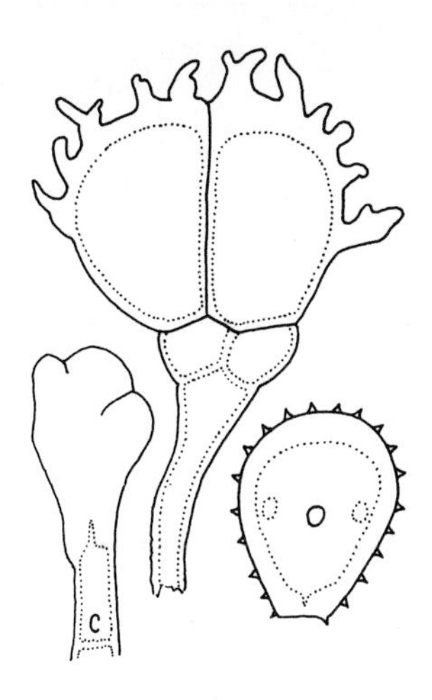

1. *DICHEIRINIA BINATA* (Berk. & Curt.) Arth. N. Amer. Flora 7:147. 1907.
   Triphragmium binatum Berk. & Curt. Proc. Amer. Acad. 4: 125. 1859.

Spermogonia amphigenous and on petioles. Aecia around the spermogonia, becoming confluent, uredinoid, cinnamon brown, with paraphyses few or numerous; spores essentially as the urediniospores. Uredinia mostly on abaxial leaf surface, about cinnamon brown, with numerous intersperced, colorless, refractive paraphyses, the head irregular and knoblike, to 30 μm wide, solid or nearly so; spores (28-)32-35(-38) x (22-)25-29(-31) μm, asymmetrical but obovoid as usually seen, wall (2.5-)3-3.5(-4) μm thick at sides, commonly thicker apically, strongly echinulate, golden brown, pores 3, rarely 4, with 2 in the rounded side side and 1 in the flattened side (a view not commonly seen). Telia on abaxial surface, exposed, blackish, with paraphyses as in the uredinia; spores 2 celled, (30-)33-38(-42) μm high, (38-)45-55(-60) μm wide when both cells show equally, wall (1.5-)2-3 μm thick at sides, 4-6 μm apically, chestnut brown, with numerous simple or branched projections to 9 μm long and 6μm wide at base, pedicel colorless, the 2 apical cells distinct, to 50 μm long but usually broken near the apical cells.

Hosts and distribution: *Erythrina glauca* Willd., *E.* sp.: Guatemala to Panama; also in Cuba, the West Indies and South America.

Type: on leaves of *Lecythea pezizaeformis*, Nicaragua, Chas. Wright (K; isotype PUR 6516).

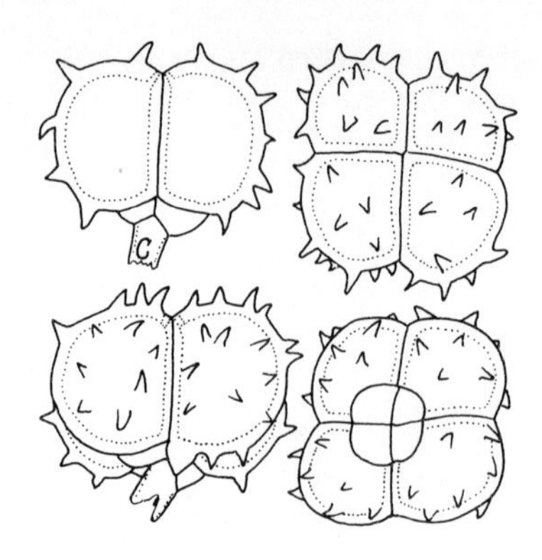

2. *DICHEIRINIA SPINULOSA* (J. W. Baxt.) Hennen & Cumm. Rept. Tottori Mycol. Inst. 10:170. 1973.
Diorchidium spinulosum J. W. Baxt. Mycologia 56:287. 1964.

Spermogonia, aecia and uredinia unknown. Telia amphigenous and on pods, early exposed, pulverulent, chocolate brown; spores 2 to 8 celled, mostly 2 or 4 celled, 26-42(-48) µm wide by 20-32(-36) µm high, wall uniformly (1.5-)2-2.5(-3) µm thick, chestnut brown, with scattered, broad based spines 2-4 µm long, pedicel colorless, with 1·inconspicuous apical cell for each cell of the spore head, commonly broken at or near the spore.

Hosts and distribution: *Cassia* or *Leucaena sp.:* coastal region of Colima, Mexico.

Type: on fruits of *Cassia* from Mexico (locality unknown), intercepted at the Plant Quarantine Station, Nogales, Arizona by Kaiser and Noel (PUR).

Hennen and Cummins (*loc. cit.*) suggested that the host of the type and of two foliar specimens collected by Cummins north of Manzanillo, Colima may be of the genus *Leucaena*.

10. DIABOLE Arthur

Bull. Torrey Bot. Club 49:194. 1922.

Spermogonia subcuticular, conical, type 7 (16). Aecia and uredinia unknown. Telia subcuticular in origin, erumpent; spores 1 celled, borne in pairs on nonhygroscopic apical cells of the pedicel, usually 2 or 3 pairs on each pedicel, pedicel simple below the apical cells, spore wall pigmented; basidium doubtless external, germ pores uncertain.

Type species: *Diabole cubensis* (Arth. & J. R. John.) Arth.

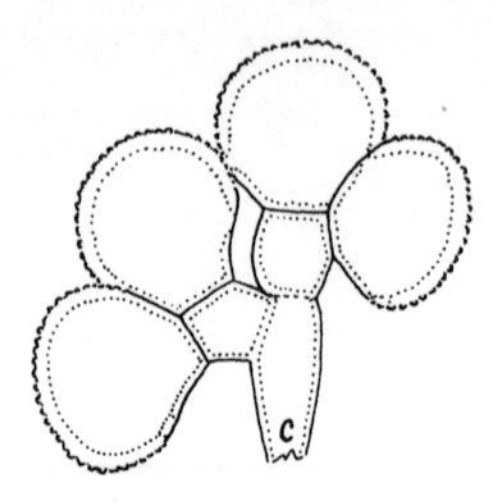

1. *DIABOLE CUBENSIS* (Arth. & J. R. John.) Arth. Bull. Torrey Bot. Club 49:194. 1922.
   Uromycladium (?) *cubense* Arth. & J. R. John. Mem. Torrey Bot. Club 17:119. 1918.

Spermogonia few on adaxial leaf surface. Telia amphigenous or mostly on abaxial surface, few in a group and becoming confluent, early exposed, blackish brown, pulverulent; spores 15-20 µm diam but commonly slightly wider than high, globoid, depressed globoid or broadly obovoid, borne in pairs on common brownish, short apical cells of a rather long, thin walled pedicel, usually 2 or 3 pairs on each pedicel, wall of spore pale basally and smooth, 0.5-1 µm thick, chestnut brown, 1.5-2 µm thick and verrucose on upper 2/3 or 3/4 of spore, the verrucae small and spaced about 0.5-1 µm or often in lines or tending to unite in series, pores obscure as to location and number but probably 2 or 3 at the junction of the pale thin part of the wall and the pigmented, verrucose part.

Hosts and distribution: *Mimosa* sp.: Nayarit, Mexico and El Salvador; also in Cuba.

Type: on *Mimosa pigra* L., Soledad Cienfuegos, Cuba, Johnston No. 191 (PUR 6530).

11. SPHAEROPHRAGMIUM Magnus

Ber. Dtsch. Bot. Ges. 9:121. 1891.

Spermogonia and aecia unknown. Uredinia subepidermal in origin, erumpent, with paraphyses; spores borne singly on pedicels. Telia subepidermal in origin, erumpent; spores of 4 to several cells by both horizontal and vertical septa, pigmented, ornamented with simple or commonly apically furcate projections, borne singly on simple pedicels, germ pores uncertain; basidium external.

Type species: *Sphaerophragmium acaciae* (Cooke) Magn.

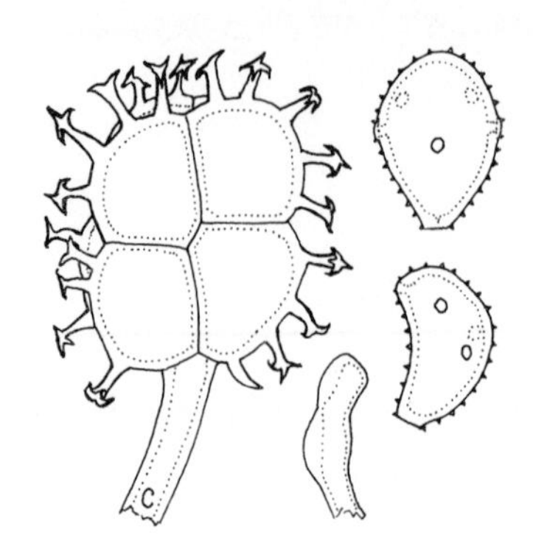

1. *SPHAEROPHRAGMIUM ACACIAE* (Cooke) Magn. Ber. Dtsch. Bot. Ges. 9:121. 1891.
Triphragmium acaciae Cooke, Grevillea 8:94. 1880.

Spermogonia and aecia unknown. Uredinia on the abaxial leaf surface, yellowish brown, with inconspicuous, peripheral, incurved, dorsally thick walled paraphyses, seldom seen except in sections; spores (18-)22-27(-30) μm long, asymmetrical, mostly obovoid in one view and (15-)17-20(-22) μm wide, mostly reniform when rotated 45 degrees and 11-15 μm wide, wall 1.5-2 μm thick, about cinnamon brown or golden, finely echinulate, pores 4 or 5, without definite arrangement but tending to be in or above the equatorial zone, with slight or no caps. Telia as the uredinia but dark brown; spores (30-)35-45(-55) x (22-)26-35(-40) μm, globoid or nearly so, composed of 6-10, commonly 8, cells, the wall uniformly 2-2.5 μm thick, chestnut brown, each cell with 6-10 apically furcate, brownish projections mostly 6-10 μm long and 2-4 μm wide basally; pedicel brownish apically, relatively thick walled, terete, the basal 1/3 to 1/2 often rugose, to 110 μm long.

Hosts and distribution: *Albizia lebbek* Benth.: southern Florida; also widely distributed in tropical regions.

Type: on *Acacia lebbek*, near Belgaum, India, Hobson (K).

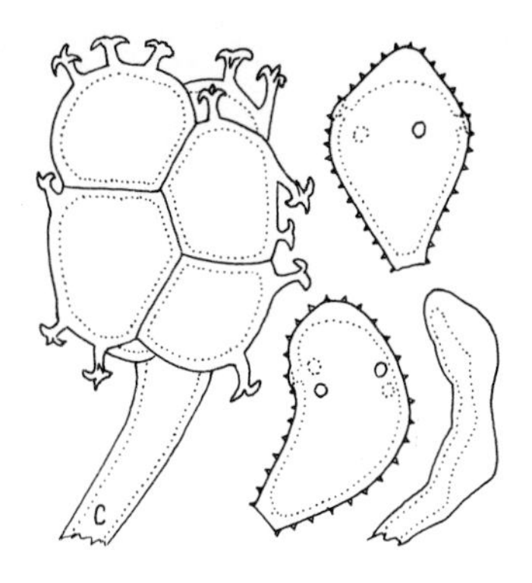

2. *SPHAEROPHRAGMIUM FIMBRIATUM* Mains, Carnegie Inst. Wash. Publ. 461:96-97. 1935.

Spermogonia and aecia unknown. Uredinia on the abaxial surface of leaves in small groups, about cinnamon brown, with abundant pale yellowish, peripheral, incurved, dorsally thick walled paraphyses; spores (24-)28-37(-40) x (15-)16-20(-22) µm, mostly asymmetrical, obovoid in one view or reniform when rotated 45 degrees, wall 1.5 µm thick at sides, (2-)4-7(-9) µm at apex, cinnamon brown or golden, uniformly echinulate, pores 4 or 5(6), equatorial or often near the thickened apex, with slight or no caps. Telia as the uredinia but chocolate brown; spores (32-)34-40(-42) x (28-)30-38(-40) µm, more or less globoid, mostly composed of 8 cells, the wall uniformly 2-2.5 µm thick, each cell bearing 5-8 apically furcate, brownish projections 4-6(-10) µm long and 2-3 µm wide; pedicel terete, brownish apically, to 100 µm long.

Type: on *Dalbergia glabra* (Mill.) Standl., Uaxactun, Peten, Guatemala, Bartlett No. 12429A (MICH; isotype PUR 47921). Known otherwise from one collection in Nicaragua on *Dalbergia*.

12. CHRYSELLA H. Sydow

Ann. Mycol. 24:292. 1926.

Spermogonia subepidermal, globoid, type 4 (16). Aecia and uredinia lacking. Telia subepidermal in origin, erumpent, waxy when moist; spores borne singly on pedicels, 1 celled, wall thin and pale, germination occurs without dormancy; basidium external.

Type species: *Chrysella mikaniae* Syd.

1. *CHRYSELLA MIKANIAE* H. Syd. Ann. Mycol. 24:292. 1926.

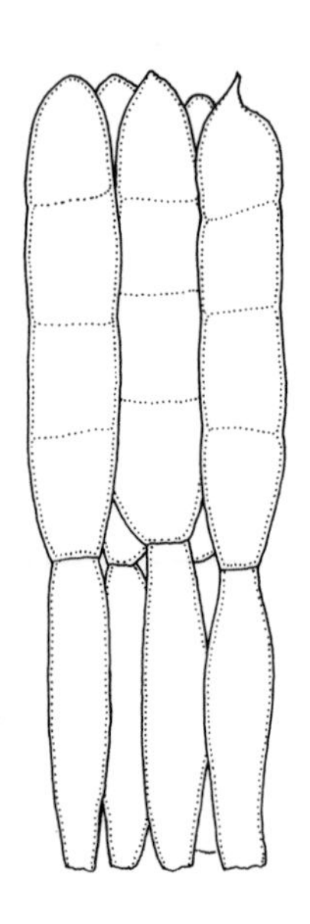

Spermogonia on the adaxial leaf surface in small groups. Aecia and uredinia lacking. Telia opposite the spermogonia, closely grouped, early exposed, waxy, bright orange yellow when fresh, pale yellowish when dry; spores 50-80 x 11-14 µm, cylindrical, wall 0.5 µm thick, colorless, smooth, germinating without dormancy by an internal basidium, pedicel colorless, 8-11 µm wide, 70-100 µm long.

Type: on *Mikania hirsutissima* DC., Angeles de San Ramon, Costa Rica, Sydow (holotype destroyed; isotypes Sydow, F. exot. exsic. No. 606). Not otherwise known.

The above description is adapted from the original. Photographs in the Arthur Herbarium with an isotype (PUR 64912) suggest that the species may belong in *Chrysocyclus* and, if so, *Chrysella* becomes a synonym of *Chrysocyclus*.

13. CHACONIA Juel

Bihang. K. Svenska Vet. Akad. Handl. 23, Afd. 3, No. 10:12. 1897.

Spermogonia subcuticular, conical, type 7 (16). Aecia subepidermal in origin, erumpent, uredinoid; spores borne singly on pedicels. Uredinia subepidermal in origin; spores borne singly on pedicels, similar to the aeciospores. Telia subepidermal in origin, erumpent; spores 1 celled, sessile and grouped on sporogenous basal cells, wall pale and thin, germ pore apical if differentiated, germinating without dormancy; basidium external.

Type species: *Chaconia alutacea* Juel.

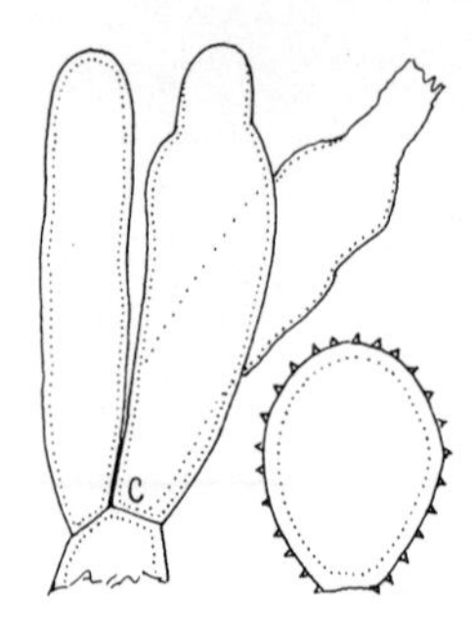

1. *CHACONIA ALUTACEA* Juel, Bihang. Svenska Vet. Akad. Handl. 23, Afd. 3, No. 10:12. 1897.

Spermogonia amphigenous in small groups. Aecia on the abaxial leaf surface around the spermogonia, uredinoid, cinnamon brown; spores 22-27(-30) x 16-20 µm, ellipsoid or obovoid, wall 2 µm thick, cinnamon brown or golden brown, echinulate, pores not seen. Uredinia, if formed, presumably as the aecia but not associated with the spermogonia. Telia on the abaxial surface in small groups, early exposed, white when old and dry, probably bright yellow or orange when fresh; spores 40-70 x 10-18 µm, variable in size, wall uniformly 0.5 µm thick, smooth, colorless; the spores germinate without dormancy.

Hosts and distribution: *Pithecellobium recordii* Britt. & Rose: British Honduras (now Belize); also in South America.

Type: on *Pithecellobium divaricatum* (Borg.) Benth., Asuncion, Gran Chaco, Paraguay, Lindman (S; isotypes Vestergren Micromy. rar. sel. No. 755).

2. *CHACONIA INGAE* (Syd.) Cumm. Mycologia 48:602. 1956.
Maravalia ingae H. Syd. Mycologia 17:257. 1925.
Bitzea ingae (Syd.) Mains, Mycologia 31:38. 1939.

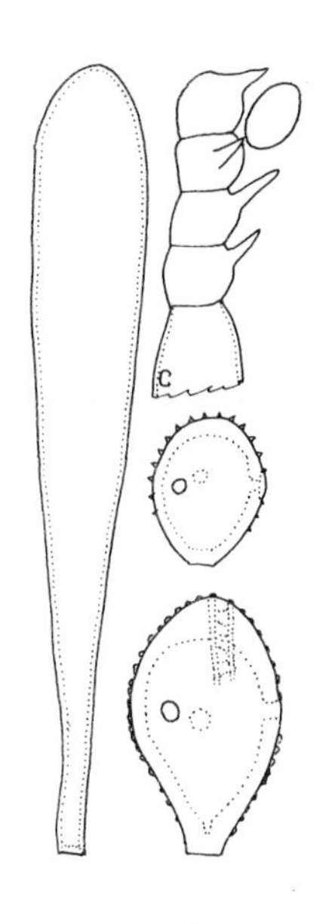

Spermogonia mostly on adaxial leaf surface in small groups. Aecia amphigenous, uredinoid, in a more or less complete circle around the spermogonia, deep seated and opening by a slit in the epidermis; spores (23-)26-45(-55) x (14-)18-24(-30) µm, variable, mostly broadly ellipsoid or obovoid, the base mostly narrowed and truncate, wall (2-)2.5-4(-4.5) µm thick at sides, 5-9 µm at apex, with longitudinal, narrow ridges spaced (1.5-)2-3(-3.5) µm and with finer, more closely spaced cross connections, thus producing a striately reticulate pattern, pores 3(4), equatorial. Uredinia amphigenous, cinnamon brown, in large confluent groups; spores (17-) 20-24(-26) x (13-)15-19(-21) µm, mostly obovoid, wall uniformly 1.5-3 µ thick or thickened apically to 5 µm, strongly echinulate, especially apically, pale golden, pores 3, equatorial. Telia on the abaxial surface in small circinate groups, white at least when old and dry, feltlike; spores 70-140 x 12-20 µm, elongately clavate, gradually tapering basally, wall uniformly 0.5 µm thick, colorless, smooth, germinating without dormancy, the basidia short and forming a tangled mass on the collapsed spores.

Hosts and distribution: *Inga* spp.: southern Mexico to Costa Rica; also in the West Indies and South America.

Type: on *Inga* sp., Vreedon Hoor, British Guiana, Stevens No. 715 (holotype destroyed; isotype PUR F18302).

Mains (*loc. cit.*) considered *Uredo ingae* P. Henn., which is the basis of the above description, to be distinct from *C. ingae*. Inoculations or continuous field observations will be needed to decide the matter.

14. CHRYSOCELIS Lagerheim & Dietel in Mayor

Mem. Soc. Neuch. Sci. Nat. 5:542. 1913.

Spermogonia subepidermal, globoid, type 4 (16). Aecia subepidermal, opening by a pore in the epidermis, without an organized peridium but with some peridial cells; spores catenulate. Uredinia unknown. Telia subepidermal in origin erumpent, waxy when moist; spores sessile, 1 celled, thin walled, germinating without dormancy, germ pore apical if differentiated; basidium external.

Type species: *Chrysocelis lupini* Lagh. & Diet.

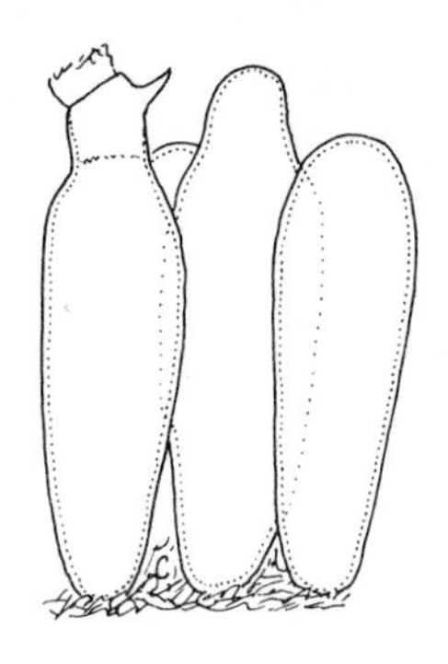

1. *CHRYSOCELIS LUPINI* Lagh. & Diet. ex Mayor, Mem. Soc. Neuch. Sci. Nat. 5:542. 1913.

Spermogonia amphigenous in small groups. Aecia on the abaxial leaf surface around the spermogonia, opening by a pore in the epidermis, yellowish when dry, without an organized peridium but with peridioid coarsely rugose cells among the spores and with a hyphal layer peripherally, spores catenulate or apparently so; spores (20-)24-32(-35) x (16-)18-23 (-26) μm, mostly broadly ellipsoid, wall 1.5-2 μm thick or in the coarsely rugose cells 3 μm thick, finely verrucose with discrete wartlets or these merging in rugose patterns and merging in characters with the putative peridial cells. Uredinia lacking. Telia on abaxial surface, closely grouped, early exposed, more or less compact, pale yellowish brown; spores 40-60 x (10-)13-18(-20) μm, cylindrical or clavate, wall uniformly 0.5 μm thick, colorless, smooth, germinating without dormancy, the basidium 10-12 μm wide.

Hosts and distribution: *Lupinus aschenbornii* Schau., *L. clarkii* Oerst.: Costa Rica; also in South America.

Lectotype: on *Lupinus* sp., Paramo Cruz Verde above Bogota, Colombia, Mayor No. 95. Lectotype designated here following Arthur's (1) citation of the above as the "Type Locality."

15. COLEOSPORIUM Léveille

Ann. Sci. Nat. ser. III, 8:373. 1847.

Spermogonia subepidermal, indeterminate, type 8 (16). Aecia subepidermal in origin, peridermioid, with strongly developed peridium; spores catenulate, verrucose with rods or columns which tend to merge in various patterns. (The aecial stage mostly on needles of *Pinus* but some species autoecious). Uredinia subepidermal in origin, erumpent; spores catenulate, verrucose as the aeciospores. Telia (basidiosori) subepidermal in origin, erumpent, waxy in appearance, hard when dry but gelatinous when wet; spores 1 celled, in 1 layered crusts, or pseudocatenulate, or catenulate, wall thin and pale but usually thick and gelatinous apically; basidium internal (actually the "teliospore" is a basidium).

Lectotype species: *Coleosporium campanulae* (Strauss) Tul. Lectotype designated by Laundon, Mycotaxon 3:154. 1975.

KEY TO SPECIES OF COLEOSPORIUM ON COMPOSITAE

1. Basidia catenulate, the sori 3 or more basidia deep; basidiospores obovoid 2
1. Basidia not catenulate but new basidia may grow up between older ones; basidiospores various 5

    2. Basidia mostly less than 50 µm long; the species is autoecious .... *viguierae* (1)
    2. Basidia mostly exceeding 50 µm long; life cycles unknown 3

3. Basidia 60-75(-80) µm long; other spores forms unknown .... *durangense* (2)
3. Basidia mostly 45-65 µm long; uredinia known 4

4. Urediniospores mostly 21-26 µm long.. *arizonicum* (3)
4. Urediniospores mostly 24-30 µm long ... *pereziae* (4)

5. Uredinia lacking; species autoecious 6
5. Uredinia produced; species presumably heteroecious .. 7

6. Aecial peridium short, inconspicuous; basidia mostly 58-66 µm long .... *incompletum* (5)
6. Aecial peridium conspicuous; basidia mostly 60-90 µm long .... *reichei* (6)

7. Uredinia with peripheral paraphyses; urediniospores mostly 28-50 µm long .... *paraphysatum* (7)
7. Uredinia lacking paraphyses; urediniospores shorter 8

8. Urediniospore wall uniformly verrucose or nearly so 9
8. Urediniospore wall with smooth or smoothish area on one side and/or base 16

9. Urediniospores mostly 28-36 x 22-26 µm ... *pacificum* (8)
9. Urediniospores mostly smaller 10

10. Urediniospores mostly 25-33 µm long 11
10. Urediniospores mostly smaller 12

11. Basidiospores mostly 26-30 µm long ..... *laciniariae* (9)
11. Basidiospores 19-22 µm long .... *senecionis* (10)

12. Urediniospores mostly 22-30 µm long 13
12. Urediniospores smaller 15

13. Basidiospores 23-27 x 14-17 µm .... *asterum* (11)
13. Basidiospores smaller 14

14. Basidia mostly 70-95 µm long .... *helianthi* (12)
14. Basidia mostly 55-70 µm long .... *sonchi* (13)

15. Basidia mostly 50-66 µm long; basidiospores 20-26 x 12-14 µm .... *vernoniae* (14)
15. Basidia mostly 65-80 µm long; basidiospores 19-24 x 13-15 µm; aecia and aeciospores differ
.... *delicatulum* (15)
.... *inconspicuum* (16)

16. Basidiospores obovoid, 25 µm or less long 17
16. Basidiospores oblong ellipsoid, 23-28 µm long ... 18

17. Basidiospores 21-25 x 14-18 µm ....*dahliae* (17)
17. Basidiospore 16-19 x 12-16 µm .... *verbesinae* (18)

18. Urediniospores 27-38 µm long ...... *longisporum* (19)
18. Urediniospores mostly less than 30 µm long 19

19. Urediniospores mostly 22-29 x 18-24 µm; basidia mostly 65-85 x 18-21 µm .... *anceps* (20)
19. Urediniospores mostly 21-25 x 17-22 µm; basidia mostly 60-78 x 19-24 µm .... *steviae* (21)

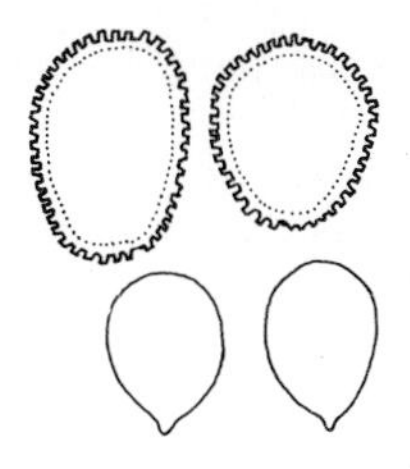

1. *COLEOSPORIUM VIGUIERAE* Diet. & Holw. in Holway, Bot. Gaz. 24:34. 1897.

Spermogonia not produced. Aecia on the abaxial surface of leaves, sometimes few but usually densely distributed over the entire leaf surface and commonly on most leaves of a plant, but probably not from systemic mycelium, orange when fresh fading to yellowish, the peridium columnar or tongue shape, relatively persistent, 0.5-1 mm long, 0.3-1 mm wide, peridial cells 35-45 x 22-30 µm face view, the outer wall 1.5-2 µm thick, punctate, the inner wall 5-9 µm thick, verrucose much as the aeciospores, the side walls striately ridged; spores (20-)22-28(-30) x (15-)17-20(-22) µm, mostly oblong ellipsoid or broadly ellipsoid, wall 2.5-3.5 µm thick including verrucae, the inner wall 1.5 µm thick, the verrucae 2-2.5 µm long, 1-2 µm wide or when elongate 2.5-3 µm, sometimes merging in small groups, mostly columnar with flat top, somewhat deciduous but the spores lack a smooth area. Uredinia preceding or among the aecia but not so abundant; spores (19-)21-26 x (16-)17-21 µm mostly broadly ellipsoid or ellipsoid, wall (1.5-)2-2.5 µm thick including verrucae, the verrucae mostly 1-1.5 µm long, not easily detached, irregular in outline (0.5-)1-2 µm wide or merged in ridges to 3 µm long, verrucae columnar with flat top, the spores lack smooth areas. Telia on the abaxial surface, among the aecia and uredinia, brownish orange, pulvinate, hard when dry, 3-5 spores deep, catenulate, the apical gelatinous layer 4-8 µm thick; spores (34-)36-48(-54) x (16-)17-22(-24) µm, mostly oblong ellipsoid, wall 0.5-1 µm thick, colorless, smooth, the spores germinate without dormancy by horizontal septa or in the uppermost layer by septa in various planes; basidiospores 14-18 x 11-13 µm, oval or ellispoid.

Hosts and distribution: *Viguiera dentata* (Cav.) Spreng. var. *dentata* and var. *lancifolia* Blake, perhaps also *Verbesina* spp., *Zexmenia helianthoides* Gray: the southwestern United States southward to southern Mexico.

Type: on *Viguiera helianthoides* H.B.K. (= *V. dentata*), near Tula, Mex., Mexico (S; isotype PUR 1933).

This rust fungus is common in the mountains of southern Arizona where the aecial infections are most conspicuous. Inoculations have not been made but the species obviously is autoecious.

2. *COLEOSPORIUM DURANGENSE* Cumm. Mycotaxon 5:399. 1977.

Spermogonia, aecia and uredinia unknown. Telia (basidiosori) on the abaxial leaf surface, early exposed, separate or in small group, orange color when fresh fading to honey color, hard when dry, mostly 3 or 4 basidia deep, the basidia catenulate, 60-75(-80) x 18-22(-24) µm, mostly cylindrical, the upper layers become crushed and disarranged as they germinate, the apical gelatinous layer (8-)10-14 µm thick, not conspicuous; basidiospores (18-)20-22(-24) x (15-)16-18(-20) µm, broadly obovoid or broadly ellipsoid.

Type: on *Eupatorium calaminthaefolium* H.B.K., Mex. hgw 40, km 1015, W of Durango, Mexico, Cummins, No. 63-577 (PUR 62369). Not otherwise known.

Neither uredinia nor urediniospores adhering to gelatinized basidiosori could be found.

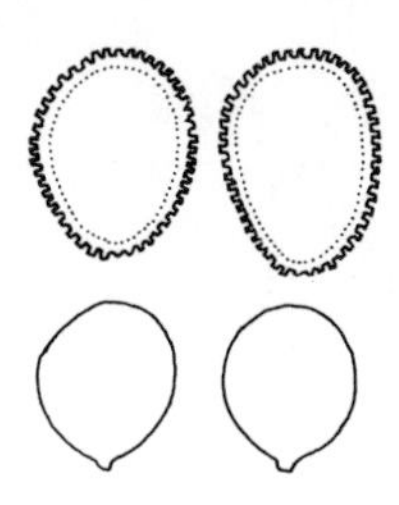

3. *COLEOSPORIUM ARIZONICUM* Cumm. Mycotaxon 5:399. 1977.
Coleosporium aridum H. S. Jack. in Arthur, Bull. Torrey Bot. Club 51:52. 1924 (based on uredinia).

Spermogonia and aecia unknown. Uredinia on the abaxial surface of leaves, yellow when fresh becoming nearly colorless with age; spores (18-)21-26(-29) x (17-)20-24 µm, broadly ellipsoid or broadly obovoid, wall uniformly verrucose (no smooth area) with rods about 1.5 µm high or these merged in short ridges, the inner wall 1 µm thick. Telia (basidiosori) on abaxial surface, early exposed, pale orange or honey color, hard when dry, 2-4 basidia deep, the basidia catenulate (40-)45-60(-65) x (17-)19-23(-26) µm, cylindrical, the apical gelatinous layer (5-)10-15 µm thick; basidiospores 15-20 x 12-16 µm, broadly obovoid or slightly napiform.

Hosts and distribution: *Brickellia californica* Torr. & Gray, *Eupatorium herbaceum* (Gray) Greene: New Mexico, Arizona and California.

Type: on *Brickellia californica*, Oak Creek Canyon, Sedona, Arizona, Cummins No. 57-72 (PUR 56550; isotypes Solheim & Cummins Mycof. Saximont. Exsic. No. 902).

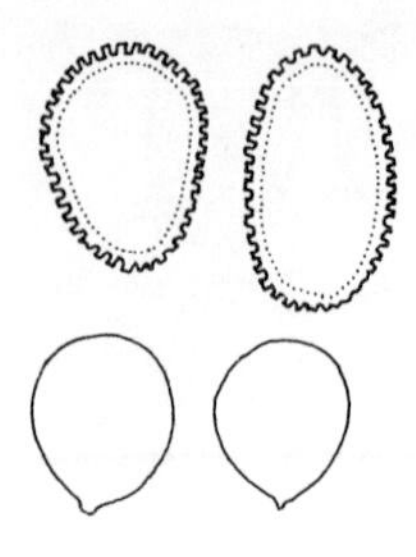

4. *COLEOSPORIUM PEREZIAE* Cumm. Mycotaxon 5:401. 1977.

Spermogonia and aecia unknown. Uredinia on the abaxial leaf surface, dull yellowish when dry, doubtless bright yellow when fresh; spores (22-)24-30(-32) x (15-)18-22(-24) µm, mostly oblong ellipsoid or broadly ellipsoid, wall nearly uniformly verrucose with rods or very short ridges 1.5-2 µm high, sometimes less prominent on one side of the spore but no smoothish area is present, the wall without the verrucae about 1 µm thick. Telia (basidiosori) on the abaxial surface, dull orange when dry, early exposed, hard when dry, 3 or more basidia deep, the basidia obviously catenulate, 44-70 x 17-23 µm, cylindrical, horizontally septate, apical gelatinous layer 15 µm or less thick, not conspicuous; basidiospores 16-18 x 14-16 µm, broadly obovoid or somewhat napiform.

Hosts and distribution: *Perezia* spp.: Nuevo Leon to Durango and Jalisco, Mexico.

Type: on *Perezia thurberi* Gray, near km 177, Mex hgw 45, S. of Durango, Dgo., Cummins No. 71-209 (PUR 65365).

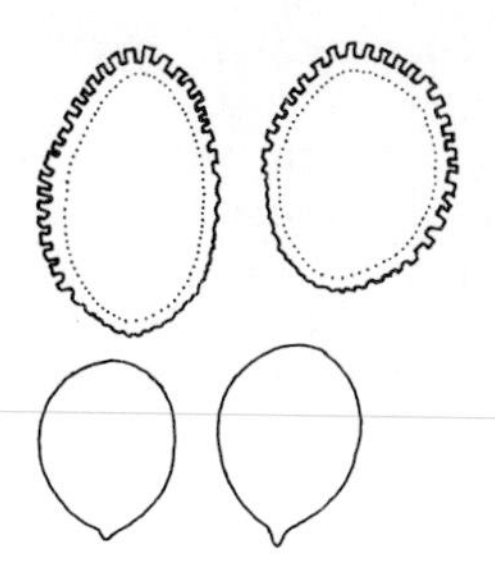

5. *COLEOSPORIUM INCOMPLETUM* Cumm. Mycotaxon 5:400. 1977.

Spermogonia lacking. Aecia on the abaxial leaf surface, scattered singly or few in a group, the peridium short, peridial cells nearly square in face view, 20-35 µm long, the same width or slightly narrower, inner wall 4-6 µm thick, verrucose as the spores but more coarsely so, outer wall 1 µm thick, finely rugosely verrucose; spores (21-)23-30(-33) x (16-)18-22(-24) µm, ellipsoid, oblong ellipsoid or less commonly globoid, verrucose with mostly irregularly shaped rods or blocks, these often branching or merging in pseudoreticulate patterns, 2-2.5 µm high on one side of spore decreasing to minutely verrucose or rugose on the opposite side, inner wall about 1 µm thick, colorless. Uredinia lacking. Telia (basidiosori) on abaxial surface in loose groups, early exposed, orange color, hard when dry, 1 or 2 basidia deep but the basidia not catenulate, the new ones pushing up between the older ones, apical gelatinous layer 18-25 µm thick; basidia (50-)58-66(-70) x 17-22(-26) µm, mostly cylindrical; basidiospores 18-20 x 16-18 µm, oval or broadly obovoid.

Type: on *Stevia berlandieri* Gray, Mex. hgw 40, km 184, near Espinosa del Diablo, Durango near the border with Sinaloa, Mexico, Cummins No. 71-544 (PUR 64124). Not otherwise known.

6. *COLEOSPORIUM REICHEI* Diet. Ann. Mycol. 21:341. 1923.

Spermogonia unknown. Aecia on the abaxial surface of leaves, often associated with veins, mostly in groups of 3-8, peridermioid, to about 1 mm long, peridium usually tubular with apical rupture, pale brownish yellow, peridial cells cuboidal, oblong, oblong ellipsoid or more or less broadly ellipsoid in face view, to 50 µm long and 25-30 µm wide, the inner wall 6-8 µm thick and closely verrucose with rods or ridges, the outer wall 2 µm thick and finely verrucose, the side walls striately ridged; spores 22-27(-30) x (18-)20-24(-26) µm, globoid or slightly depressed globoid, the upper 2/3 or so closely beset with rods to 4 µm long, these discrete or sometimes fused in pseudoreticulate patterns, decreasing rather abruptly basally so that the lower 1/3 becomes almost smooth, the wall (minus verrucae) about 1 µm thick, colorless. Uredinia lacking. Telia (basidiosori) on abaxial surface, early exposed, about orange color, hard when dry, 1 basidium deep; basidia 60-90 x 20-33 µm, cylindrical or clavate, transversely septate, the apical gelatinous layer mostly 18-25 µm thick, the side wall thin, the basidiospores not seen.

Type: on *Stevia* sp., Tres Marias, between Mexico City and Cuernavaca, Reiche (S; isotype PUR 2004). Not otherwise known.

7. *COLEOSPORIUM PARAPHYSATUM* Diet. & Holw. in Holway, Bot. Gaz. 31:337. 1901.

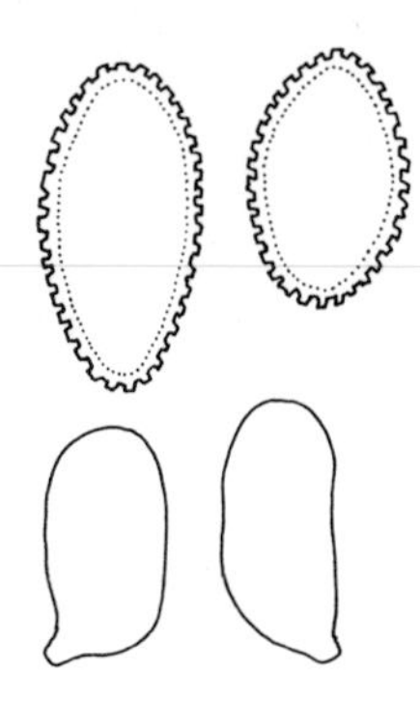

Spermogonia and aecia unknown. Uredinia on abaxial leaf surface, pale yellowish (dry), doubtless bright yellow or orange when fresh, with a peridium, cellular basally, the apical cells free as short paraphyses but difficult to observe in the usually tomentose leaves; spores (22-)28-50(-65) x (9-)12-20(-22) μm, mostly narrowly ellipsoid but some globoid and some (the long ones) narrowly cylindrical or fusiform, the latter sometimes adhere in a column, wall colorless, beset with rods or ridges or irregularly shaped verrucae, isodiametric or to 4 μm long, these frequently tending to be oriented parallel to the long axis of the spore, sometimes forming pseudoreticulate patterns, without a smooth area, 1.5-2 μm high, the wall minus verrucae 1-1.5 μm thick. Telia (basidiosori) on abaxial surface, pale orange when dry, early exposed, 1 basidium deep, the apical gelatinous layer 12-25 μm thick, basidia (36-)55-66(-78) x (17-)19-27(-30) μm, cylindrical or clavate, septa all horizontal or the uppermost often vertical; basidiospores (20-)22-30(-32) x (9-)11-15 (-17) μm, oblong ellipsoid or slightly allantoid.

Hosts and distribution: *Liabum* spp.: southern Mexico to Belize, Guatemala and Costa Rica.

Type: *Liabum discolor* Benth. & Hook., Chapala, Jal., Mexico, Holway No. 3483 (S; isotype PUR 1980).

In some specimens the urediniospores mostly are shorter than 44 μm, in others they range much longer and some are very narrow and one wonders if they are viable. Both types may occur on the same leaf, usually in separate sori although the long type may also occur with spores of the more typical size.

An undescribed *Peridermium* on needles of *Pinus* sp. from Chiapas, Mexico has spores 44-62 x 24-32 μm but there are no field observations to indicate its relationship.

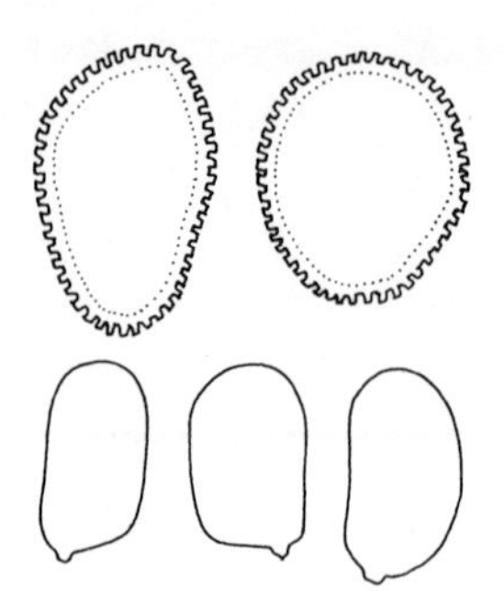

8. *COLEOSPORIUM PACIFICUM* Cumm. Mycotaxon 5:401. 1977.
Coleosporium madiae Cooke Grevillea 7:102. 1879 (based on uredinia).
Stichopsora madiae Syd. Ann. Mycol. 2:30. 1904.
Coleosporium arnicale Arth. N. Amer. Flora 7:94. 1907 (based on uredinia).

Spermogonia amphigenous, conspicuous. Aecia amphigenous on leaves of *Pinus radiata* D. Don in California, tongue shape, peridermioid; spores 40-45 x 25-29 µm, broadly ellipsoid, wall 3-4.5 µm thick, coarsely verrucose with rods. Uredinia on the abaxial leaf surface, bright orange when fresh fading to nearly colorless with age; spores (24-)28-36 (-40) x (18-)22-26(-28) µm, mostly broadly ellipsoid, nearly or quite uniformly verrucose with small, irregularly shaped, mostly discrete rods 1-1.5 µm high. Telia (basidiosori) on abaxial surface, early exposed, dull orange color, hard when dry, the apical gelatinous layer 20-30 µm thick, the sori mostly 1 basidium deep but with young spores pushing up a among the older ones, basidia (50-)55-75(-85) x (18-)20-23 µm, mostly cylindrical; basidiospores (18-)21-25(-27) x 13-16 µm, oblong ellipsoid.

Hosts and distribution: *Hemizonia* spp., *Holozonia filipes* (Hook. & Arn.) Greene, *Madia* spp. (of tribe Madieae), *Gaillardia aristata* Pursh, *Tagetes erecta* L., *T. patula* L. (of tribe Helenieae): southern British Columbia south to California.

Type: on *Madia sativa* Molina, California, Copeland (S; not seen).

Whether *Coleosporium arnicale* belongs here is uncertain, as is the identity of the host.

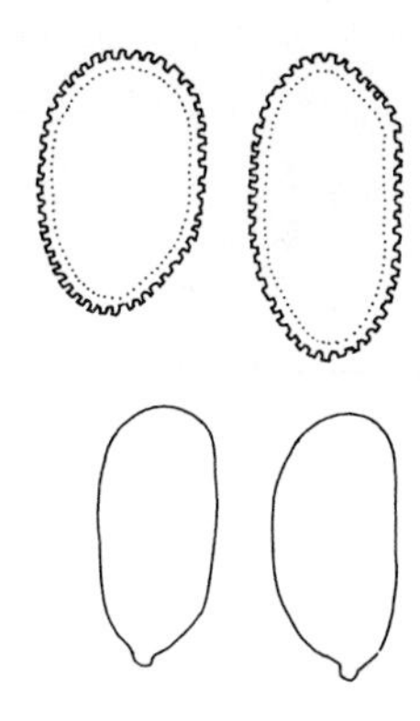

9. *COLEOSPORIUM LACINIARIAE* Arth. N. Amer. Flora 7:90. 1907.

Spermogonia amphigenous. Aecia on abaxial leaf surface of species of *Pinus*, the peridium rather fragile, laterally flattened, peridermioid; spores 25-34 x 18-22 µm, ovoid or ellipsoid, wall 2-3 µm thick, verrucose as the urediniospores. Uredinia amphigenous and on stems, bright yellow when fresh fading to dull yellowish with age; spores (20-) 25-33(-38) x (16-)17-21(-23) µm, mostly oblong ellipsoid or broadly ellipsoid, wall more or less uniformly verrucose with rods or short ridges 1-1.5 µm high or these discrete or often merged into pseudoreticulate patterns, wall minus verrucae about 1 µm thick. Telia (basidiosori) amphigenous and on stems, early exposed, dull orange color, hard when dry, mostly 2 basidia deep with young spores pushing in between the older ones, the apical gelatinous layer 20-35 µm thick; basidia (60-)66-85(-90) x (17-)20-24 µm, cylindrical, basidiospores 26-30 x 14-16 µm, oblong ellipsoid with a slight tendency to be alantoid.

Hosts and distribution: *Liatris* spp.: New Jersey to Arkansas and Florida.

Type: on *Laciniaria graminifolia* (= *Liatris graminifolia* (Walt.) Pursh, Auburn, Alabama, Oct. 1895, Underwood (PUR 904).

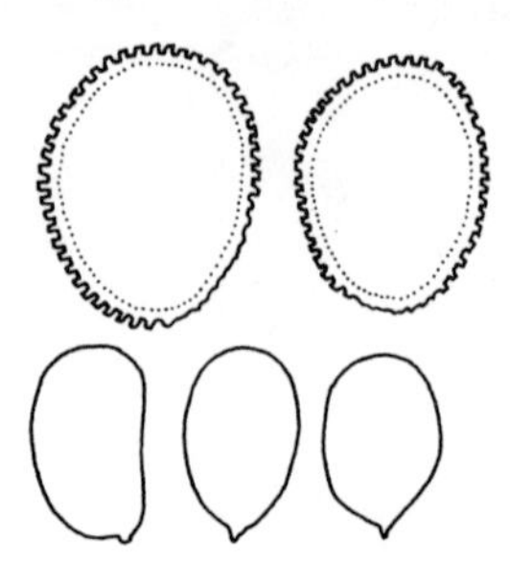

10. *COLEOSPORIUM SENECIONIS* Kichx Flora Flandres 2:53. 1867.
Coleosporium occidentale Arth. N. Amer. Flora 7:94. 1907.

Spermogonia amphigenous, conspicuous. Aecia amphigenous on leaves of *Pinus* spp. but its occurrence in North America not confirmed, tongue shape, peridermioid; spores 28-36 x 17-24 µm, oblong ellipsoid or broadly ellipsoid, wall 3-4 µm thick, coarsely verrucose with rod-like verrucae. Uredinia on the abaxial surface and often on stems, bright yellow when fresh fading to dull yellowish with age; spores (20-)24-33(-38) x (16-)19-23(-26) µm, broadly ellipsoid or oblong ellipsoid, rather finely and nearly or quite uniformly verrucose with small, irregularly shaped, mostly discrete rods 1-1.5 µm high or these occasionally reduced on one side of spore toward the base, wall minus verrucae about 1.5 µm thick. Telia (basidiosori) on abaxial surface, early exposed, orange color, hard when dry, the gelatinous apical layer 15-25 µm thick, the sori 1 basidium deep; basidia (55-)64-72(-80) x (18-)20-24 µm, cylindrical, basidiospores 19-22 x 11-14 µm, oblong ellipsoid or ellipsoid.

Hosts and distribution: *Raillardella pringlei* Greene, *Senecio* spp.: Rhode Island and Colorado to Washington and California; rare or rarely collected; also in Europe, Asia and South America.

Lectotype: on *Senecio vulgaris* L., Flanders, Belgium (not seen). Lectotype designated by Hylander, Joerstad and Nannfeldt (17).

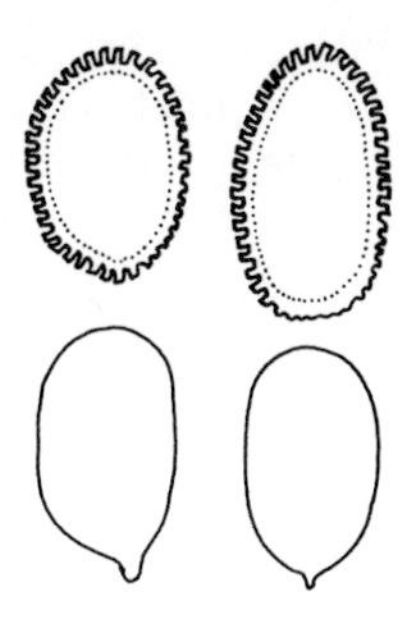

11. *COLEOSPORIUM ASTERUM* (Diet.) Syd. Ann. Mycol. 12:109. 1914.
Coleosporium solidaginis Thuem. Bull. Torrey Bot. Club 6:216. 1878 (based on uredinia).
Stichopsora asterum Diet. Bot. Jahrb. 27:566. 1899.

Spermogonia on both leaf surfaces. Aecia on both surfaces of leaves of *Pinus* sp., peridermioid, tongue shape and prominent but shattering easily; spores (20-)26-35(-38) x (16-)18-22(-24) µm, mostly ellipsoid or oblong ellipsoid, wall including verrucae 2-4 µm thick, the verrucae as in the urediniospores. Uredinia on the abaxial surface of leaves, bright orange when fresh fading to colorless with age, spores (19-)22-30(-33) x (14-)17-21(-23) µm, variable in size and shape, mostly ellipsoid or oblong ellipsoid, wall 2-2.5(-3) µm thick, verrucose with rods or ridges or irregularly shaped verrucae or even pseudoreticulate, usually somewhat reduced on one side. Telia (basidiosori) on the abaxial surface, early exposed, orange or honey color, hard when dry, the apical gelatinous layer 20-35 µm thick, 1 or 2 basidia deep with young basidia pushing up among the older ones; basidia (50-)55-75(-80) x (16-)19-23(-25) µm, cylindrical; basidiospores 18-23 x 15-17 µm, broadly ellipsoid or obovoid.

Hosts and distribution: especially common on species of *Aster* and *Solidago* but also recorded on species of *Callistephus*, *Doellingeria*, *Erigeron*, (*Gaillardia*?), *Grindelia* *Haplopappus*, *Heterotheca*, and *Macaeranthera*: southern Alaska and Canada south into the northern half of Mexico; also in China and Japan.

Lectotype: on *Aster tartaricus* L., Japan, Hort. Botan., Oct. 1898, Kusano No. 30 (S). Lectotype designated here.

C. asterum is heterogeneous. The urediniospores vary in the coarseness of the wall sculpture and the basidiospores vary in size. As to basidiospores, Dietel (*loc. cit.*) wrote "Die Sporidien sind gross, gleich denen von *Coleosporium*, ca. 20 μm lang und 16 μm breit, am unteren Ende oft citronenförmig zugespitzt." and this is true of Japanese specimens that I have seen. It is true of some American specimens but others have basidiospores mostly 23-28 x 14-17 μm and oblong ellipsoid or slightly alantoid.

12. *COLEOSPORIUM HELIANTHI* (Schw.) Arth. N. Amer. Flora 7: 93. 1907.
Caeoma (*Uredo*) *helianthi* Schw. Trans. Amer. Phil. Soc. II. 4:291. 1832 (Telia described).
Coleosporium terebinthinaceae Arth. N. Amer. Flora 7: 93. 1907.

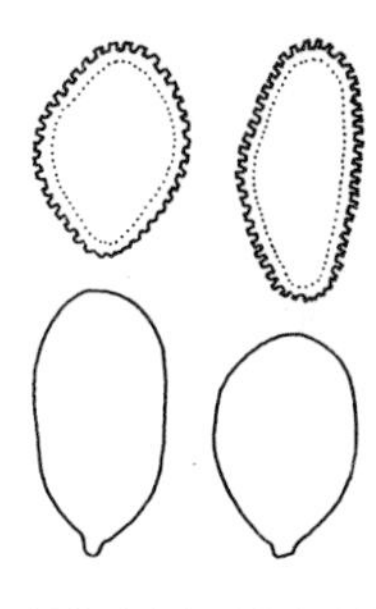

Spermogonia amphigenous, few. Aecia on species of *Pinus* amphigenous, peridermioid, mostly long and tongue shape; spores (20-)24-32(-35) x (15-)17-22(-24) μm, variable but mostly ellipsoid or oblong ellipsoid, when broadly ellipsoid the spores usually shorter, verrucose as the urediniospores but more coarsely so and with a smooth side more obvious. Uredinia on the abaxial leaf surface, bright yellow when fresh fading to nearly colorless; spores (18-)22-30(-35) x (12-)14-18(-22) μm, variable, mostly oblong, oblong ellipsoid or ellipsoid, more or less uniformly verrucose with small rods or short ridges 1-1.5 μm high, these usually discrete on one side of spores becoming more pseudoreticulate or striolate and slightly finer on opposite side. Telia (basidiosori) on abaxial surface, early exposed, hard when dry, dull orange color, 1 basidium deep, the apical gelatinous layer 15-30 μm thick; basidia (60-)70-95(-115) x (17-)19-24(-26) μm, cylindrical, basidiospores (20-)22-25(-27) x 13-16 μm, oblong ellipsoid.

Hosts and distribution: species of *Helianthus*, *Parthenium* and *Silphium*; New York and Wisconsin south to Oklahoma and Georgia.

Lectotype: on *Helianthus giganteus* L., Bethlehem, Pennsylvania (PH). Lectotype designated here. It is obvious that Schweinitz described telia although placing the species in *Caeoma*.

Arthur and Bisby (3) considered that the material in the Schweinitz collections was part *H. giganteus* and part possibly *H. strumosus* L. Because Schweinitz designated *H. giganteus* as the host plant, I am selecting that part, indicated by Arthur and Bisby as on *H. giganteus*, as lectotype.

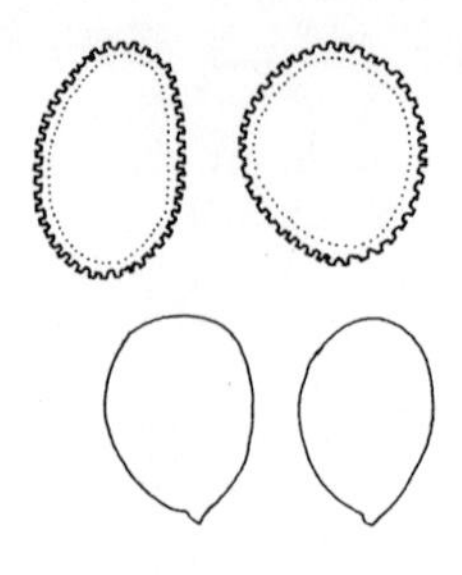

13. *COLEOSPORIUM SONCHI* (Strauss) Tul. Ann. Sci. Nat. Bot. IV, 2:190. 1854.
    Uredo tremellosa y *sonchi* Strauss, Ann. Wetter. Ges. 2: 90. 1810 (based on telia).

Spermogonia mostly on the adaxial leaf surface, obvious. Aecia, on *Pinus sylvestris* L. in Wisconsin, on both leaf surfaces, conspicuous, flattened laterally, peridermioid; spores (22-)25-30(-35) x (18-)21-25 µm, mostly broadly ellipsoid, wall 2-3 µm thick, surface sculpture as the urediniospores but more prominent. Uredinia on the abaxial leaf surface, doubtless bright yellow when fresh, dull yellowish with age; spores (18-)20-29(-33) x (13-)14-19(-21) µm, variable but mostly broadly ellipsoid or ellipsoid, wall more or less uniformly pseudoreticulate with narrow, variously merging ridges or some ridges or rods discrete, ridges 1(-1.5) µm high, without a smoothish area. Telia (basidiosori) on abaxial surface, early exposed, hard when dry, orange or honey color, 1 basidium deep, apical gelatinous layer 10-25 µm thick; basidia (50-)55-70 x (17-)19-23(-26) µm, cylindrical, basidiospores 18-22 x 15-17 µm, ellipsoid.

Hosts and distribution: *Sonchus asper* (L.) Hill: Wisconsin; also in the West Indies and Europe.

Type: on *Sonchus oleraceus* L., Germany (not seen).

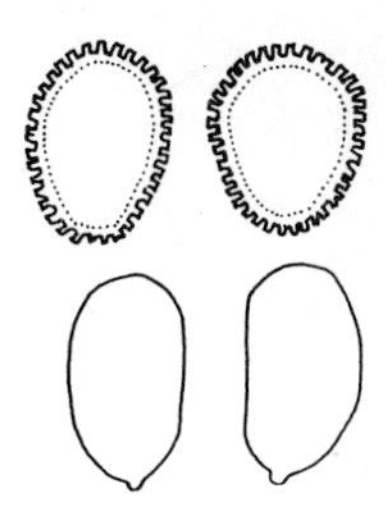

14. *COLEOSPORIUM VERNONIAE* Berk. & Curt. in Berkeley, Grevillea 3:57. 1874.
Coleosporium elephantopodis Thuem. Myc. Univ. No. 953. 1878.

Spermogonia mostly on the upper leaf surface, in 1 or 2 rows. Aecia on both surfaces, large and conspicuous, flattened laterally, peridermioid, peridium firm; spores (20-)25-34(-38) x (14-)17-23(-25) μm, mostly obovoid or ellipsoid, wall uniformly 3-6(-9) μm thick or often thicker above, verrucose as the urediniospores but more coarsely so. The aecia are on species of *Pinus*. Uredinia on both leaf surfaces or usually only on the abaxial surface, yellow when fresh but nearly white when old and dry; spores (17-)20-26 (-30) x (14-)16-20(-22) μm, mostly ellipsoid or broadly so, nearly uniformly verrucose with rods or ridges mostly 1.5-2 μm high, these commonly united in irregular or pseudoreticulate patterns, the wall minus verrucae about 1 μm thick, colorless. Telia (basidiosori) on the abaxial surface, early exposed, hard when dry, about orange color or paler, mostly 2 or 3 basidia deep but the basidia not catenulate, (44-)50-66(-70) x (16-)18-22(-24) μm, cylindrical or elongately clavate, mostly transversely septate but some collections commonly have some cruciately septate basidia, the apical gelatinous layer (20-)25-35(-40) μm thick; basidiospores (few seen) 20-26 x 12-14 μm, oblong ellipsoid or slightly alantoid.

Hosts and distribution: species of *Elephantopus* and *Vernonia*: Massachusetts to Nebraska south to Guatemala; also in West Indies and South America.

Type: on *Vernonia* sp., Alabama, U.S.A., Beaumont No. 4643 (not seen).

C. vernoniae and *C. elephantopodis* have traditionally been kept as species but they are separable only by the hosts.

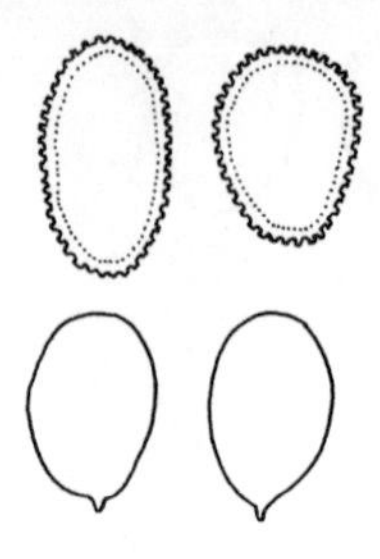

15. *COLEOSPORIUM DELICATULUM* Arth. N. Amer. Flora 7:657. 1924.
Coleosporium delicatulum Hedgc. & Long Phytopathology 3:250. 1913, *nomen nudum.*

Spermogonia amphigenous, conspicuous. Aecia on both leaf surfaces of species of *Pinus*, inconspicuous, peridermioid but the peridium delicate and not conspicuous; spores (17-)20-26(-28) x (14-)17-20(-22) µm, mostly ellipsoid, wall conspicuously verrucose with rods or short ridges 2-3 µm high which may be discrete or merged. Uredinia amphigenous or sometimes mostly on the abaxial leaf surface, bright yellow fading to pale yellowish; spores (16-)18-25(-28) x (12-) 13-17 µm, mostly ellipsoid or oblong ellipsoid, wall mostly uniformly verrucose with small rods or very short ridges, both usually discrete or at least not merging in a pseudoreticulate pattern, the rods about 0.5-1 µm high, there is no obvious smoothish area. Telia (basidiosori) amphigenous, early exposed, hard when dry, dull orange color, 1 basidium deep, the apical gelatinous layer 18-25 µm thick; basidia (55-)65-80(-85) x (16-)19-24(-26) µm, cylindrical, basidiospores 19-22 x 13-14 (few seen), ellipsoid or obovoid.

Hosts and distribution: *Euthamia* spp.: Maine to eastern Kansas, eastern Texas and Florida.

Lectotype: on *Euthamia graminifolia* (L.) Nutt., Sylvan Beach, New York, H. D. House, Barth. N. Amer. Ured. No. 2302 (PUR 1794). Lectotype designated here.

16. *COLEOSPORIUM INCONSPICUUM* Arth. N. Amer. Flora 7:659. 1924.
Coleosporium inconspicuum Hedgc. & Long, Phytopathology 3:250. 1913, *nom. nudum*.

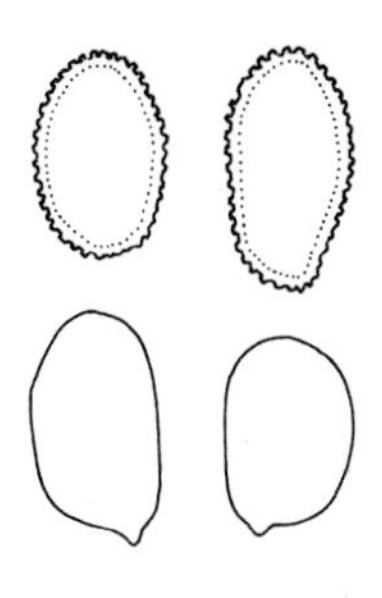

Spermogonia mostly on the abaxial leaf surface. Aecia on both sides of leaves of *Pinus* spp., peridermioid, mostly long and narrowly tongue shape; spores (21-)25-35 (-38) x 13-18 µm, mostly narrowly ellipsoid or oblong ellipsoid, wall finely verrucose as the urediniospores. Uredinia on the abaxial leaf surface, bright yellow when fresh fading to nearly colorless; spores (18-)20-26(-29) x 14-18(-20) µm, variable, ellipsoid, oblong ellipsoid or globoid, wall more or less uniformly verrucose with mostly discrete, small rods and short ridges 0.5-1 µm high on one side of spores becoming finer or pseudoreticulate on opposite side. Telia (basidiosori) on abaxial surface, early exposed, hard when dry, dull orange, 1 basidium deep, the apical gelatinous layer 20-30 µm thick; basidia (60-)65-75(-80) x 21-23 µm, cylindrical, basidiospores (few seen) 20-24 x 13-15 µm, oblong ellipsoid.

Hosts and distribution: *Coreopsis* spp.: Maryland and Ohio south to Georgia.

Lectotype: on *Coreopsis major* Walt. var. *oemleri* (Ell.) Britt., Seneca, South Carolina, Bartholomew; Barth. N. Amer. Ured. No. 2109 (PUR 1900). Lectotype designated here.

This species is similar to *C. helianthi* but has shorter basidia, slightly smaller urediniospores and more finely verrucose urediniospores and, especially, aeciospores. The aecia, whether the basidial stage is on *Coreopsis*, *Helianthus* or *Silphium*, are similar and are longer and narrower than in most species of *Coleosporium*. Basidiospores also are similar from these hosts.

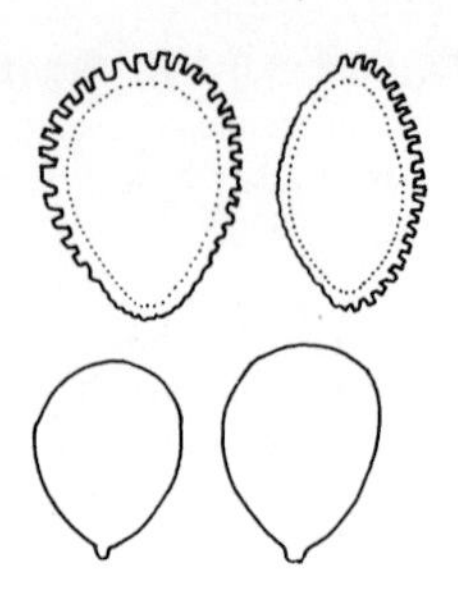

17. *COLEOSPORIUM DAHLIAE* Arth. Bot. Gaz. 40:197. 1905.

Spermogonia and aecia unknown. Uredinia on the abaxial leaf surface, almost colorless when dry, doubtless bright yellow when fresh, spores (22-)24-30(-35) x (15-)16-22(-24) µm, variable but mostly ellipsoid, wall verrucose on one side with rods 2 µm high or elongate ridges to 3 µm long or these sometimes branched, grading to finely verrucose rugose or nearly smooth on the opposite side, the wall without verrucae about 1 µm thick. Telia (basidiosori) on abaxial surface, orange when fresh fading to dull orange, early exposed, hard when dry, 1 or 2 basidia deep, when 2 apparently catenulate, basidia (50-)57-75(-80) x 20-28(-33) µm, cylindrical, horizontally septate or sometimes the upper septum vertical, apical gelatinous layer mostly less than 12 µm thick; basidiospores 21-25(-28) x (13-)14-18 µm, mostly obovoid or somewhat napiform.

Hosts and distribution: *Dahlia* spp., Veracruz, Durango and Jalisco, Mexico.

Type: on *Dahlia variabilis* (Willd.) Desf., Guadalajara, Jal., Mexico Holway No. 5121 (PUR 1937).

Arthur described the gelatinous layer as 10-20 µm thick but in sections it is scarcely discernible.

18. *COLEOSPORIUM VERBESINAE* Diet. & Holw. in Holway, Bot. Gaz. 31:337. 1901.

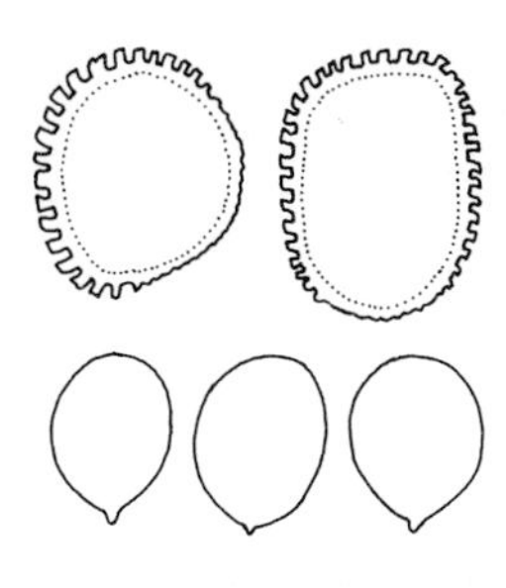

Spermogonia and aecia unknown. Uredinia on the abaxial leaf surface, nearly colorless in age, presumably bright yellow when fresh; spores (24-) 26-31(-33) x (19-)21-24(-26) µm, mostly oblong ellipsoid or broadly ellipsoid, wall on one side of spore verrucose with mostly discrete rods or ridges 2-3 µm high, these decreasing, usually abruptly, to fine verrucae or rugosity on the opposite side, the wall minus verrucae 1-1.5 µm thick. Telia (basidiosori) on abaxial surface, scattered, early exposed, orange or honey color, hard when dry, usually 1 basidium deep with new spores growing up between the old ones, the apical gelatinous layer 25-35 µm thick, the basidia (55-)60-78(-80) x 16 -20 µm, mostly cylindrical; basidiospores ellipsoid or slightly napiform, 16-19 x 12-16 µm.

Hosts and distribution: *Verbesina* spp.: southern Florida to southern Mexico and Guatemala.

Lectotype: on *Verbesina* sp., Cuernavaca, Mor., Mexico, Holway No. 3542 (S; isotype MIN). Lectotype designation made here.

In the original publication, *Verbesina virgata* Cav. and *V.* sp. were listed as hosts. *V. virgata* bears only a few telia of *Coleosporium* but abundant telia of a *Puccinia*. The *V.* sp. bears abundant telia of the *Coleosporium* and none of the *Puccinia*. This specimen was cited, and is in the Holway Herbarium (MIN), as Holway No. 3542. Apparently, Holway sent part of the collection to Dietel, but without the number because the packet in the Dietel Herbarium (S) bears no number. It is to be noted that the manuscript was written by Holway, including the translation from German to English of the diagnoses provided by Dietel.

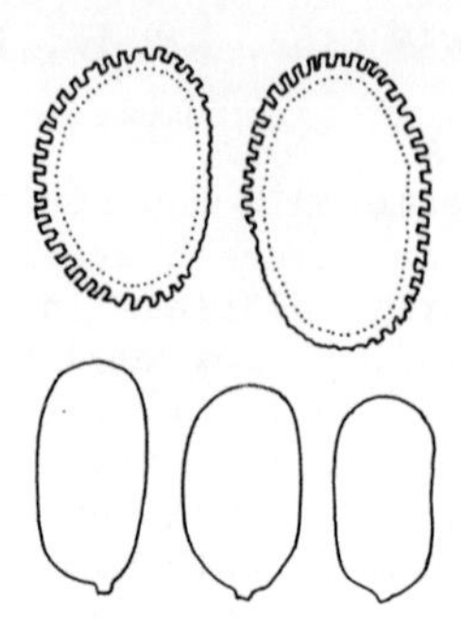

19. *COLEOSPORIUM LONGISPORUM* Cumm. Mycotaxon 5:400. 1977.
Aecidium carphochaetes Syd. Ann. Mycol. 1:20. 1903.

Spermogonia and aecia unknown. Uredinia on the abaxial leaf surface, orange yellow when fresh fading to pale yellowish or nearly colorless with age, sometimes appearing aecidioid due to adherence of the spores; spores variable in size and shape, (23-)27-38(-44) x (16-)18-24(-27) µm, oblong ellipsoid, ellipsoid or broadly ellipsoid, verrucose with rods, ridges or irregularly merged verrucae that are 2-3 µm high and 1.5-2 µm wide on the coarsely verrucose side decreasing to finely rugose or pseudoreticulate on the opposite side toward the base, wall without the verrucae 1-1.5 µm thick. Telia (basidiosorus) on abaxial surface, early exposed, pale orange, hard when dry, the apical gelatinous layer about 20 µm thick, 1 basidium deep but with young basidia pushing up between the older germinating ones, (55-) 60-70(-80) x (16-)18-22 µm, cylindrical; basidiospores (20-) 24-28(-30) x 11-14 µm, oblong ellipsoid or with a tendency to be slightly allantoid.

Hosts and distribution: *Carphochaete grahami* Gray, *Stevia lucida* Lag., *S. salicifolia* Cav.: central to south central Mexico.

Type: on *Carphochaete grahami*, Mex hgw No. 40, km 1016 west of Durango, Dgo., Cummins No. 63-581 (PUR 62353).

Arthur (1), transferred *A. carphochaetes*, based on uredinia and telia, to the synonymy of *C. steviae*, recognizing that it was not an *Aecidium*.

20. *COLEOSPORIUM ANCEPS* Diet. & Holw. in Holway, Bot. Gaz. 31:337. 1901.

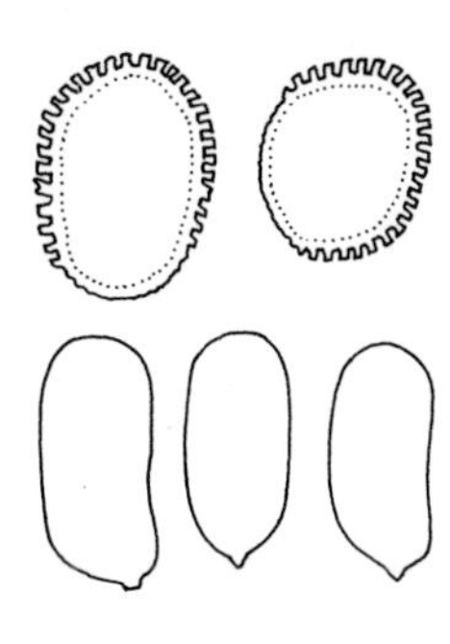

Spermogonia and aecia unknown. Uredinia mostly on the adaxial leaf surface, conspicuous, often concentrically arranged; in chlorotic becoming necrotic areas, nearly white when dry; spores (20-)22-29(-32) x (16-)18-24 μm, mostly broadly ellipsoid but variable, beset with rods 2-2.5 μm long and 1-1.5 μm wide or these sometimes slightly elongated as short ridges, these decreasing, sometimes abruptly, to a nearly smooth area at base or along one side toward the base, the wall minus verrucae 1.5 μm thick. Telia (basidiosori) mostly on the abaxial surface opposite the uredinia, typically in a circle, pale orange when dry, early exposed, hard, 1 basidium deep but newer basidia push up between the older ones, the apical gelatinous layer 25 μm or thicker, basidia 65-85(-90) x (16-)18-21(-23) μm, mostly cylindrical, horizontally septate, the lower most septum separates a basal stalk like part of varying length; basidiospores 23-28 x 12-14 μm, oblong ellipsoid or tending to be slightly allantoid.

Hosts and distribution: species of *Verbesina* and *Zexmenia helianthoides* (H.B.K.) Jacks. & Hook. f.: southern Florida to southern Mexico and Costa Rica.

Lectotype: on *Verbesina sphaerocephala* Gray, Chapala, Jal., Mexico, Holway No. 3492 (S; isotype PUR 1857, MIN). Lectotype designation made here.

The original publication states that No. 3492 is "mostly Uredo" and No. 3501 is "only teleutospores" but the reverse is true of the specimens that I have seen. The basidiospore size was given as 24-30 x 12-15 μm. They are abundant in the lectotype.

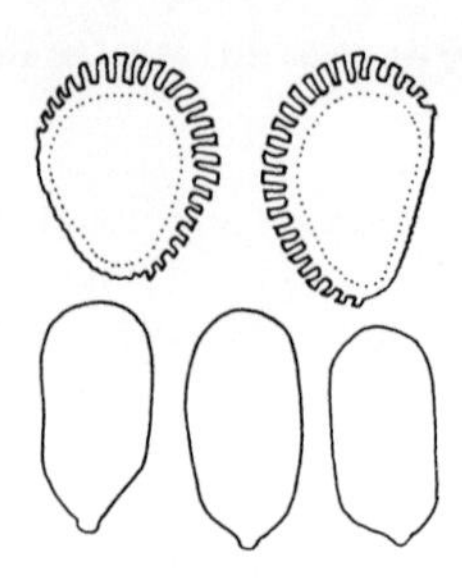

21. *COLEOSPORIUM STEVIAE* Arth. Bot. Gaz. 40:197. 1905.
Coleosporium eupatorii Arth. Bull. Torrey Bot. Club 33: 31. 1906, based on uredinia.
Coleosporium eupatorii Cumm. Mycologia 48:603. 1956, not Hirat. f. 1927.

Spermogonia and aecia not recognized in North America; recorded in Japan as *C. eupatorii* Hirat. f. but it is not certain that this is synonymous with *C. steviae*. Uredinia mostly or only on the abaxial leaf surface, yellow when fresh fading to colorless; spores (19-)21-25(-30) x (15-) 17-22(-24) µm, mostly broadly ellipsoid, the inner wall 0.5 -1 µm thick, the verrucose with rods or ridges 2.5-4 µm long on apex and upper sides, decreasing abruptly to a conspicuous smooth area at base and/or one side toward base, the verrucae often merging to pseudoreticulate pattern especially toward the smooth area. Telia (basidiosori) on abaxial surface, early exposed, orange color, compact and hard when dry, 1 basidium deep, basidia (50-)60-78(-88) x (15-)19-24 µm, cylindrical, 4 celled, the septa mostly horizontal but may be oblique or even cruciate, the side wall thin, the apex gelatinous and mostly 20-35 µm thick; basidiospores (20-)22-26(-30) x (14-)17-23 µm, oblong ellipsoid or tending to be slightly allantoid.

Hosts and distribution: species of *Eupatorium*, *Stevia* and perhaps *Brickellia*: northern Mexico south to Costa Rica.

Type: on *Stevia trachelioides* (DC.) Hook. (= *monardaefolia* H.B.K.), Nevada de Toluca, Mex., Mexico, Holway No. 5159 (PUR 888).

16. PHAKOPSORA Dietel

Ber. Dtsch. Bot. Ges. 13:333. 1895.

Spermogonia subcuticular, conical, type 7 (16). Aecia subepidermal in origin, erumpent, uredinoid; spores borne singly on pedicels. Uredinia subepidermal in origin, mostly with incurved, basally septate paraphyses; spores borne singly on pedicels. Telia subepidermal, not erumpent, consisting of crusts 3 or more spores deep; spores sessile, 1 celled, irregularly arranged, not catenulate, with a single, obscure germ pore, wall pigmented; basidium external.

Type species: *Phakopsora punctiformis* (Diet. & Barcl.) Diet.

The aecia described here are those of rust fungi often treated as *Bubakia*. Aecia are not known for *Phakopsora sensu stricto*, hence for any of the species on legumes.

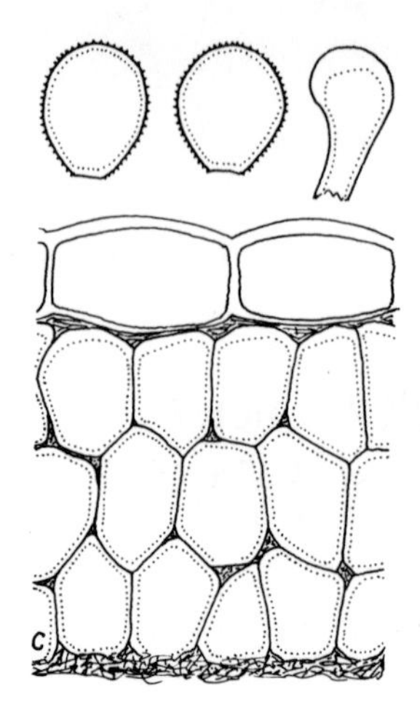

1. *PHAKOPSORA DIEHLII* Cumm. Mycologia 66:892. 1974.
   Physopella aeschynomenis Arth. N. Amer. Flora 7: 1907 (based on uredinia).
   Phakopsora aeschynomenis Arth. ex Cast. N. Giorn. Bot. Ital. 49:18, 1942 (telia described, the name a hymonym).

Spermogonia and aecia unknown. Uredinia mostly on abaxial leaf surface, with abundant basally united, multiple series of peripheral paraphyses, incurved, 8-14 µm wide at apex, wall brownish and 3-7 µm thick apically and more or less dorsally; spores (16-)18-20(-23) x (12-)14-16(-18) µm, mostly ellipsoid or broadly so, wall 1 µm thick, finely echinulate, pale yellowish, pores obscure, probably 6 or 7, scattered. Telia on abaxial surface with the uredinia, covered by epidermis, in irregular layers of (2)3(4) spores; spores (12-)15-18(-20) x (6-)8-10(-12) µm, ovoid or oblong ellipsoid, wall uniformly 1-1.5 µm thick or 2(-2.5) µm at apex, golden brown, smooth.

Hosts and distribution: *Aeschynomene americana* L.: southern Mexico to Guatemala; also in South America and perhaps Ethiopia.

Type: on *Aeschynomene americana*, Sierra Madre, states of Michoacan and Guerrero, Mexico. Langlasse No. 758 ex US 385549 (BPI 6279).

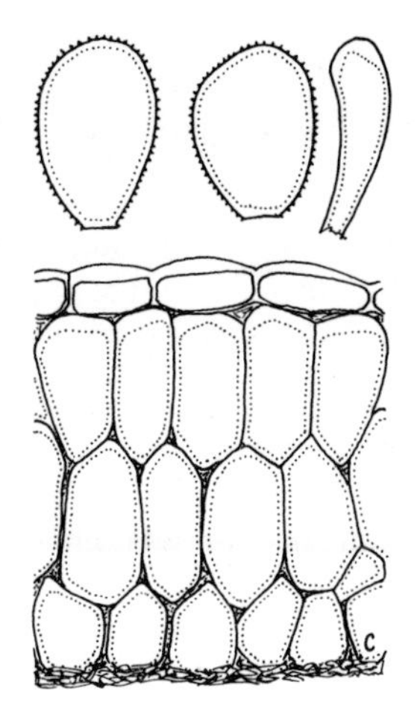

2. *PHAKOPSORA PACHYRHIZI* H. Syd. & P. Syd. Ann. Mycol. 12: 108. 1914.
   Physopella meibomiae Arth. Mycologia 9:59. 1917.
   Physopella concors Arth. Mycologia 9:60. 1917 (based on uredinia).
   Phakopsora vignae Arth. Bull. Torrey Bot. Club 44:509 1917. (based on uredinia).

Spermogonia and aecia unknown. Uredinia on abaxial leaf surface, usually grouped, small, opening by a pore, with incurved, basally united, multiseriate paraphyses (5-)6-8(-13) µm wide at apex, the wall colorless and thin below, 2-5 µm thick and nearly colorless at apex; spores (18-)21-28(-30) x (13-)15-19(-22) µm, mostly obovoid or ellipsoid, wall 1(-1.5) µm thick, finely echinulate, yellowish to pale brown, pores obscure, probably about 8 and scattered, with no or inconspicuous caps. Telia on abaxial surface among the uredinia, reddish brown, covered by epidermis, crustose, with 2-5 irregular layers of spores; spores (14-)18-24(28) x (5-)7-11(-13) µm, mostly more or less oblong but variable, wall uniformly 1 µm thick in lower spores and pale yellowish, uniform or to 5 µm thick and near chestnut brown at apex of outer spores, smooth.

Hosts and distribution: *Canavalia villosa* Benth., *Desmodium* spp., *Erythrina berteroana* Urban, *Glycine max* (L.) Merr., *Phaseolus Macrolepis* Piper, *Vigna repens* (L.) Kuntze: southern Mexico to Costa Rica; also in the West Indies, South America, the Orient and Africa.

Type: on *Pachyrhizus angulatus* Rich., Taihoku, Taiwan, Fujikuro No. 37 (S).

The genus needs much study and most material is now inadequate to permit conclusions as to specific limitations. I am following Hiratsuka (15) and grouping the above "species." The fungus also occurs on *Dolichos*, *Pueraria* and probably others. Cummins (7) described telia on *Canavalia villosa* under the invalid name *Phakopsora vignae*.

Uredo erythrinae P. Henn. is assigned here but there is no assurance that this correct.

17. BAEODROMUS Arthur

Ann. Mycol. 3:19. 1905.

Spermogonia subepidermal, globoid, type 4 (16), Aecia and uredinia unknown. Telia subepidermal in origin, erumpent, compact; spores 1 celled, catenulate in chains of relatively few spores, wall lightly pigmented, germ pore 1; basidia external.

Type species: *Baeodromus holwayi* Arth.

1. *BAEODROMUS CALIFORNICUS* Arth. Ann. Mycol. 3:19. 1905.

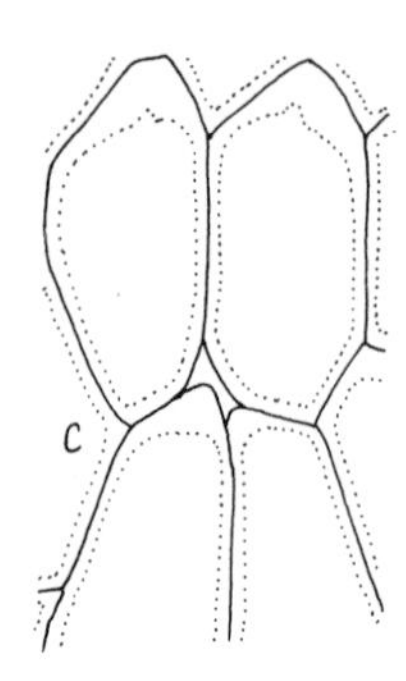

Spermogonia not seen. Aecia and uredinia lacking. Telia amphigenous and on stems, crowded in groups of various sizes, early exposed, pulvinate, compact, discrete, brownish; spores (26-)30-40(44) x (12-)16-20(21) µm, variable in size and shape, internal spores mostly ellipsoid or oblong ellipsoid, external spores the same or often obovoid, wall 1.5-2 µm thick except 4-8 µm at apex, golden brown, smooth.

Hosts and distribution: *Senecio douglasii* DC.: southern California.

Type: Lytle Creek, San Bernardino, Calif., Parish No. 2562 (PUR 5785).

2. *BAEODROMUS HOLWAYI* Arth. Ann. Mycol. 3:19. 1905.

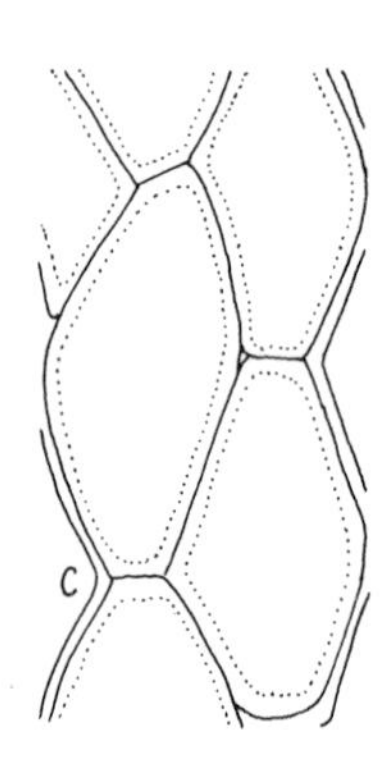

Spermogonia amphigenous in small groups. Aecia and uredinia lacking. Telia on the abaxial leaf surface, densely grouped around the spermogonia, early exposed, discrete, pulvinate, compact, brownish; spores (30-)36-48(-58) x (15-)20-28(-30) µm, variable in size and shape, internal spores mostly oblong ellipsoid, external spores more robust and often nearly globose, wall of internal spores uniformly about 1.5 µm thick or occasionally to 4 µm apically, wall of external spores usually 2.5-4 µm thick on the outer side, golden brown, smooth.

Hosts and distribution: *Senecio argutus* H. B. K., *S. cinerarioides* H.B.K., *S. warscewiczii* A. Br. & Bouche: southern Mexico and Guatemala.

Type: on *Senecio cinerarioides*, Nevada de Toluca, Mexico, Holway No.5160 (PUR 5780; isotypes Barth. N. Amer. Ured. No. 207).

3. *BAEODROMUS EUPATORII* (Arth.) Arth. N. Amer. Flora 7:125. 1907.
Dietelia eupatorii Arth. Bot. Gaz. 40:197-.98. 1905.

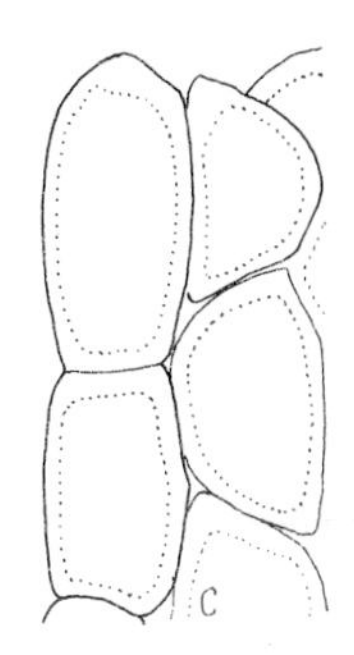

Spermogonia amphigenous, in close groups. Aecia and uredinia lacking. Telia on the abaxial surface and on petioles and stems, in close groups around the spermogonia, discrete, pale brownish, each sorus surrounded with a compact layer of hyphae and collapsed host cells; spores (24-)26-35(-38) x (16-) 18-22(-24) μm, variable in size and shape, mostly more or less oblong or shorter spores cuboidal or angularly globoid, wall uniformly (1.5-)2-2.5 μm thick or the apex sometimes to 5 μm thick, pale golden, smooth.

Hosts and distribution: *Eupatorium* spp.: Durango and Nayarit, Mexico to Guatemala.

Type: on *Eupatorium patzcuarense* H.B.K., Amecameca, Mexico, Holway No. 5205 (PUR 5777).

4. *BAEODROMUS* sp.

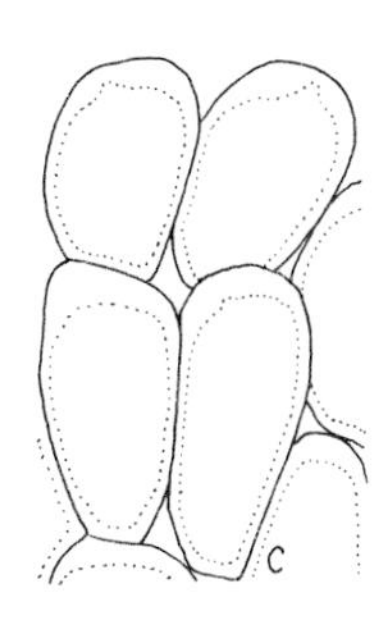

Spermogonia not seen. Aecia and uredinia lacking. Telia in close groups, early exposed, pulvinate, compact, discrete, brown becoming gray from germination; spores (20-) 26-35(-40) x (13-)15-18(-20) μm, variable in size and shape, internal spores mostly oblong ellipsoid, external spores tending to be obovoid or globoid, wall 1-1.5 μm thick at sides and 2-5(-6) μm at apex, side walls of external spores usually thicker than the internal spores, yellow to golden, smooth.

Type: on *Senecio eremophilus* Rich., Hand Hills, near Delta, Alberta, Brinkman No. 5017. Not otherwise known.

This apparently distinctive species has not been formally described but presumably will be by Hennen and Buritica.

18. CEROTELIUM Arthur

Bull. Torrey Bot. Club 33:30. 1906.

Spermogonia subcuticular, conical, type 7 (16). Aecia subepidermal in origin, erumpent, with peridium, aecidioid; spores catenulate. Uredinia subepidermal in origin, scarcely erumpent, with peridium, peripheral, basally united paraphyces, or neither.; spores borne singly on pedicels or perhaps sessile. Telia subepidermal in origin, becoming erumpent; spores 1 celled, catenulate in short chains, the chains scarcely adherent laterally, wall thin and pale, germ pore obscure if differentiated, spores germinate without dormancy; basidium external.

Type species: *Cerotelium canavaliae* Arth.

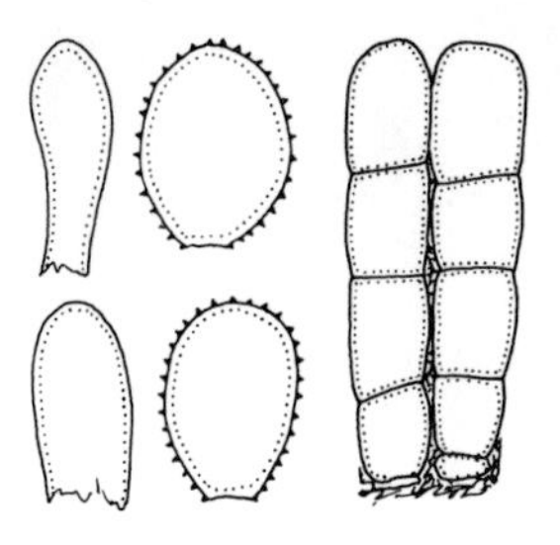

1. *CEROTELIUM TANAKAE* S. Ito in Ito & Homma, Trans. Sapporo Nat. Hist. Soc. 15:118. 1938.

Spermogonia and aecia unknown. Uredinia mostly on abaxial leaf surface, bright yellow when fresh fading to colorless, with short thin walled paraphyses united basally to peridial tissue, opening by a pore; spores (15-)17-20(-23) x (13-)15-17(-19) µm, mostly obovoid or broadly ellipsoid, wall 1 µm thick, pale yellowish or colorless, echinulate, pores obscure. Telia on abaxial surface, early exposed, essentially colorless; spores (14-)16-22(-26) x (8-)9-11(-12) µm, oblong or more or less cuboidal, delicate, wall about 0.5 µm thick, colorless, germ pore not differentiated, the upper spores germinate and collapse early, the chains thus short.

Hosts and distribution: *Amphicarpaea bracteata* (L.) Fern.: known in North America only in Brown County State Park, Indiana; also in Japan.

Type: on *Falcata comosa* Kuntze var. *japonica* Makino (= *Amphicarpaea edgworthii* Benth.), Hokkaido, Japan, Tanaka (SAPA).

The teliospores, as described by Ito, are larger than those in the Indiana material but the spores are so delicate and collapse so rapidly that accurate measurements are difficult, even in fresh specimens.

19. CIONOTHRIX Arthur

N. Amer. Flora 7:124. 1907.

Spermogonia subepidermal, globoid, type 4 (16). Aecia and uredinia unknown. Telia subepidermal in origin, becoming erumpent as long, filiform columns of spores; spores 1 celled, catenulate, germ pore obscure, perhaps not differentiated, wall pale and thin, spores germinate without dormancy; basidium external.

1. *CIONOTHRIX PRAELONGA* (Wint.) Arth. N. Amer. Flora 7:125. 1907.
   Cronartium praelongum Wint. Hedwigia 26:24. 1887.

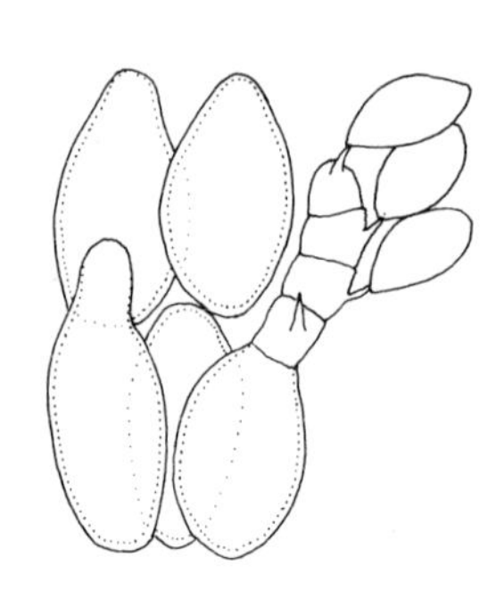

Spermogonia on adaxial leaf surface,few in a group. Aecia and uredinia lacking. Telia in groups on abaxial surface, deeply seated, flask shape, the basal region wide but narrowed to a neck region with abundant upwardly directed, cylindrical, thick walled colorless paraphyses, mostly 4-7 µm wide, these joined below, the spores extruded in filiform columns; spores (24-)28-34(-40) x (12-)16-20(-24) µm, mostly narrowly ovoid or ellipsoid, initially catenulate but not remaining in chains, without obvious intercallary cells, wall 1 µm thick, colorless.

Hosts and distribution: *Eupatorium* spp.: southeastern Mexico to Panama; also in South America.

Type: on "compositae", near Sao Francisco, Prov. St. Catherina, Brazil, Ule (isotype Rab.-Wint. F. europ. No. 3419).

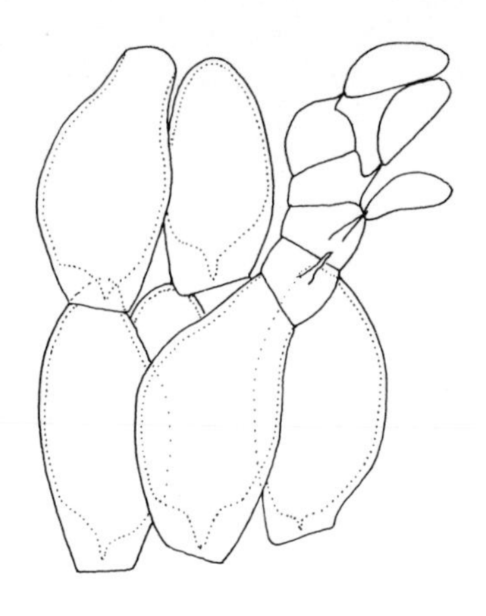

2. *CIONOTHRIX* sp.

Spermogonia on adaxial leaf surface, deeply seated, in groups. Aecia and uredinia lacking. Telia in loose groups on abaxial surface, deeply seated, more or less flask shape, the basal part large and globoid, the upper part narrowed to a pore lined with a layer of hyphae and collapsed host cells but without paraphyses, the spores extruded in long, filiform, pale yellow columns; spores 30-44(-53) x (15-) 17-22(-24) μm, oblong ellipsoid or ovoid, the base usually obtuse, initiated catenulately but apparently without intercallary cells, wall 0.5-1 μm thick except basally where 3-6(-8) μm thick, colorless or pale yellowish.

Hosts and distribution: *Eupatorium morifolium* Mill.: Guatemala City, Holway No. 688 (PUR 5682). Not otherwise known.

This species will be described by Hennen and Buritica.

20. ENDOPHYLLUM Léveille

Mem. Soc. Linn. Paris 4:208. 1825.

Spermogonia subepidermal, subcuticular, or wanting, mostly type 4 (16). Aecia and uredinia lacking. Telia erumpent, with peridium, aecidioid; spores catenulate, 1 celled, usually with intercallary cells, germinating without or after dormancy; basidium external.

Type species: *Endophyllum sempervivi* (Alb. & Schw.) DeBary.

1. *ENDOPHYLLUM PUMILIO* (Kunze) H. Syd. & P. Syd. Ann. Mycol. 18:179. 1920.
   Aecidium pumilio Kunze in Weigelt Exsic. no No. 1827.
   Endophyllum decoloratum (Schw.) Whet. & Olive, Amer. J. Bot. 4:49. 1917.

Spermogonia unknown. Telia on abaxial leaf surface, in groups, peridium short cylindrical or evanescent, the margin incised or erose; spores 15-20 x 11-15 µm, mostly broadly ellipsoid or globoid, wall about 1 µm thick, colorless, finely verrucose.

Hosts and distribution: *Clibadium* spp., *Wedelia trilobata* (L.) Hitch.: southern Mexico to Panama; also in the West Indies and South America.

Type: on Urticaceae or *Baillera aspera* (considered by the Sydows to be *Clibadium surinamense* L. var. *asperum* (Aubl.) Baker), Surinam, Weigelt (not seen).

The Weigelt exsiccata is accompanied by a Latin description of the species.

21. ENDOPHYLLOIDES Whetzel & Olive

Amer. J. Bot. 4:50. 1917.

Spermogonia subepidermal, globoid, type 4 (16), or wanting. Aecia and uredinia lacking. Telia subepidermal in origin, erumpent as short, compact columns, aecidioid, with peridium often adherent; spores 1 celled, catenulate in more or less adherent chains, with intercallary cells, spores germinate without dormancy, wall pale and thin; basidium external.

Type species: *Endophylloides portoricensis* Whet. & Olive.

1. *ENDOPHYLLOIDES PORTORICENSIS* Whet. & Olive in Olive & Whetzel, Amer. J. Bot. 4:51. 1917.

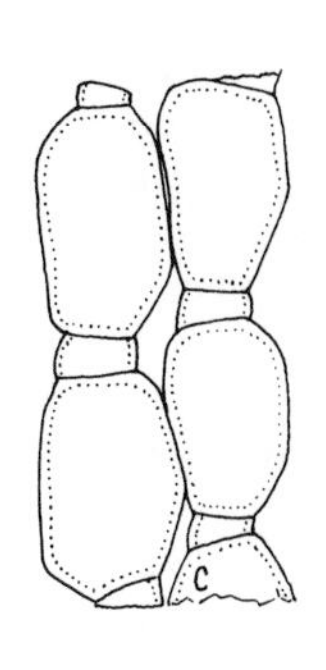

Spermogonia, aecia and uredinia lacking. Telia on the abaxial leaf surface and sometimes on petioles and stems, aecidioid, pale yellowish, in close groups, the individual sori short columnar, with poorly formed peridium; spores (15-)18-26(-28) x (12-)14-17(-20) µm, mostly ellipsoid or oblong ellipsoid, catenulate, separated by intercallary cells, wall 1-1.5 µm thick, colorless, smooth.

Hosts and distribution: *Mikania* spp.: Guatemala to Panama; also in the West Indies and South America.

Type: on *Mikania cordifolia* (L. *f.*) Willd., Puerto Rico, Whetzel & Olive No. 83 (CUP; isotype PUR 5724).

22. PUCCINIOSIRA Lagerheim

Ber. Dtsch. Bot. Ges. 9:344. 1891 (issued 1892).

Spermogonia subepidermal, globoid, type 4 (16). Aecia and uredinia lacking. Telia subepidermal in origin, erumpent, with peridium, aecidioid; spores 2 celled by horizontal septum, the cells often separating easily, catenulate in loosely or not adherent chains, with intercallary cells, germ pores obscure, spores germinate without dormancy; basidium external.

Type species: *Pucciniosira triumfettae* Lagh.

1. *PUCCINIOSIRA BRICKELLIAE* Diet. & Holw. in Holway, Bot. Gaz. 24:34. 1897.

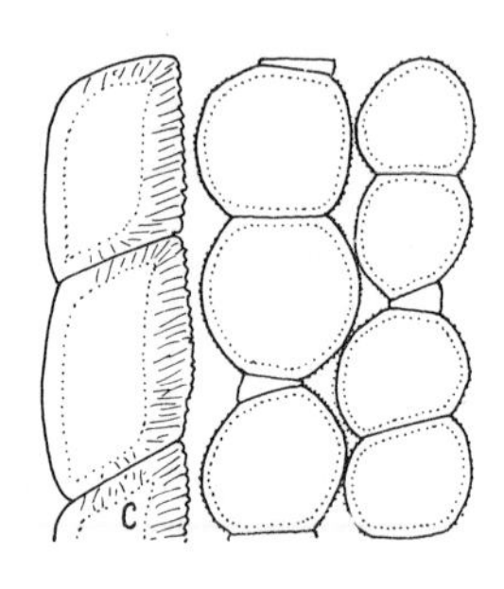

Spermogonia on the adaxial leaf surface. Telia on abaxial leaf surface, commonly along veins, and on stems, often causing slight hypertrophy, pale yellowish, aecidioid, peridium cylindrical, rupturing variously; spores (24-)27-35 (-40) x (11-)17-21(-24) µm, variable in size, mostly oblong ellipsoid but constricted at the septum, separating easily into 2 cells, wall about 1 µm thick, essentially colorless, rather indistinctly verrucose, intercallary cells usually obvious.

Hosts and distribution: *Brickellia* spp. and *Montanoa* sp.: Baja California Sur to Durango, Mexico and south to El Salvador and Guatemala.

Type: on *Brickelia* sp., Rio Hondo, near Mexico City, Holway (S).

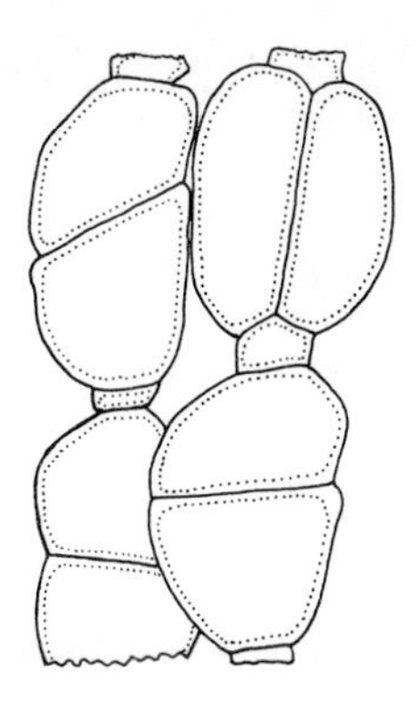

2. *PUCCINIOSIRA* sp.

Spermogonia lacking. Telia on the abaxial leaf surface in groups, pale yellowish, cylindrical, aecidioid, peridium loosely organized; spores (26-)30-38(-40) x (16-)18-24 (-28) µm, oblong or oblong ellipsoid, slightly or no constricted at the septum which often is oblique or nearly vertical; wall 1 µm thick, colorless, smooth, intercallary cells obvious.

Hosts and distribution: *Eupatorium* sp.: Mexico, Guatemala and Honduras.

This species will be described by Hennen and Buritica.

UREDO

1. *UREDO ADENOCAULONIS* Cumm. Mycologia 48:607. 1956. *Coleosporium adenocaulonis* H. S. Jack. Brooklyn Bot. Gard. Mem. 1:202. 1918. Based on uredinia.

Uredinia on abaxial leaf surface in yellowish spots, exposed, orange yellow fading to whitish, pulverulent, the spores catenulate; spores 23-26 x 18-24 µm, globoid or broadly ellipsoid, the wall verrucose with rods or ridges which may be discrete or somewhat merged, 1.5-2 µm high, the wall minus verrucae about 1 µm thick, colorless, pores not seen.

Type: on *Adenocaulon bicolor* Hook., Corvallis, Oregon, Jackson. Not otherwise known.

The fungus doubtless will prove to belong in *Coleosporium*, as Jackson anticipated.

2. *UREDO GARCILASSAE* P. Henn. Hedwigia 43:160. 1904.

Uredinia on the abaxial leaf surface, cinnamon brown; spores (18-)20-22(-24) x (16-)19-22 µm, mostly broadly ellipsoid or broadly obovoid with pores face view, obovoid or triangularly so with pores lateral, wall 1-1.5 µm thick, finely echinulate, golden or cinnamon brown, pores 2, subequatorial,often near the hilum, with slight, smooth caps.

Hosts and distribution: *Garcilassa rivularis* Poep. & Endl.: Costa Rica and Panama; also in South America.

Type: on *Garcilassa rivularis*, Terapoto, Peru, Ule No. 3168.

The spores are similar to those of *Puccinia gymnolomiae* Arth.

3. *UREDO WILSONII* Arth. Bull. Torrey Bot. Club 37:577. 1910.

Uredinia on the abaxial leaf surface, chocolate brown, with abundant, peripheral, straight or slightly incurved, cylindrical, pale brownish, uniformly thin walled, 1 or more septate, 9-15 μm wide paraphyses, these joined basally; spores (24-)27-30(-32) x (21-)23-27(-29) μm, globoid or broadly ellipsoid, wall 2(-2.5) μm thick, nearly chestnut brown, uniformly echinulate verrucose with round cones 1-1.5 μm wide and about 1.5-2 μm long, spaced about 2-2.5 μm, pores 2, conspicuous, equatorial in the slightly flattened sides, with low caps.

Hosts and distribution: *Gochnatia magna* M. C. Johns.: southern Tamaulipas, Mexico; known otherwise in the Bahamas.

Type: on *Gochnatia bahamensis* (Urb.) Howard & Dunb. *(Anastraphia bahamensis)*, Hanna Hill, Long Cay, Bahamas, Brace No. 4029.

4. *UREDO ACACIAE-BURSARIAE* Cumm. Mycologia 48:607. 1956.
Ravenelia inquirenda Arth. & Holw. Amer. J. Bot. 5:423. 1918. Not *Uredo inquirenda* Arth. 1907.

Uredinia amphigenous, pale cinnamon brown, with intermixed cylindrical or clavate, 7-10 μm wide, brownish, uniformly thin walled paraphyses; spores (22-)26-31(-35) x (16-) 17-20(-23) μm, mostly ellipsoid or obovoid, wall 1-1.5 μm thick, pale golden, echinulate, pores (3)4 or 5(6), equatorial, without obvious caps.

Type: on *ACACIA bursaria* Schr., Laguna, Lake Amatitlán, Guatemala, Holway No. 196. There is one other collection, from Nayarit, Mexico on *Acacia standleyi* Saff.

5. *UREDO HOFFMANSEGGIAE* Cumm. Mycologia 48:608. 1956.
Ravenelia hoffmanseggiae Long, Bot. Gaz. 64:57. 1917, based on uredinia.

Uredinia amphigenous, subepidermal, erumpent, about cinnamon brown; spores (23-)25-29(-31) x (16-)18-21(-23) μm, variable in shape, broadly ellipsoid, ellipsoid or obovoid, wall 1.5-(-2) μm thick, cinnamon brown or golden, finely echinulate, pores (4)5-7(8), scattered, with obvious caps.

Type: on *Hoffmanseggia oxycarpa* Gray, Del Rio, Texas, Long No. 6082. Not otherwise known.

6. *UREDO HYMENAEAE* Mayor, Mém. Soc. Neuch. Sci. Nat. 5:585. 1913.

Uredinia on the abaxial leaf surface, subepidermal, round, opening by a pore, with abundant, peripheral, cylindrical or clavate, colorless, thin walled paraphyses; spores (22-)24-31(-34) x (13-)15-18 μm, mostly ellipsoid or elongately obovoid, wall 1-1.5 μm thick at sides, 2-3.5(-5) μm at apex, clear chestnut brown or slightly darker at the apex, echinulate, pores 2 or 3 near the apex.

Hosts and distribution: *Hymenaea courbaril* L.: Chiapis, Mexico; also in the West Indies and South America.

Type: on *Hymenaea* sp., near Titiribi, Dept. Antioquia, Colombia, Mayor No. 149.

7. *UREDO IERENSIS* Dale, Commonw. Mycol. Inst. Kew Mycol. Papers 59:8. 1955.

Uredinia amphigenous, subepidermal, in groups to 1 cm across, cinnamon brown; spores (20-)22-27(-30) x (16-)18-22 (-25) μm, broadly ellipsoid or obovoid, wall (1.5-)2-2.5(-3) μm thick, cinnamon brown, strongly echinulate, pores 3, equatorial, with slight or no caps.

Hosts and distribution: *Lonchocarpus salvadorensis* Pitt. *L.* sp.: El Salvador and Guatemala; also in Brazil and the West Indies.

Type: on *Lonchocarpus latifolius* H.B.K., Trinidad, Dale.

8. *UREDO MACHERIICOLA* Cumm. Bull. Torrey Bot. Club 70:79. 1943.

Uredinia mostly on abaxial leaf surface, subepidermal, grouped, yellowish brown, with colorless, incurved or sinuous, cylindrical, 4-7 μm wide, thick walled paraphyses, these united basally; spores (14-)15-18(-21) x (11-)12-15 (-16) μm, broadly ellipsoid or obovoid, wall 1-1.5(-2) μm thick, echinulate, usually smooth at base, pores obscure.

Type: on *Macherium biovulatum* Mich., vicinity of Retalhuleu, Dept. Retalhuleu, Guatemala, Standley No. 88557. Not otherwise known.

9. *UREDO MEXICENSIS* Cumm. Mycotaxon 5:407. 1977.

Uredinia amphigenous, subcuticular, in often circinate groups, sometimes associated with veins, cinnamon brown, with peripheral, capitate or clavate capitate paraphyses, (10-) 14-20(-25) μm wide in the head, wall brown apically and 2-3.5(-4) μm thick, 1.5 μm in the stalk which is nearly colorless; spores (20-)23-29(-33) x (11-)13-17(-19) μm, mostly obovoid, wall at sides 1.5-2(-2.5) μm thick, 2.5-4(-5) μm at apex, golden brown apically, paler below, uniformly echinulate, pores (3)4-6, equatorial.

Type: on *Leucaena macrocarpa* Rose, east of Malpica, Sin., Mexico, Cummins No. 71-568. Also in Colima and Jalisco.

10. *UREDO QUICHENSIS* Cumm. Bull. Torrey Bot. Club 70:80. 1943.

Uredinia on the adaxial leaf surface, subcuticular, yellowish brown, with abundant, mostly capitate paraphyses, 10-16 μm wide, nearly uniformly thin walled, golden brown; spores 25-30 x 16-19 μm, ellipsoid or more or less oblong ellipsoid, wall uniformly 1.5(-2) μ thick or the apex to 3 μm, about cinnamon brown, closely verrucose echinulate with small cones, pores bizonate with 4 or 5 in each zone.

Type: on *Calliandra conzattiana* (Britt. & Rose) Standl., Aguacatán, Dept. El Quiche, Guatemala, Standley No. 81393. Not otherwise known.

11. *UREDO RAMONENSIS* H. Syd. Ann. Mycol. 23:325. 1925.

Uredinia on abaxial leaf surface, subepidermal, in small groups or singly, cinnamon brown; spores (22-)25-32 (-35) x (17-)18-22(-24) μm, variable in shape and size, oblong ellipsoid or more commonly triangularly obovoid, wall 1-1.5 μm thick, about cinnamon brown except often paler basally, echinulate, pores 3, equatorial in the angles, perhaps rarely 4.

Type: on *Cassia bacillaris* L. *f.*, Cerro de San Isidoro, pr. San Ramon, Costa Rica. Not otherwise known.

12. *UREDO YUCATANENSIS* Mains, Contrib. Univ. Michigan Herb. 1:17. 1939.

Uredinia amphigenous, subepidermal, pale yellowish (dry), probably bright yellow when fresh, with peripheral, cylindrical or clavate, nearly colorless, 8-10 μm wide, thin walled paraphyses which are united basally; spores (18-)20-24 x (15-)17-20 μm, mostly broadly ellipsoid or obovoid, wall 2(-2.5) μm thick, colorless or essentially so, echinulate, pores obscure, scattered, 7 or 8.

Type: on *Mimosa albida* Humb. & Bonpl., San Agustin, British Honduras (now Belize), Mains No. 3889. Not otherwise known.

EXCLUDED SPECIES

UREDO POSITA J. J. Davis ex Arth. Torreya 34:46. 1934.

Type: on *Aeschynomene virginiana* (L.) B.S.P., Crowley, Louisiana, A. L. Smith.

There are spores, as described, present but it is probable that they are contaminants. Until additionally collection proves otherwise, the species is best excluded.

AECIDIUM

1. *AECIDIUM AMPLIATUM* H. S. Jack. & Holw. in Arthur, Mycologia 10:148. 1918.

Spermogonia amphigenous, numerous. Aecia on the abaxial leaf surface, often along the veins, cupulate, peridium white or pale yellowish, margin recurved, erose or lacerate; spores 27-30 x 23-27 µm, globoid or broadly ellipsoid, wall 1 µm thick, finely verrucose, colorless.

Type: on *Eupatorium* sp., El Alto, near Cartago, Costa Rica, Holway No. 434. Also known in Guatemala.

2. *AECIDIUM ARCHIBACCHARIDIS* Cumm. Bull. Torrey Bot. Club 68:471. 1941.

Spermogonia mostly on the abaxial leaf surface. Aecia on abaxial surface, in spots to 1 cm across, peridium cupulate, margin recurved; spores 18-25 x 18-23 µm, globoid, wall 1 µm thick at sides, 7-13 µm apically, colorless or pale yellowish, loosely verrucose with rather large and somewhat deciduous warts.

Type: on *Archibaccharis serratifolia* (H.B.K.) Blake, Mauchen, Guatemala, Johnston No. 79. There is another collection from the same locality.

3. *AECIDIUM BATESII* Arth. Bull. Torrey Bot. Club 47:479. 1920.

Spermogonia grouped on adaxial leaf surface. Aecia amphigenous, grouped in spots to 8 mm diam, peridium yellow, tardily exposed, the margin finally recurved, lacerate; spores 24-27 x 19-23 µm, angularly globoid, wall 1-1.5 µm thick, finely verrucose.

Type: on *Rudbeckia hirta* L., Callaway, Nebr., Bates.

4. *AECIDIUM BORRICHIAE* H. Syd. & P. Syd. Hedwigia 40(Beibl.) p. 129. 1901.

Spermogonia amphigenous in small groups. Aecia on abaxial leaf surface in groups to 5 mm across, peridium cupulate, pale yellowish, the margin recurved; spores 26-32 x 24-28 µm, angularly globoid or ellipsoid, wall 2.5-3.5 µm thick, colorless, minutely verrucose.

Type: on *Borrichia frutescens* (L.) DC., Fort Morgan, Alabama, Tracy. Not otherwise known.

5. *AECIDIUM COLUMBIENSE* Ellis & Ever. Erythea 1:206. 1893.

Spermogonia scattered over entire leaves from systemic mycelium. Aecia on abaxial leaf surface, systemic, peridium cupulate, pale yellowish, the margin incurved and erose; spores 18-21 x 16-20 µm, globoid, wall 1-1.5 µm thick, colorless, minutely verrucose.

Hosts and distribution: *Hieracium* spp.: British Columbia to northern California and Long Island, New York.

Type: on *Hieracium* sp., British Columbia, Macoun.

6. *AECIDIUM DAHLIAE* H. Syd. & P. Syd. Ann. Mycol. 18:155. 1920.

Spermogonia amphigenous in small groups. Aecia on abaxial leaf surface in close groups to 8 mm across, peridium cupulate, the margin recurved and lacerate; spores 16-20 x 15-18 µm, angularly globoid, wall 1-1.5 µm thick, nearly colorless, minutely verrucose.

Type: on *Dahlia variabilis* (= *D. rosea* Cav.), Pedregal, Mex., Mexico, Reiche. Not otherwise known.

7. *AECIDIUM DAHLIAE-MAXONII* Cumm. Bull. Torrey Bot. Club 70:69. 1943.

Spermogonia lacking. Aecia on abaxial leaf surface, few in loose groups, peridium cupulate, yellowish, margin recurved; spores 17-20 x 14-18 µm, broadly ellipsoid or globoid, wall 0.5-1 µm thick, colorless, minutely verrucose.

Type: on *Dahlia maxonii* Saff., Las Calderas, Dept. Chimaltenango, Guatemala, Johnston No. 1875. Not otherwise known.

8. *AECIDIUM HUALTATINUM* Speg. Bol. Acad. Cien. Cordoba 11: 184. 1888.
Aecidium herrerianum Arth. Bull. Torrey Bot. Club 33: 520. 1906.

Aecia on abaxial leaf surface in small groups, almost wholly immersed. Aecia on abaxial surface, crowded in groups to 5 mm across, peridium cylindrical, to 1.5 mm long, pale yellow, the margin erect, deeply lacerate; spores 28-34 x 23-27 µm, mostly globoid, wall 3-4 µm thick, pale cinnamon brown, finely verrucose.

Hosts and distribution: *Senecio salignus* DC.: Hidalgo State, Mexico; also in South America.

Type: on *Senecio*, Gabel Island, Patagonia.

9. *AECIDIUM IVAE* H. S. Jack. Proc. Indiana Acad. Sci. 1917: 373. 1918.

Spermogonia amphigenous in yellowish spots. Aecia on abaxial leaf surface, peridium cupulate, brownish yellow, the margin erose and recurved; spores 26-33 x 21-29 µm, globoid or broadly ellipsoid, wall 2-3 µm thick, colorless or pale yellowish, finely verrucose.

Hosts and distribution: *Iva frutescens* L.: coastal areas from Delaware to Louisiana.

Type: on *Iva ovaria* = error for *oraria* (= *I. frutescens* var. *oraria* (Bartl.) Fern & Grisc.), Lewes, Delaware, Jackson No. 1676.

10. *AECIDIUM KEERLIAE* Arth. Bull. Torrey Bot. Club 45:154. 1918.

Spermogonia amphigenous in small groups. Aecia in spots to 8 mm across on the abaxial leaf surface, peridium short cupulate, the margin erose and slightly recurved; spores 15-20 x 12-18 µm, globoid or ellipsoid, wall 1 µm thick except 3-6 µm apically, finely verrucose.

Type: on *Keerlia mexicana* Gray, Guadalajara, Jal., Mexico. Not otherwise known.

Possibly this is an aecial stage of *Puccinia cyperi* Arth.

11. *AECIDIUM LIABI* Mayor, Mém. Soc. Neuch. Sci. Nat. 5:576. 1913.
Aecidium liabi Arth. Bull. Torrey Bot. Club 47:479. 1920.

Spermogonia in loose groups on adaxial leaf surface. Aecia on abaxial surface in groups to 8 mm across, peridium cupulate, pale yellowish, fragile and lacerating early; spores 26-31 x 21-26 µm, angularly globoid, wall 1.5-2.5 µm thick, colorless, distinctly verrucose.

Hosts and distribution: *Liabum* sp.: Jalapa, Veracruz, Mexico and in Nicaragua; also in South America.
Type: on *Liabum igniarium* Humb. & Bonpl., Alto Don Elias, Dept. Antioquia, Colombia, Mayor No. 197.

12. *AECIDIUM MESADENIAE* Arth. Bull. Torrey Bot. Club 47: 479. 1920.

Spermogonia on adaxial leaf surface in small groups. Aecia mostly on abaxial surface, in groups to 6 mm across, peridium cupulate, the margin recurved, lacerate, pale yellowish; spores 15-19 x 12-16 µm, globoid or ellipsoid, wall 1 µm or less thick, finely verrucose.

Hosts and distribution: *Cacalia atriplicifolia* L., *C. reniformis* Muhl.: Wisconsin, Kansas and Missouri.

Type: on *Mesadenia reniformis* (= *C. reniformis*), Somers, Wisconsin, Davis.

13. *AECIDIUM MIKANIAE* P. Henn. Hedwigia 35:261. 1896.

Spermogonia not seen. Aecia on the abaxial leaf surface, in groups 1-8 mm across, peridium cupulate, pale yellowish, the margin erect, erose; spores 16-20 µm diam, globoid, wall 1 µm or less thick, very finely verrucose.

Hosts and distribution: *Mikania houstoniana* (L.) B. L. Rob.: Honduras; also in South America and Africa.

Type: on *Mikania confertifolia* Sch. Bip., St. Cathar. pr. Blumenau, Brazil, Ule No. 911.

14. *AECIDIUM PEREZIAE* Arth. Bull. Torrey Bot. Club 45:153. 1918.

Spermogonia mostly on adaxial leaf surface, closely grouped. Aecia grouped in yellowish areas to 15 mm across, peridium cupulate, the margin erose, usually not recurved; spores 16-19 x 13-18 µm, globoid, wall 1-1.5 µm thick,mintely verrucose, colorless.

Hosts and distribution: *Perezia* sp.: Durango and Jalisco, Mexico.

Type: on *Perezia* sp., Barranca, Jal., Pringle.

15. *AECIDIUM POASENSIS* H. Syd. Ann. Mycol. 23:324. 1925.

Spermogonia not seen. Aecia amphigenous, deeply seated in thickened spots 2-5 mm across, peridium scarcely protruding; spores 17-23 x 16-18 µm, globoid or broadly ellipsoid, wall 1 µm thick, colorless, minutely verrucose.

Type: on *Otopappus verbesinoides* Benth., monte Poas, pr. Grecia, Costa Rica, Sydow No. 315.

16. *AECIDIUM PRAECIPUUM* Arth. Bull. Torrey Bot. Club 47:480. 1920.

Spermogonia amphigenous in small groups. Aecia on abaxial leaf surface, in close groups to 10 mm across, peridium short cylindrical, pale yellowish, the margin erose, erect; spores 19-26 x 16-19 µm, angularly globoid or ellipsoid, wall 2.5-3.5 µm thick, colorless, inconspicuously verrucose.

Type: on *Senecio praecox* DC.: Sierra de Pachuca, Hidalgo, Mexico, Rose, Painter and Rose No. 8791. Also in Mexico State on the same host.

17. *AECIDIUM STEVIICOLA* Arth. Bull. Torrey Bot. Club 45:154. 1918.

Spermogonia amphigenous, numerous. Aecia on the abaxial leaf surface, loosely grouped in spots to 15 mm across, peridium cupulate, the margin somewhat recurved and erose; spores 30-40 x 24-32 µm, globoid, wall 1.5 µm thick at sides, 6-9 µ m at apex, finely verrucose, colorless.

Type: on *Stevia* sp., Popo Park, D. F., Mexico, Hitchcock. Not otherwise known.

18. *AECIDIUM WEDELIAE-HISPIDAE* Diet. Ann. Mycol. 20:294. 1922.

Spermogonia on adaxial leaf surface, few in groups. Aecia mostly on abaxial surface in small groups, peridium white, cylindrical, the margin lobed; spores 24-40 x 13-23 µm, oblong or ellipsoid, wall 1-1.5 µm thick, colorless, finely verrucose.

Type: on *Wedelia hispida* H.B.K., Mexico State, Mexico, Reiche. Not otherwise known.

Reference Materials

Literature Cited

1. Arthur, J. C. 1907-1927. Order Uredinales. N. Amer. Flora 7:83-848.
2. Arthur, J. C. 1934. *Manual of the rusts in United States and Canada*. Purdue Res. Found., Lafayette, IN. 438 pp.
3. Arthur, J. C. & Bisby, G. R. 1918. An annotated translation of the part of Schweinitz's two papers giving the rusts of North America. Proc. Amer. Phil. Soc. 57:173-292.
4. Baxter, J. W. 1959. A monograph of the genus *Uropyxis*. Mycologia 51:210-221.
5. Baxter, J. W. 1965. Studies of North American species of *Ravenelia*. Mycologia 57:77-84.
6. Bromfield, K. R. 1971. Peanut rust: a review of literature. J. Amer. Peanut Res. 3:111-121.
7. Cummins, G. B. 1943. Descriptions of tropical rusts -V. Bull. Torrey Bot. Club 70:68-81.
8. Cummins, G. B. 1975. Two nomenclatural changes in *Ravenelia*. Mycologia 67:1042-1043.
9. Cummins, G. B., Britton, M. P. & Baxter, J. W. 1969. The autoecious species of *Puccinia* on North American Eupatorieae. Mycologia 61:924-944.
10. Cummins, G. B. & Stevenson, J. A. 1956. A check list of NorthAmerican rust fungi (Uredinales). Pl. Dis. Reptr. Suppl. 240:109-193.
11. Davis, H. B. 1936. *Life and works of Cyrus Guernsey Pringle*. Univ. Vermont. Burlington. 756 pp.
12. Fromme, F. D. 1924. The rust of cowpea. Phytopathology 14:67-79.
13. Harkness, H. W. & Moore, J. P. 1880. Catalogue of the Pacific Coast fungi. Calif. Acad. Sci. 46. pp.
14. Hennen, J. F. & Cummins, G. B. 1969. The autoecious species of *Puccinia* and *Uromyces* on North American Senecioneae. Mycologia 61:340-356.

15. Hiratsuka, N, 1935. *Phakopsora* of Japan. I. Bot. Mag. Tokyo 49:781-788.
16. Hiratsuka, Y. & Cummins, G. B. 1963. Morphology of the spermogonia of the rust fungi. Mycologia 55:487-507.
17. Hylander, H., Joerstad, I. & Nannfeldt, J. A. Enumeratio Uredinearum Scandinavicarum. Opera Bot. I:1-102.
18. Joerstad, I. 1958. The genera *Aecidium, Uredo* and *Puccinia* of Persoon. Blumea 9:1-20.
19. Lawrence, G. H. M. 1951. *Taxonomy of vascular plants*. Macmillan, New York, 823 pp.
20. Leon Gallegos, H. M. 1962. El chauhuixtle del garbanzo; una nueva enfermedade en México. Agr. Tech. Mex. 2:71.
21. Leppik, E. E. 1972. Evolutionary specialization of rust fungi (Uredinales) on the Leguminosae. Ann. Bot. Fenn. 9:135-148.
22. Lindquist, J. C. 1958. Las royas de "Baccharis." Rev. Fac. Agron. Univ. Nac. La Plata 34:1-79.
23. Long, W. H. The *Ravenelias* of the United States and Mexico. Bot. Gaz. 35:111-133.
24. Parmelee, J. A. 1962. *Uromyces striatus* Schroet. in Ontario. Can. J. Bot. 40:491-510.
25. Parmelee, J. A. 1967. The autoecious species of *Puccinia* on Heliantheae in North America. Can. J. Bot. 45:2267-2327.
26. Parmelee, J. A. 1969. The autoecious species of *Puccinia* on Heliantheae ('Ambrosiaceae') in North America. Can. J. Bot. 47:1391-1402.
27. Parmelee, J. A. 1972. Additions to the autoecious species of *Puccinia* on Heliantheae in North America. Can. J. Bot. 50:1457-1459.
28. Robinson, B. L. 1917. A monograph of the genus *Brickellia*. Mem. Gray Herb. 1:1-151.
29. Savile, D. B. O. 1970. Autoecious *Puccinia* species attacking Cardueae in North America. Can J. Bot. 48:1567-1584.
30. Urban, Z. 1973. The autoecious species of *Puccinia* on Vernonieae in North America. Acta Univ. Carol. Biol.-1971:1-84.
31. Watson, S. 1886. Contributions to American botany. Proc. Amer. Acad. Arts Sci. 21:414-468.

Index of Fungus Names

Index of Host Plant Names